Advanced Materials Design and Manufacturing Technologies of Nonferrous Metals

Advanced Materials Design and Manufacturing Technologies of Nonferrous Metals

Editors

Yilong Dai
Youwen Yang
Deqiao Xie

Basel • Beijing • Wuhan • Barcelona • Belgrade • Novi Sad • Cluj • Manchester

Editors

Yilong Dai
School of Materials Science
and Engineering
Xiangtan University
Xiangtan
China

Youwen Yang
School of Mechanical and
Electrical Engineering
Jiangxi University of Science
and Technology
Ganzhou
China

Deqiao Xie
School of Astronautics
Nanjing University of
Aeronautics and Astronautics
Nanjing
China

Editorial Office
MDPI
St. Alban-Anlage 66
4052 Basel, Switzerland

This is a reprint of articles from the Special Issue published online in the open access journal *Materials* (ISSN 1996-1944) (available at: www.mdpi.com/journal/materials/special_issues/ AMDMT_nonferrous_metals).

For citation purposes, cite each article independently as indicated on the article page online and as indicated below:

Lastname, A.A.; Lastname, B.B. Article Title. *Journal Name* **Year**, *Volume Number*, Page Range.

ISBN 978-3-7258-0854-0 (Hbk)
ISBN 978-3-7258-0853-3 (PDF)
doi.org/10.3390/books978-3-7258-0853-3

© 2024 by the authors. Articles in this book are Open Access and distributed under the Creative Commons Attribution (CC BY) license. The book as a whole is distributed by MDPI under the terms and conditions of the Creative Commons Attribution-NonCommercial-NoDerivs (CC BY-NC-ND) license.

Contents

About the Editors vii

Preface ix

Yun Zhou, Kai Zhang, Yaru Liang, Jun Cheng and Yilong Dai
Selective Laser Melted Magnesium Alloys: Fabrication, Microstructure and Property
Reprinted from: *Materials* 2022, 15, 7049, doi:10.3390/ma15207049 1

Zaure Karshyga, Albina Yersaiynova, Azamat Yessengaziyev, Bauyrzhan Orynbayev, Marina Kvyatkovskaya and Igor Silachyov
Synthesis of Manganese Oxide Sorbent for the Extraction of Lithium from Hydromineral Raw Materials
Reprinted from: *Materials* 2023, 16, 7548, doi:10.3390/ma16247548 24

Jarosław Kalabis, Aleksander Kowalski and Santina Topolska
Properties of Padding Welds Made of CuAl2 Multiwire and CuAl7 Wire in TIG Process
Reprinted from: *Materials* 2023, 16, 6199, doi:10.3390/ma16186199 48

Puli Cao, Chengbo Li, Daibo Zhu, Cai Zhao, Bo Xiao and Guilan Xie
Effect of Grain Structure and Quenching Rate on the Susceptibility to Exfoliation Corrosion in 7085 Alloy
Reprinted from: *Materials* 2023, 16, 5934, doi:10.3390/ma16175934 64

Ergen Liu, Qinglin Pan, Bing Liu, Ji Ye and Weiyi Wang
Microstructure Evolution of the Near-Surface Deformed Layer and Corrosion Behavior of Hot Rolled AA7050 Aluminum Alloy
Reprinted from: *Materials* 2023, 16, 4632, doi:10.3390/ma16134632 78

Daibo Zhu, Na Wu, Yang Liu, Xiaojin Liu, Chaohua Jiang and Yanbin Jiang et al.
The Portevin–Le Chatelier Effect of Cu-2.0Be Alloy during Hot Compression
Reprinted from: *Materials* 2023, 16, 4455, doi:10.3390/ma16124455 96

Guilan Xie, Zhihao Kuang, Jingxin Li, Yating Zhang, Shilei Han and Chengbo Li et al.
Thermal Deformation Behavior and Dynamic Softening Mechanisms of Zn-2.0Cu-0.15Ti Alloy: An Investigation of Hot Processing Conditions and Flow Stress Behavior
Reprinted from: *Materials* 2023, 16, 4431, doi:10.3390/ma16124431 111

Mingyu Hu, Xuemei Ouyang, Fucheng Yin, Xu Zhao, Zuchuan Zhang and Xinming Wang
Effect of Boronizing on the Microstructure and Mechanical Properties of CoCrFeNiMn High-Entropy Alloy
Reprinted from: *Materials* 2023, 16, 3754, doi:10.3390/ma16103754 125

Shujuan Yan, Caihong Hou, Angui Zhang and Fugang Qi
Influence of Al Addition on the Microstructure and Mechanical Properties of Mg-Zn-Sn-Mn-Ca Alloys
Reprinted from: *Materials* 2023, 16, 3664, doi:10.3390/ma16103664 138

Xu Zheng, Yi Yang, Jianguo Tang, Baoshuai Han, Yanjin Xu and Yuansong Zeng et al.
Influence of Retrogression Time on the Fatigue Crack Growth Behavior of a Modified AA7475 Aluminum Alloy
Reprinted from: *Materials* 2023, 16, 2733, doi:10.3390/ma16072733 155

Yufang Zhang, Zhu Xiao, Xiangpeng Meng, Lairong Xiao, Yongjun Pei and Xueping Gan
Experimental and Numerical Studies on Hot Compressive Deformation Behavior of a Cu–Ni–Sn–Mn–Zn Alloy
Reprinted from: *Materials* **2023**, *16*, 1445, doi:10.3390/ma16041445 169

Hang Zheng, Ruixiang Zhang, Qin Xu, Xiangqing Kong, Wanting Sun and Ying Fu et al.
Fabrication of Cu/Al/Cu Laminated Composites Reinforced with Graphene by Hot Pressing and Evaluation of Their Electrical Conductivity
Reprinted from: *Materials* **2023**, *16*, 622, doi:10.3390/ma16020622 188

Songsong Hu, Yunsong Zhao, Weimin Bai, Yilong Dai, Zhenyu Yang and Fucheng Yin et al.
Orientation Control for Nickel-Based Single Crystal Superalloys: Grain Selection Method Assisted by Directional Columnar Grains
Reprinted from: *Materials* **2022**, *15*, 4463, doi:10.3390/ma15134463 200

Yongming Yan, Yanjun Xue, Wenchao Yu, Ke Liu, Maoqiu Wang and Xinming Wang et al.
Predictions and Experiments on the Distortion of the 20Cr2Ni4A C-ring during Carburizing and Quenching Process
Reprinted from: *Materials* **2022**, *15*, 4345, doi:10.3390/ma15124345 210

Jing Qin, Haibin Zhao, Dongsheng Wang, Songlin Wang and Youwen Yang
Effect of Y on Recrystallization Behavior in Non-Oriented 4.5 wt% Si Steel Sheets
Reprinted from: *Materials* **2022**, *15*, 4227, doi:10.3390/ma15124227 223

Bo Deng, Yilong Dai, Jianguo Lin and Dechuang Zhang
Effect of Rolling Treatment on Microstructure, Mechanical Properties, and Corrosion Properties of WE43 Alloy
Reprinted from: *Materials* **2022**, *15*, 3985, doi:10.3390/ma15113985 234

About the Editors

Yilong Dai

Yilong Dai, Ph.D., Postdoctoral Fellow, Associate Professor, Master Supervisor of the School of Materials Science and Engineering, Xiangtan University, He received his bachelor's, master's, and doctoral degrees from Central South University. He is a reviewer of more than 50 SCI journals, such as *Advanced Science, Bioactive Materials, Acta Biomaterialia, Journal of Materials Science and Technology, Corrosion Science*, etc., and an excellent reviewer of the *Journal of Magnesium and Alloys*. He is the managing editor-in-chief of "Smart Materials in Manufacturing" and the young editorial board of three journals: "Advanced Powder Materials", "Rare Metals", and the "Journal of Central South University". He is mainly engaged in the research of light metal preparation and processing, metal medical materials, metal anode materials, etc. He has published more than 60 SCI papers, including 26 as the first author or corresponding author, with an H-index of 20.

Youwen Yang

Youwen Yang, a Ph.D., Associate Professor, and Master Supervisor at the School of Mechanical and Electrical Engineering, Jiangxi University of Science and Technology, specializes in laser 3D printing and functional design of regenerative structures. His groundbreaking work includes the development of technology for the spheroidization of biodegradable polymer powders, the elucidation of laser shaping and macro-microstructural regulation mechanisms in regenerative structures, and the creation of physical stimuli and photodynamic nano-systems for imparting osteogenic and antibacterial functions to regenerative tissues. As a prolific researcher, Dr. Yang has authored 53 papers in esteemed journals like *Bioact Mater* (IF= 18.9) as the first or corresponding author, including 3 ESI hot papers and 9 highly cited papers. His contributions have garnered over 4800 citations from experts worldwide, with 29 papers published in top-tier journals. Notably, he holds 14 authorized invention patents and has led research projects funded by the National Natural Science Foundation of China and the Jiangxi Provincial Natural Science Foundation. Dr. Yang's excellence is recognized through awards such as the first prize in provincial natural science (ranked third). He is esteemed as a leading academic and technical figure in the Jiangxi province, an outstanding young talent at Jiangxi University of Science and Technology, and serves as a young editorial board member for journals like *JMA* (IF= 17.6) and *IJEM* (IF= 14.7).

Deqiao Xie

Deqiao Xie graduated with a Bachelor's degree in 2011, a Master's degree in 2014, and a Ph.D. in 2019 from Nanjing University of Aeronautics and Astronautics, where he currently serves as a research assistant. His primary research interests lie in additive manufacturing technology and advanced structural design, including structural distortion and residual stress, innovative design of porous metallic/ceramic structures, advanced manufacturing techniques, and evaluation of mechanical properties. He has led or participated in over 10 key projects, published more than 60 academic papers, and has been granted over 20 invention patents.

Preface

The design and manufacturing technology of advanced materials for non-ferrous metals have become an important driving force behind the development of modern industry. This reprint aims to provide readers with a comprehensive, systematic, and in-depth reference that elaborates on the principles of design, manufacturing technology, and application areas of advanced materials for non-ferrous metals in order to contribute to material science research and industrial development. Non-ferrous metals play an irreplaceable role in various fields such as aerospace, electronics information, new energy, and transportation due to their unique physical, chemical, and mechanical properties. With continuous technological advancements, the design and manufacturing technology of advanced materials for non-ferrous metals have made significant progress. A series of high-performance and high-value-added new materials have emerged that provide strong support for economic and social development.

This reprint collects a total of sixteen Special Issue papers, including one review article and fifteen research papers. It covers the microstructure, performance characteristics, and preparation processes of materials. Based on this foundation, it focuses on introducing the principles and methods for designing advanced materials for non-ferrous metals. Additionally, the book also provides detailed explanations about the manufacturing technologies involved in these advanced materials, including smelting, casting, plastic processing, heat treatment, surface treatment, etc., along with their technical principles and practical applications. In addition, this reprint also focuses on the application areas and development trends of advanced materials in non-ferrous metals. Through case studies in industries such as aerospace, electronics, and new energy, it demonstrates the enormous potential of advanced materials in non-ferrous metals to promote industrial upgrading, improve product quality, and reduce production costs.

Finally, we would like to express our gratitude to all the authors who have provided support and assistance in writing this reprint. Their hard work and selfless dedication have enabled this reprint to be published smoothly and meet the needs of a wide range of readers.

Yilong Dai, Youwen Yang, and Deqiao Xie
Editors

Review

Selective Laser Melted Magnesium Alloys: Fabrication, Microstructure and Property

Yun Zhou [1,2], Kai Zhang [1], Yaru Liang [2], Jun Cheng [3] and Yilong Dai [2,*]

[1] Department of Automotive Engineering, Hunan Industry Polytechnic, Changsha 410208, China
[2] School of Materials Science and Engineering, Xiangtan University, Xiangtan 411105, China
[3] Key Laboratory of Biomedical Metal Materials, Northwest Institute for Nonferrous Metal Research, Xi'an 710016, China
* Correspondence: daiyilong@xtu.edu.cn

Abstract: As the lightest metal structural material, magnesium and its alloys have the characteristics of low density, high specific strength and good biocompatibility, which gives magnesium alloys broad application prospects in fields of biomedicine, transportation, and aerospace. Laser selective melting technology has the advantages of manufacturing complex structural parts, high precision and high degree of freedom. However, due to some disadvantages of magnesium alloy, such as low boiling point and high vapor pressure, the application of it in laser selective melting was relatively undeveloped compared with other alloys. In this paper, the fabrication, microstructure, mechanical performance and corrosion resistance property of magnesium alloys were summarized, and the potential applications and the development direction of selective laser melting magnesium alloys in the future are prospected.

Keywords: selective laser melting; magnesium alloys; microstructure; mechanical property; corrosion behavior

Citation: Zhou, Y.; Zhang, K.; Liang, Y.; Cheng, J.; Dai, Y. Selective Laser Melted Magnesium Alloys: Fabrication, Microstructure and Property. *Materials* 2022, 15, 7049. https://doi.org/10.3390/ma15207049

Academic Editors: Hajo Dieringa and Gregory N. Haidemenopoulos

Received: 29 July 2022
Accepted: 24 September 2022
Published: 11 October 2022

Publisher's Note: MDPI stays neutral with regard to jurisdictional claims in published maps and institutional affiliations.

Copyright: © 2022 by the authors. Licensee MDPI, Basel, Switzerland. This article is an open access article distributed under the terms and conditions of the Creative Commons Attribution (CC BY) license (https://creativecommons.org/licenses/by/4.0/).

1. Introduction

With the advantages of low density, high specific strength, good thermal conductivity and good biocompatibility, magnesium and its alloys are the lightest metal structural materials in practical applications [1–3], which are applied in transportation, communication electronics, aerospace, biomedicine and other fields. However, the Mg alloy workpieces manufactured by traditional technology has some shortcomings, such as poor microstructure uniformity, long processing cycle, low material utilization and low efficiency [4]. Thus, the development and application of high-performance magnesium alloys were restricted. Selective laser melting (SLM) is a kind of additive manufacturing technology that uses high-energy laser beam to selectively sinter and stack powder layer by layer. It has the advantages of manufacturing complex structural parts, high precision, high degree of freedom and refining alloy structure [5]. Owing to that, the selective laser melting process has become an important method to prepare high-performance magnesium alloy workpieces. The principle of SLM technology is shown in Figure 1 [6]. According to the geometric data model in 3D software, the metal powder layer is selectively melted by a high-energy laser beam, forming a large number of molten pools. After that, the produced molten pool solidifies rapidly (10^3–10^6 k/s), and finally the parts with the required shape are obtained [7–9]. SLM technology has the following characteristics: (1) Suitable for a wide range of processing materials, including refractory metals, high reflectivity materials and low melting boiling point metals. (2) The forming accuracy is high. After grinding, sandblasting and other subsequent treatment, the surface of workpieces can meet the accuracy requirements. (3) It can process structures with complex shapes, such as spatial curved porous structure, light lattice sandwich structure, special-shaped complex cavity structure, etc., [10]. SLM technology is developed and widely used in iron [11], titanium [12,13], nickel base superalloys [14,15], aluminum alloys [16], and so on [17]. However, due to its active chemical properties, high affinity with oxygen, low melting boiling point and high vapor pressure, the application and development of magnesium alloys in SLM technology are relatively behind. In this review, the characteristics of the forming process of magnesium alloy by SLM technology were summarized, how parameters in the forming process affected the microstructure and properties of magnesium alloys reported in recent research were reviewed, and the possible applications of SLM forming magnesium alloys in the future are prospected.

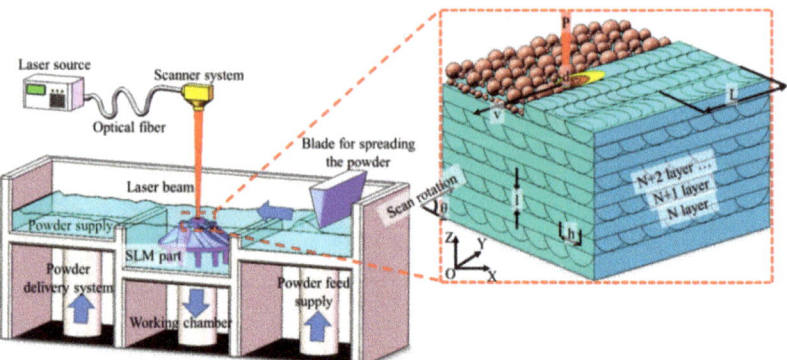

Figure 1. Schematic diagram of selective laser melting. Reprinted with permission from ref. [6]. Copyright 2020 Springer Nature.

2. Fabrication of Magnesium Alloy with Selective Laser Melting

2.1. Influence of Magnesium Alloy Powder

Since metal or alloy powders are used as raw materials in SLM process, the characteristics and quality of powders have a great impact on the stability of SLM process and the performance of final samples. The quality of powder is determined by its size, shape, surface morphology, composition and internal porosity. In this paper, several factors related to product performance will be reviewed.

2.1.1. Alloying Elements of Mg in SLM

Due to the low melting point of magnesium (923 K), it is easy to burn in the SLM process. Thus, addition of alloy elements is beneficial to broaden the melting point and boiling point range of magnesium alloys, and further limit the selective vaporization of magnesium elements [18–21]. In addition, the research shows that the oxidation of magnesium oxide alloy powder can be prevented [22,23]. For example, in Mg–Be and Mg–Al alloys, active metal elements added to the alloy are required to react before Mg elements, so as to prevent large-scale oxidation of Mg elements [24–27]. Therefore, Mg Al, AZ (Mg–Al–Zn), ZK (Mg–Zn–Zr), AM (Mg–Al–Mn) and WE (Mg–Re) series are the most commonly used alloy systems in SLM process [23,28,29].

2.1.2. Shape and Size of Alloy Powder

Spherical powder is usually used in SLM processing because it helps to improve the fluidity of the powder and obtain high precision. The research shows that because the irregular shaped powder is not easy to flow and has a strong tendency to aggregate, the use of non-spherical powder particles in SLM processing will make a negative impact on the uniform deposition of powder [30]. The shape of magnesium powder plays an important role in fluidity and laser reflection behavior [31]. Irregular shaped powder will significantly reduce the fluidity of the powder, cause uneven distribution of the powder and affect the quality after forming. Instead, the spherical powder with a uniform and smooth surface can improve the fluidity of the powder and the formability of magnesium alloy [32]. In particular, the powders should be without defects such as satellite powder and caking, which will lead to insufficiency fusion between the particles, thereby affecting the densification process [33].

The behavior of powders with different sizes is different in the laser processing process. SLMed workpieces that used powders with large particle size always showed poor adhesion between each layer and low density, due to the poor penetrating ability of it. Although relatively fine particles and high laser energy density can be used to prepare high-density components with better surface, fine powder particles are easy to be blown away by the protective air flow, thus affecting the deposition process. Furthermore, powder with small size will evaporate when comes to the situation of high laser power, the consequent generation of smoke and dust in the print bin will affect the output of laser energy and make the formed sample performed a poor surface [34]. In the recently work of Wang [35], the influence of powder particle size on smoke and dust generation in the process of forming AZ91D Magnesium Alloy by SLM was studied. The fine powders are easier to melt than the coarse powders, and more likely to experience the overburning phenomenon under the same laser energy density. Screening the fine powder below 20 μm can effectively reduce the amount of smoke and dust, and better surface quality and higher tensile strength can be obtained. Dong et al. [36] studied the size effect of Mg powder in SLM process. As shown in Figure 2, the sample fabricated by 400 mesh powder showed a rough surface with un-melted particles (Figure 2a,b), while the sample prepared with 250 mesh powder was well consolidated (Figure 2c,d). The reason for this result can be expressed as: with higher temperature and the bigger molten pool, some un-melted or melted powders were blown away by the air flow when fine powder particles were used. These powders adhered to the substrate and made 400 mesh powders exhibit more severe balling and agglomeration than that of 250 mesh powder. Aside from the size effect of powder, due to the different melt speed between larger size powders and smaller size powders, the uneven distribution of powder particles on the construction platform should be avoided. When the powders adopted show a wide size distribution mixed with both large-sized particles and small-sized particles, more attention should be paid to the selection of process parameters to avoid uneven melting of the powders. Spherical powders with a narrow range of particle sizes can help to improve the thermal conductivity of the powders, resulting in an increase in the density of SLM formed parts [37].

2.2. Influence of SLM Process Parameters

The main advantage of SLM technology is that it can produce metal parts with high density. However, since the absence of mechanical pressure, fluid dynamics is mainly driven by gravity, capillary action and thermal effect, so it is not easy to achieve this goal. In addition, the lack of mechanical pressure during the SLM process may reduce the solubility of some elements during solidification, resulting in discontinuous melting to form pores and surface irregularities. Therefore, appropriate parameters (include power density, scanning distance and speed, and layer thickness etc.) need to be carefully selected to obtain parts with good quality. Taking WE43 as an example, Figure 3 shows the influence of SLM process parameters on the formability of magnesium alloy. When specimens were produced in the low energy input zone, they exhibited poor mechanical strength due to the extensive presence of process-induced pores and lack of fusion. At high energy inputs, a loss of the Mg content occurred due to its low boiling point. An appropriate volume energy density (E_V) can be obtained by adjusting process parameters [38]. Volume energy density E_V is defined as the laser energy per volume and can be calculated by the following equation:

$$E_V = P/vdh \tag{1}$$

where P (W) is the laser power, V (mm/s) is the scanning speed, D (mm) is the scanning distance, h (mm) is the layer thickness. Under proper laser parameters, the SLM prepared can achieve a negligible fraction of metallurgical and process-induced defects. However, in the recent work of Deng [39], the depth and width of molten pool and the porosity vary significantly despite using the same LED value but different combinations of p and V values. When p or V is changed alone, it is reliable to use LED as a design parameter, but when p and V are changed at the same time, LED has limitations.

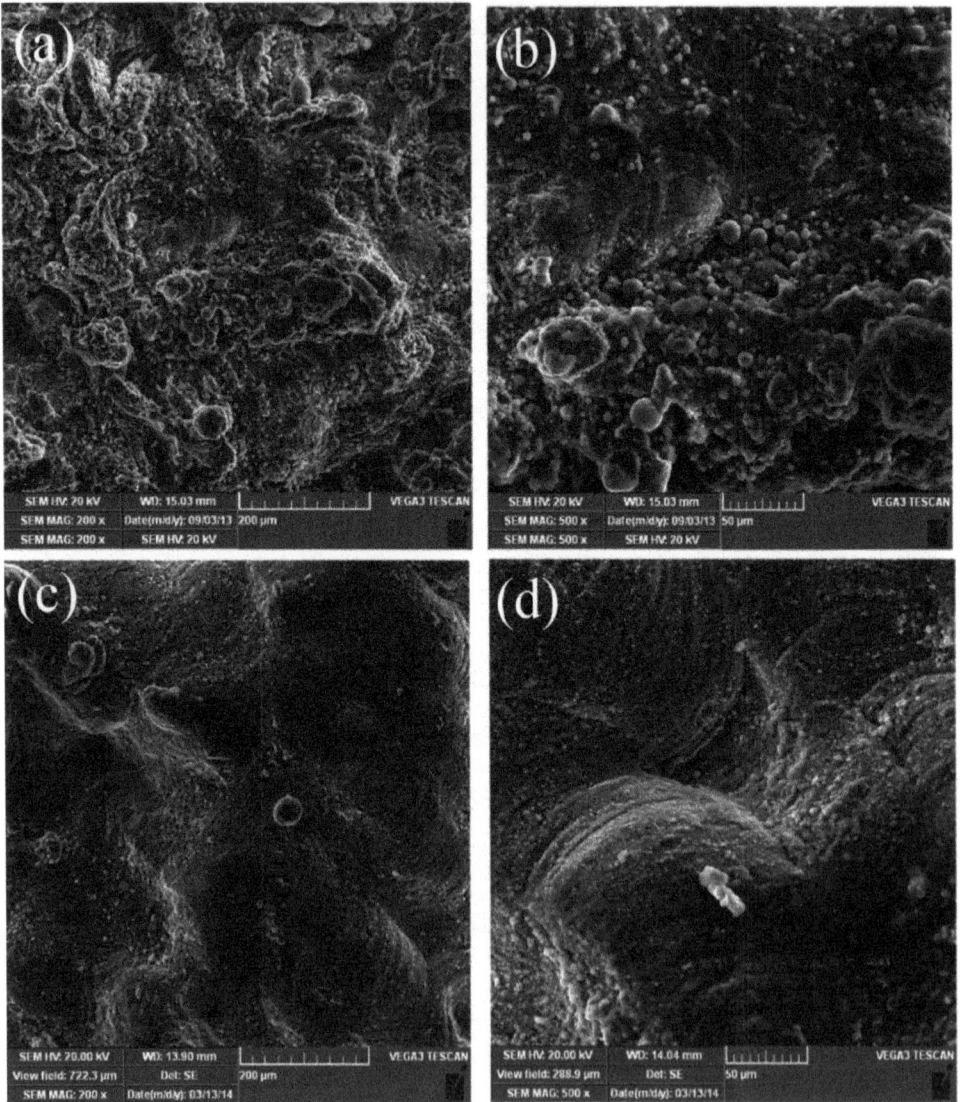

Figure 2. SEM topology under different magnifications of specimens fabricated using magnesium powders with granularity of (**a,b**) 400 mesh and (**c,d**) 250 mesh. Reprinted with permission from ref. [36]. Copyright 2015 Taylor and Francis Group.

2.2.1. Laser Energy Density

In the SLM forming process, the densification process of the sample is as follows: first, the metal powder is melted by a single laser beam, and then its melting trajectory is overlapped with the adjacent melting trajectory. When the multilayer is formed, the laser beam irradiates the powder layer to melt it and weld it with the previous powder layer to form a solid interlayer bond. Olakanmi et al. [41] observed that under the appropriate laser power and scanning speed, the metal powder can be completely melted, and there is good solid–liquid interface wettability between the powder particles and the melt, so that the formed sample can be nearly completely dense. The complete melting of metal powder can not only enhance the adhesion between powders and form complex structural parts, but also facilitate the discharge of gas, reduce the formation of pores in the sample, obtain more dense formed parts and obtain excellent comprehensive properties. Shown in Figure 4, the processing parameter influence on porosity was summarized in Oliveira's work [17]. Low E_V is not enough to melt magnesium alloy powder completely, and it is difficult to obtain dense formed parts. Too high E_V will cause surface spheroidization and burning of alloy elements [42]. For very high energy densities, Mg vapor will generate back pressure on the molten pool, and the huge temperature gradient of the molten pool causes a strong Marangoni convection effect in the molten pool, and the depth of the molten pool is much greater than the width of the molten pool. However, the solidification rate of SLM process is very fast, and more Mg vapor has no time to escape from the molten pool, so it exists in the molten pool in the form of pores, which will make the workpiece more porous. Thus, a depression filled by steam and external gas was formed, and its morphology is called keyhole, which may degrade the fatigue life of the part by acting as a crack initiator [43,44]. Increasing the canning speed is a simple way to reduce the formation of keyholes, but at high energy density, too large scanning speed will also bring about "balling". In order to address defects such as keyhole porosity, lack-of-fusion porosity, and "balling", suitable laser power and scanning speed need to be investigated.

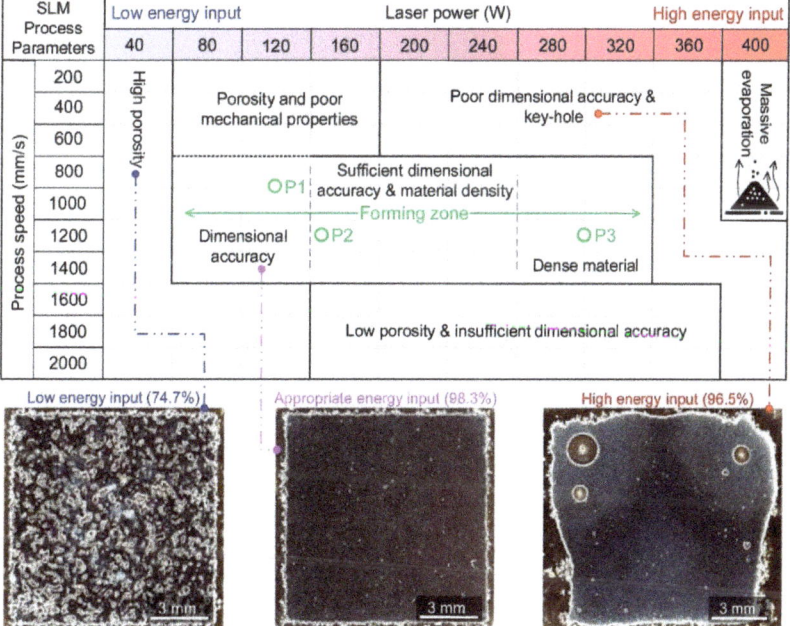

Figure 3. Effect of SLM process parameters on the formability of WE43. Reprinted with permission from ref. [40]. Copyright 2020 Elsevier.

In the study of Wang with Mg–Y–Sm–Zn–Zr alloy [45], it was found that the size of molten pool was positively correlated with laser power. As the laser power increased from 40 W to 80 W, the length and depth increased from 71.14 μm and 21.62 μm to 110.8 μm and 31.9 μm, respectively. Zhang et al. [46] studied the effect of laser energy density on the density of Mg–9%Al alloy samples by adjusting laser power and laser scanning speed. The results show that at a low laser energy density, the powder melts incompletely, resulting in discontinuous scanning tracks and spheroidization, which leads to more pore defects in the sample. With the increase of laser energy density, the powder melts better, so that more liquid phase can flow and penetrate into the gap between particles. The relative density of the sample is higher, and the surface is relatively smooth. However, with the further increase of laser energy density, the powder will melt completely, and defects such as spheroidization and scum will appear in the molten pool, resulting in reduced density and poor surface finish. Wei et al. [47] observed the formability of AZ91D alloy under different bulk energy densities. When the bulk energy density is 83–167 J/mm^3, the sample has no obvious macroscopic defects and has a high density. When the bulk energy density is higher than 214 J/mm^3, the alloy elements volatilize and burn seriously, and the sample cannot be deposited and formed. In Yang's research [48], the SLM printing process shows that with a low laser energy density, the powders were in a discrete state, and there was no fusion between the powders (Figure 5a). By increasing the laser energy density, the powders can be partially melted and sintered together forming sintering neck as a weak bonding (Figure 5d). It is only when the laser density is large enough to melt the powder completely that smooth and continuous trajectory can be obtained, as shown in Figure 5f. With the further increase of laser energy density, the powder evaporates. In Ng's study of magnesium [49], the laser energy density also shown an effect on the average grain size of α-Mg grown. It changed from 2.30 μm to 4.87 μm with the laser energy density increasing from 1.27×10^9 J/mm^2 to 7.84×10^9 J/mm^2. It is attributed to the high energy density which keeps the Mg-melt at a relative high temperature state, affected the cooling process of the melt and resulted in grain growth.

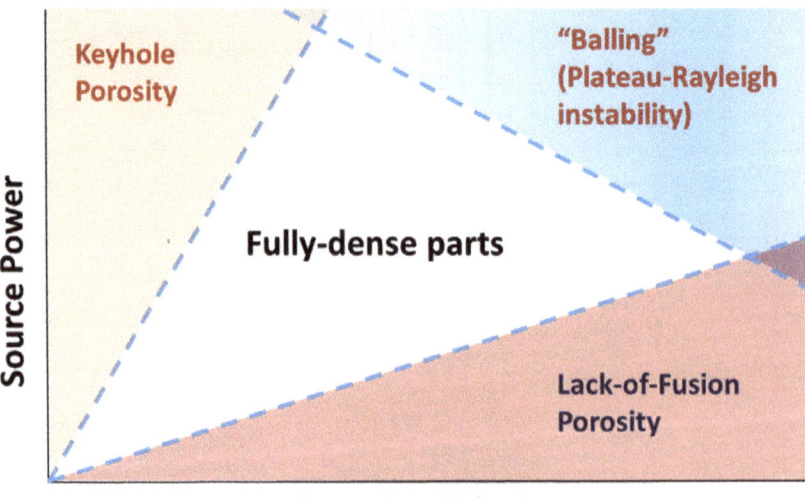

Figure 4. Illustration of processing parameter influence on porosity. Reprinted with permission from ref. [17]. Copyright 2020 Elsevier.

Although the SLM process is usually based on the complete melting, partial melting of powder caused by low energy density is not necessarily bad. The molten metal on the surface of partially melted particles can make the powders adhere to each other and leave pores between the powders, which can be effectively used to produce porous structures with complex shapes [50], whereas a loose structure formed at energy inputs below 77 J/mm^3 due to "balling effect" and incomplete melt of powders. It is worth noting that the specific value of energy density is meaningful only in specific cases, because different overlapping distance and powder layer thickness will affect the energy density, and the same energy density does not necessarily ensure the stability of the quality of the formed sample [51], so the best E_V value depends on the specific powder material composition and laser beam scanning forming strategy [52].

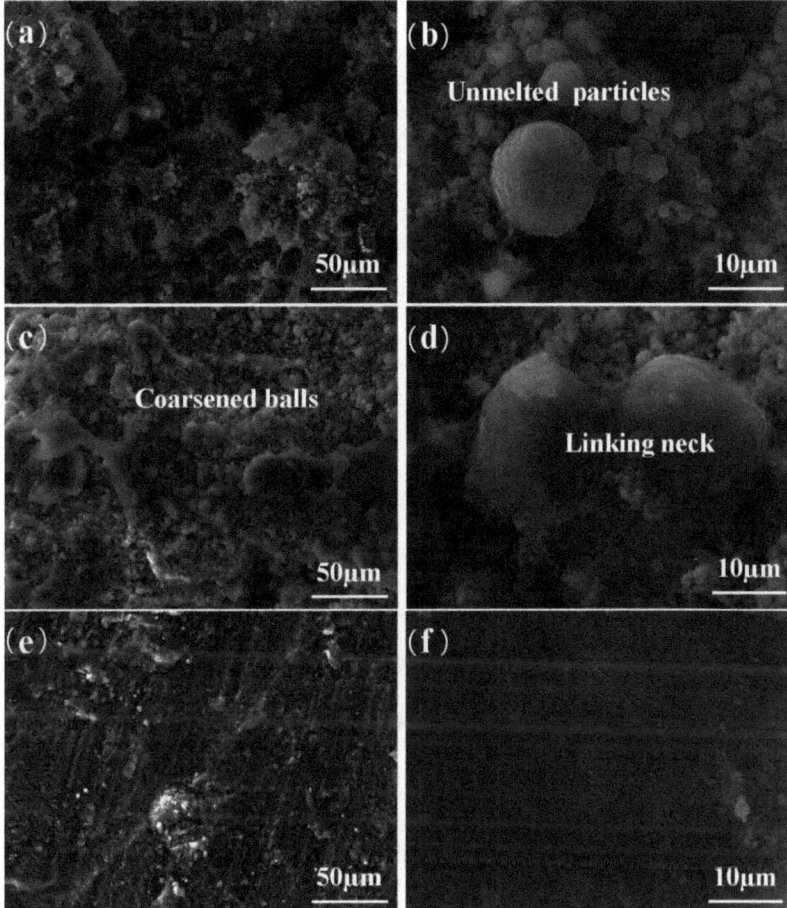

Figure 5. Surface of SLM tracks at various laser energy density: (**a**) 2.0 J/mm, (**c**) 5.0 J/mm, (**e**) 10.0 J/mm, (**b,d,f**) are high resolution of (**a,c,e**), respectively. Reprinted with permission from ref. [48]. Copyright 2016 Taylor and Francis Group.

2.2.2. Scanning Speed and Spacing

At a constant laser power, the scanning speed is related to the residence time of the laser beam on the surface of the molten pool. Thereby, by reducing the scanning speed, the energy density will be increased, and higher workpiece density will be obtained. The scanning distance (also known as the spacing distance) is another important parameter that affects the relative density of the alloy. It determines the degree of overlap of laser points when a new laser line sweeps through the previously scanned line. Ng et al. [53] studied the single pass experiment of magnesium alloy at different scanning speeds, and proved the feasibility of magnesium alloy powder in SLM manufacturing. With too fast scanning speed, powder splashing can be observed, which will cause instability of molten pool and increase thermal stress of the production. On the contrary, when the scanning speed is too low, it will lead to the recrystallization of magnesium alloy samples during solidification, which will affect the microstructure and morphology of the alloy. In Wei's work [47], the effects of scanning speed and scanning distance on the relative density of AZ91D alloy prepared by SLM were summarized. The relative density of the sample decreases with the increase of scanning speed and hatch spacings. Similar results were also obtained for ZK60 alloy [54]. As the scanning speed increased, the relative density of the sample reached a peak of 94.05% at 300 mm/s. When the scanning speed was 100 mm/s, serious vaporization and metal powder burning were observed, leaving ablative pits on the surface of the substrate, resulting in the termination of the molding process. When the scanning speed is higher than 500 mm/s, the powder particles cannot be completely melted, and pores will be formed between unmelted powders, resulting in a sharp decline in the relative density of the sample. The influence of scanning spacing on the overlap area between adjacent scanning passes was studied by Deng [43]. It can be seen from Figure 6 that suitable hatch spacing will make the height of the overlap region and the height of the unlapped region tend to be uniform, which will result in even powder distribution in the next layer. With a width of the molten pool approximately 140 to 200 μm, the corresponding overlap ratio is about 30–50% when the hatch spacing is 100 μm.

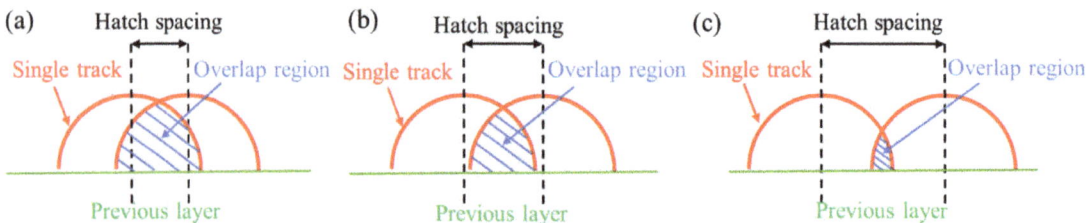

Figure 6. Schematic diagram illustrating the effect of hatch spacing on area of overlap region: (**a**) low hatch spacing; (**b**) suitable hatch spacing; (**c**) high hatch spacing. Reprinted with permission from ref. [43]. Copyright 2020 Elsevier.

2.2.3. Thickness of Powder Layer

The thickness of powder layer is another important parameter which has an important impact on porosity and interlayer adhesion, and also affects the tensile strength, hardness and dimensional accuracy in the construction direction. A thicker powder layer will reduce the thermal penetration depth of the laser beam, so the bottom powder cannot be effectively melted, and the adhesion of each layer is reduced, resulting in local stress concentration, uneven chemical composition, pores, microcracks and incomplete fusion areas. These defects greatly reduce the fracture toughness and tensile ductility of the alloy. A thicker powder layer will lead to non-fusion between particles, and the same amount of laser energy will radiate more powder than a thin powder layer. The laser energy density penetrating the powder bed will not be enough to completely melt the powder particles, which will cause more pores and reduce the density of the sample. Therefore, it is necessary to establish the best layer thickness parameters, so that there is a good adhesion between layers, and reduce the defects that may occur in the SLM process, so as to obtain a higher density. Savalani et al. [55] studied the influence of powder layer thickness on single channel pure magnesium prepared by SLM. In their experiment. The layer thickness of preheated samples was adjusted from 150 to 300 μm. The results show that there is a critical value (250 μm) for the thickness of the powder layer. When the powder thickness was relatively thin, a flat surface without any surface defects could be obtained. Because the amount of material to be melted is significantly less, the heat transmitted in the molten pool has enough energy to completely melt the adjacent particles, rather than partially melt.

3. Microstructure of SLMed Samples

When the incident laser beam irradiates the metal powder layer, most of the laser energy is absorbed by the metal powder particles, resulting in rapid heating and local melting of the powder. Therefore, laser selective melting process is a process of rapid melting and solidification [56]. Thus, by adjusting the process parameters and changing the thermodynamics and dynamics of the molten pool, so as to control the size and shape of grains, as well as the content and composition of phases in the solidification process, so as to obtain the required microstructure [57]. The characteristics of SLM with high cooling rate can obtain fine microstructure, while the cooling rate of Mg alloy can reach 10^6 K/s during SLM. Compared with the traditional process, SLMed magnesium alloy is also affected by recrystallization, but it is easier to control the evolution of microstructure. In the research of Zumdick, as shown in Figure 7, due to the characteristics of powder sintering and rapid cooling in the SLM process, the magnesium alloy presents a uniform and fine microstructure (Figure 7a–c); Although the microstructure of magnesium alloy prepared by extrusion is relatively uniform and fine, it has serious anisotropy (Figure 7d,e); The microstructure grain of as cast state is relatively rough (Figure 7f) [58].

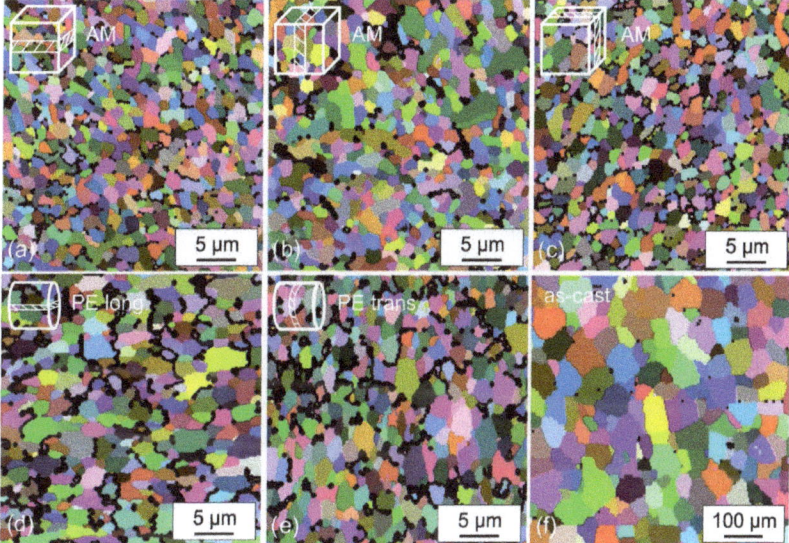

Figure 7. Electron backscatter diffraction of (**a**–**c**) selective laser melting, (**d**,**e**) of powder extruded, (**f**) of as-cast material. Reprinted with permission from ref. [58]. Copyright 2019 Elsevier.

The multiple remelting in the SLM process makes the microstructure change in different height directions, which is attributed to the different heat treatment, conduction, convection and radiation conditions between different layers. After repeated remelting, the cooling rate decreases relatively, which coarsens the grains in the remelted part. The microstructure is different due to the difference of heat affected zone between the edge and center of molten pool. In the preparation of AZ91D alloy by SLM, the grains in the scanning track area are smaller than those in the overlapping area, which is caused by the repeated remelting of the grains at the edge of the molten pool, the reduction of the cooling rate, and the existence of temperature gradient difference. Figure 8 shows the distribution of molten pool under the optical microscope of AZ91D sample under the bulk energy density of 166.7 J/mm^3, which obeys the Gaussian distribution of laser energy. The molten pool with elliptical bottom is arranged layer by layer, which is the inherent layered feature of SLM technology. The molten pool is closely stacked, forming a good metallurgical bond between two adjacent layers. [47] In addition, due to the temperature gradient difference between the edge and center of the molten pool, the transformation from columnar crystal to equiaxed crystal may occur from the edge of the molten pool to the center. Columnar α-Mg grains occupy the edge area of the molten pool, while the α-Mg grains in the center area of the molten pool show equiaxed crystal state.

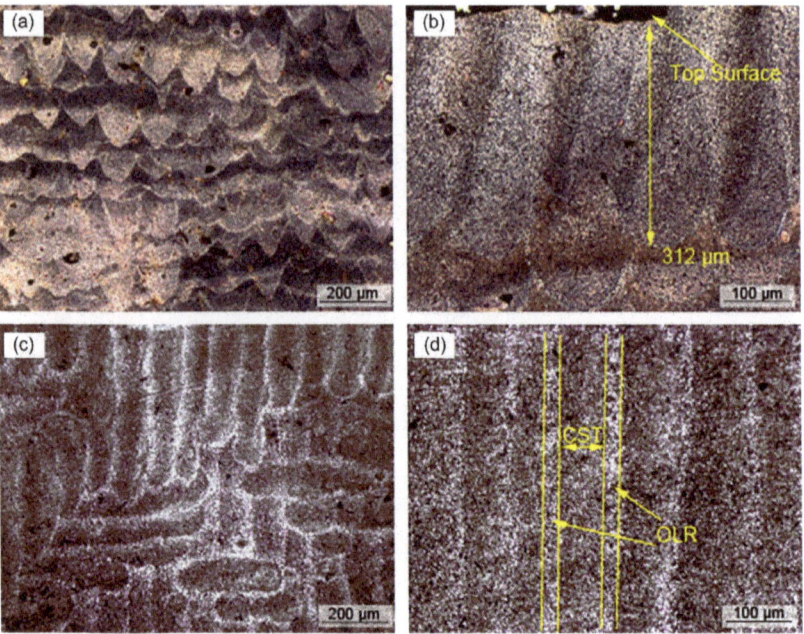

Figure 8. Optical micrographs of the AZ91D sample deposited at Ev of 166.7 J/mm^3. (**a**,**b**) The vertical section, (**c**) the cross section, and (**d**) the detail of scanning tracks shown in (**c**). Reprinted with permission from ref. [47]. Copyright 2014 Elsevier.

The microstructure of SLM samples is closely related to the preparation parameters. In the study of Mg–Zn–Zr (ZK60) alloy (Figure 9). It is found that, with the increase of laser energy density, the crystal structure changes into: columnar crystal, equiaxed crystal, and coarsened equiaxed crystal, which is caused by the increase of laser energy density and the decrease of its cooling rate [59]. In addition to the adjustment of preparation parameters, appropriate component selection and subsequent processing can effectively control the microstructure of samples. SLM will lead to changes in composition and microstructure, which is due to the high vapor pressure of Mg, Zn and other elements. During the printing process, Mg, Zn and other elements will be selectively evaporated, and Al, Zr and other elements will be enriched on the surface through solute capture effect. In the process of rapid laser melting, due to the great temperature gradient difference, it is conducive to the formation of Marangoni convection and promotes the uniform dispersion of alloy elements in the molten pool. The expansion of the solid–liquid interface contributes to the solute capture phenomenon in the α-Mg matrix, and more solute atoms are concentrated in the α-Mg matrix, which improves the solid solution limit of alloy elements in the α-Mg matrix and inhibits the nucleation of β- phase [47,59]. Therefore, the change of composition will affect the microstructure during laser selective melting. In order to obtain finer microstructure, severe plastic deformation is a very effective processing method. After extrusion and heat treatment, the mechanical properties of JDBM alloy are improved at room temperature compared with those as cast [60]. The main reason is that after extrusion, the process will undergo dynamic recrystallization, resulting in grain refinement.

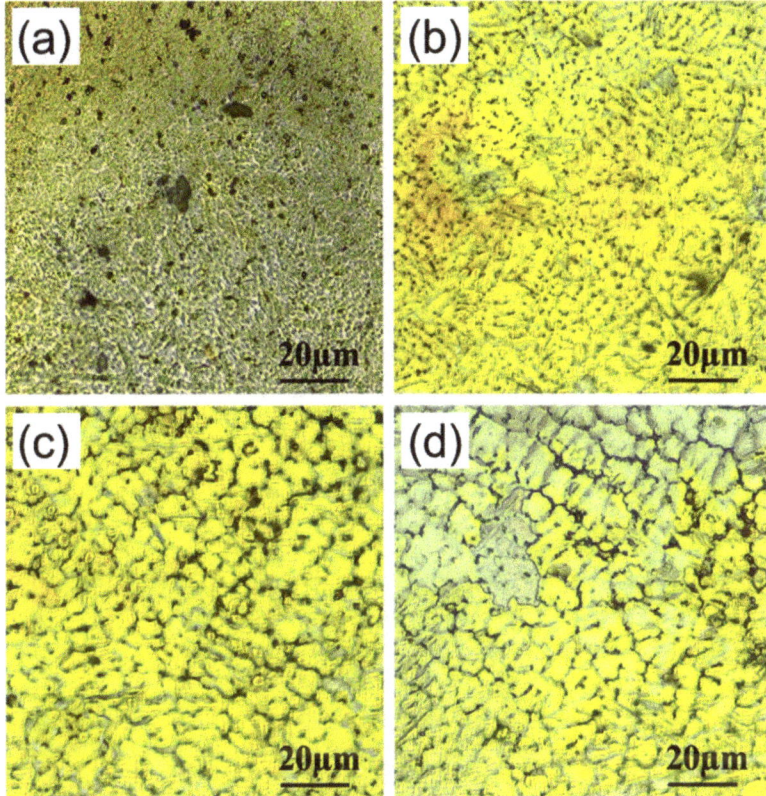

Figure 9. Microstructures of SLM ZK60 samples at different parameters, (**a**) 420 J/mm^3; (**b**) 500 J/mm^3; (**c**) 600 J/mm^3; (**d**) 750 J/mm^3. Reprinted with permission from ref. [59]. Copyright 2017 Elsevier.

4. Properties
4.1. Mechanical Property
4.1.1. Hardness

The rapid solidification effect on the parts in the SLM process leads to the grain refinement of its microstructure, which is one of the main reasons for the increase of the hardness of SLM manufactured parts. Thus, the hardness performance of samples can effectively be controlled by adjusting the processing parameters in SLM process. High cooling rate stimulated by low laser energy density can refine the grain size, consequently, increase the hardness values in the melted zone. In addition, according to the solid solution strengthening theory, high solid solubility can also improve the hardness of products [61]. Therefore, the solid solution of different elements also affects the hardness value of magnesium alloy. For example, the microhardness of Mg–Al alloy is different from that of α-Mg. In the study of Cáceres, the hardness was increasing with the aluminum content between 1%~8% in Mg–Al alloys [62]. In Table 1, the relationship between hardness values of various magnesium alloys was summarized. It can be seen from the table, beside the grain size, the hardness value of SMLed magnesium alloy is closely related to the microstructure, element composition and content. Moreover, according to Mercelis and Kruth [63], the residual stress maintained at a reasonable level in the parts manufactured by SLM can improve its hardness, provided that sufficient densification can be achieved without cracks or pores.

Table 1. Summary of the hardness and microstructural properties of different Mg alloy.

Materials System	Laser Energy Density	Grain Size (μm)	Relative Density (Maximum)	Hardness
Pure Mg [49]		2.30–4.87		0.95–0.59 (GPa)
Pure Mg [36]	300 J/mm^3	1.656–1.671	95.28–96.13	44.75–52.43 Hv
Mg–Ca [64]	600–1200 J/mm^3	5–30	81.52%	60–68 Hv
Mg–9%Al [46]	7.5–20 J/mm^2	10–20	82%	~65–85 Hv$_{25}$
AZ61 [65]		5–14	98%	69–93 Hv
AZ91D [47]	83–167 J/mm^3		99.52%	85–100 Hv
AZ31 [66]	123.81–242.86 J/mm^3			67–71 Hv
AZ61D [67]		1–3	99.2%	125 Hv
AZ61 + HIP [68]		23.9 ± 6	close to 100%	98.9 ± 5.9 Hv
ZK30 [69]		~20		80 Hv
WE43 [58]		1.0–1.1		
ZK60 [54]	138.8–416.6 J/mm^3		94.05%.	~78 Hv
ZK60 [59]	600 J/mm^3		97.3%	89.2 Hv
Mg–3.4Y–3.6Sm-2.6Zn–0.8Zr [70]		1–3 μm		105 Hv (cross section) 95 Hv (vertical section)
ZK60-Cu [71]		4.5–13.6		80.5 ± 1.9–105.2 ± 2.9
AZ61 [72]	600 J/mm^3		93.2 ± 2%	90.5 ± 0.9 Hv
AZ61–0.4Mn [72]	600 J/mm^3	11.4 ± 0.55	91.5 ± 1.8%	95.8 ± 1.2 Hv
AZ61–0.8Sn [72]	600 J/mm^3		90.3 ± 2.1%	97.2 ± 1.2 Hv
AZ61–0.4Mn–0.8Sn [72]	600 J/mm^3	4.2 ± 0.42	91.1 ± 1.5%	105 ± 1.4 Hv

4.1.2. Tensile Properties

Previous studies have shown that the elastic modulus of SLMed magnesium is related to powder thickness and energy density. In Ng's study [73], pure Mg prepared with energy density of 7.84×10^3 J/mm^2 can reach an elastic modulus of 32.83–34.97 GPa. which are higher than the conventional cast magnesium (~28 GPa). Wei et al. [47] studied the tensile properties under different volume energy densities. The results showed that, at a relatively low energy density, the ultimate tensile strength of the specimen with an energy density of 83.3 J/mm^3 was 274 MPa. The solid solubility of Mg in the sample was restricted, and the precipitation of the second phase was also relatively reduced, which further affects the tensile properties of SLM formed samples. With an increased energy density of 166.7 J/mm^3, the ultimate tensile strength of the specimen reached 296 MPa, which is superior to the die-cast AZ91D (~230 MPa). In the study of Savalani [55], the powder layer thickness also affected the SLM of magnesium. When the thicknesses of the powder layers were set to 0.15–0.20 mm and 0.25–0.30 mm, the elastic moduli of the samples were 31.88–34.28 GPA and 28.43–31.47 GPA, respectively. In the work of Peng [74], the room temperature tensile properties of typical magnesium alloys in as cast, SLM and extruded states were compared. As shown as Figure 10, the yield strength of SLM magnesium alloy is significantly higher than that of as cast alloy, close to or even higher than that of extruded alloy, which is mainly due to the fine grains in SLM state. The tensile strength of SLM magnesium alloy is mostly significantly higher than that of as cast alloy and lower than that of extruded alloy. The elongation of SLM magnesium alloy is generally low, and it does not show the trend of improvement of the plasticity which may be caused by fine grains. For example, the elongation of AZ91D alloy in SLM state is 1.24–1.83% with the laser energy density of 83.3–167.7 J/mm^3, which is lower than the elongation of die-cast AZ91D (3%) [47]. Similarly, due to the β-Mg$_{17}$Al$_{12}$ phase precipitated along the grain boundaries, the elongation of SLMed AZ61 alloy with the laser energy density of 138.89–208.33 J/mm^3 is 2.14–3.28%, which is lower than the elongation of 5.2% of AZ61 as cast [75].

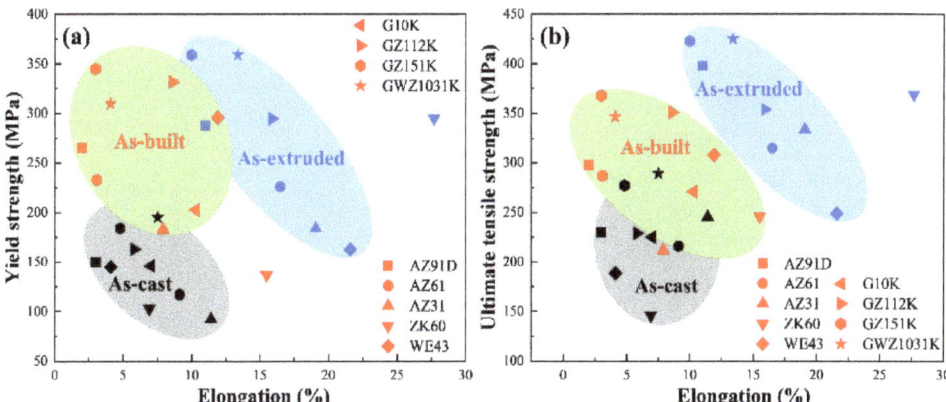

Figure 10. Comparison of room temperature tensile properties of the typical Mg alloys under the as-cast, as-built and as-extruded states (**a**) YS vs. EL, (**b**) UTS vs. EL. Reprinted with permission from ref. [74]. Copyright 2022 Chinese Academy of Sciences.

4.1.3. Fracture Behavior

The fracture behavior analysis of SLM parts shows the characteristics of ductile brittle hybrid fracture. Wei et al. [47] analyzed the fracture morphology of SLMed AZ91D alloy (Figure 11). All tensile specimens showed ductile brittle fracture characteristics. This fracture mode is mainly related to the mixing of soft α-Mg and brittle hard eutectic multiphase structure. It can be seen from Figure 11 that the dimple of the alloy is shallow, and there are defects such as pores and microcracks on the micro fracture diagram, resulting in poor plasticity of SLMed alloy. In addition, due to the layer-by-layer manufacturing method, it is observed that the construction direction of the part during SLM will affect the final tensile properties of the part. Samples deposited along the length direction of the tensile sample usually show higher tensile strength than samples deposited perpendicular to its length. Performing hot isostatic pressing (HIP) procedures after SLM can significantly reduce the anisotropic mechanical behavior of the parts by reducing manufacturing-induced pores [76]. However, the elongation at break of SLM parts is usually low, which may be attributed to micropores and oxide inclusions in the parts, which is the result of non-optimized SLM process. It is worth noting that, even at room temperature, magnesium can be oxidized under extremely low oxygen partial pressure [77,78], and local oxidation caused by residual air in the powder gap can often be observed in SLMed magnesium alloys [79,80]. Oxide films are believed to inhibit the densification mechanism and induce spheroidization, and then destroy the intergranular coalescence/wetting between the molten layers. Oxide inclusions, which usually lead to crack initiation and reduce the mechanical properties of SLM manufactured parts [81,82], need to be avoided in research.

4.1.4. Effect of Heat Treatments on the Mechanical Properties

Although the cooling rate of SLM process is very high, most of the SLM magnesium alloys still have hard brittle eutectic phase on the grain boundary and cannot form single-phase supersaturated solid solution. Therefore, it is necessary to carry out subsequent heat treatment to adjust the microstructure and improve the mechanical properties especially for the tensile properties. Room temperature tensile properties of some magnesium alloys under the state of SLMed and post-treatmented were summarized in Table 2.

Table 2. The tensile properties of the as-built and post-processed Mg alloys fabricated by SLM under optimized process parameters.

Alloys	State	YS/MPa	UTS/MPa	EL/%	Ref
ZK60	As-built	137	246	15.5	[83]
	T4	107	224	16.7	
	T6	191	287	14.1	
WE43	As-built	215	251	2.6	[84]
	T6	219	251	4.3	
G10K	As-built	203	271	10.3	[39]
	T5	285	360	2.9	
GZ112K	As-built	332	351	8.6	[85]
	T4	281	311	14.4	
	T6	343	371	4.0	
GZ151K	As-built	345	368	3.0	[86]
	T5	410	428	3.4	
GWZ1031K	As-built	310	347	4.1	[87]
	T5	365	381	0.8	
	T4	255	328	10.3	
	T6	316	400	2.2	

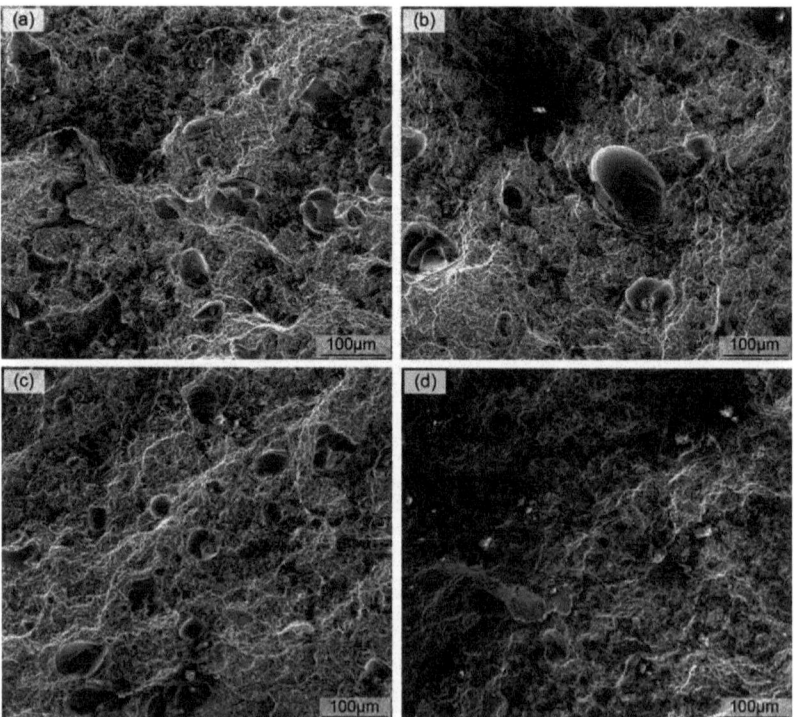

Figure 11. Fracture topography of SLMed AZ91D at different energy densities (a) 166.7 J/mm^3, (b) 142.9 J/mm^3, (c) 111.1 J/mm^3, (d) 83.3 J/mm^3. Reprinted with permission from ref. [47]. Copyright 2022 Copyright 2014 Elsevier.

Hyer et al. [84] conducted a heat treatment of solutionizing at 536 °C for 24 h and subsequent ageing at 205 °C for 48 h on SLM state WE43 alloy. High temperature and long-time solid solution treatment led to significant coarsening of grain. Meanwhile, the

$β_1$-Mg_3Nd phase was found to dissolve, and developed into plate-like precipitates. The combined action of the two causes made the tensile strength remain unchanged at 251 MPa, and the yield strength slightly increase from 215 MPa to 219 MPa, and the elongation to increase from 2.6% to 4.3%. The above T6 heat treatment has not significantly improved the mechanical properties, so it is necessary to develop a special follow-up heat treatment system for the unique rapid solidification non-equilibrium structure in SLM state. In Fu's study of GZ151K [86], after subsequent aging at 200 °C for 64 h (T5 treatment), both the strength and elongation of the SLMed-T5 GZ151K increased. The YS, UTS and elongation of the SLMed-T5 GZ151K alloy are 410 MPa, 428 MPa and 3.4%, respectively, which are higher than those of the as-fabricated (SLMed) GZ151K alloy (YS of 345 MPa, UTS of 368 MPa and elongation of 3.0%) and conventional cast-T6 alloy (YS of 288 MPa, UTS of 405 MPa and elongation of 2.9% [88]). The performance improvement brought by the subsequent aging was attributed to the YS improved by precipitation hardening and the improvement of ductility due to the probably released part of the residual stress. Recently, Liang et al. [83] studied the evolution of microstructure and mechanical properties of SLM ZK60 alloy under different heat treatment systems. As shown in Figure 12, the crystal boundary of SLM ZK60 magnesium alloy is composed of reticulated Mg_7Zn_3 phase, and there are three precipitates (Zn rich phase with Zr rich core, Zn rich phase without Zr and Zr rich particles) in the crystal. After the solution heat treatment (410 °C for 24 h), the Mg_7Zn_3 phase at the grain boundary is dissolved into the matrix, and the precipitated phase in the crystal mainly evolves into the Zn rich phase with Zn_2Zr core. The T4 heat treatment reduces the strength and slightly increases the elongation from 15.5% (SLM state) to 16.7%; Rod-like precipitates after aging heat treatment $β_1'$- MgZn precipitates can significantly improve the mechanical properties. The length and quantity of the rod-shaped precipitates of the two-stage aging (90 °C for 24 h + 180 °C for20 h) are more than that of the single-stage aging (180 °C for 20 h). Compared with the single-stage aging, the strength of the two-stage aging can be improved while maintaining a considerable plasticity. T6 heat treatment can significantly improve the yield strength and slightly reduce the elongation to 14.1%.

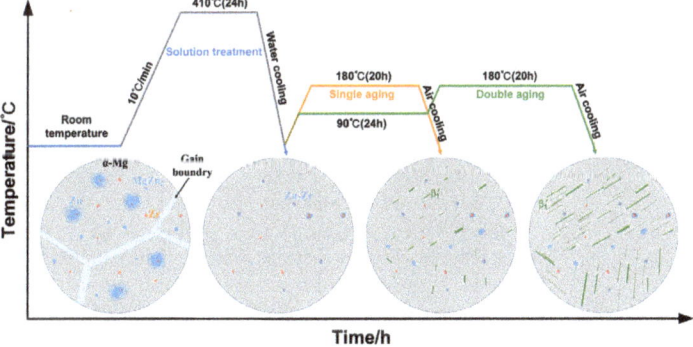

Figure 12. Schematic illustration of the microstructure evolution of LPBF ZK60 Mg alloy during different heat treatments. Reprinted with permission from ref. [83]. Copyright 2022 Elsevier.

4.1.5. Abrasion Resistance

Chang et al. [89] used additive manufacturing equipment to prepare Mg–Zn–Ca alloy, and explored the Mg–Zn–Ca alloy at wear behavior in 3.5 wt.% NaCl solution. The dry friction and wear mechanism of the alloy is adhesive wear and oxidation wear, while the main wear mechanism of wet friction is corrosion and adhesive wear. When the sliding distance is 28 m, the wet friction process first enters the stable wear stage compared with the dry friction process. The average wear width of wet friction is larger than that of dry friction, and the average wear rate of wet friction is more serious than that of dry friction.

4.2. Corrosion Resistance Properties

Generally, poor corrosion resistance of magnesium alloys caused by high chemical activity of magnesium alloys and the lack of protective passivation oxide film was considered as a major obstacle to its further application [90,91]. In addition, with a high negative standard electrode potential, magnesium and its alloys always showed a rapid corrosion feature under similar body fluid conditions [92]. The hydrogen produced by corrosion cannot be treated by the host tissue, which have prevented magnesium-based materials from being applied in practice so far. Therefore, the corrosion rate of magnesium and its alloys should be carefully controlled. The corrosion properties of magnesium alloys are closely related to their phase and alloy elements. Biodegradable Mg alloys, such as Mg–Ca AZ31, AZ91, and WE43 have been reported with improved corrosion resistance [72]. The corrosion rates of different Mg alloy fabricated by SLM technology are summarized in Table 3.

Table 3. Summary of corrosion rate of different Mg alloys.

Materials System	Tests Solution	Corrosion Rate (mm Year^{-1})	Hydrogen Evolution Rate (ml cm^{-2} h^{-1})
AZ31 [66]	0.9% NaCl solution	0.312	
	3.0% NaCl solution	1.071	
AZ61(80 W) [65]	Simulated body fluid	2.4 (after immersion for 24 h) 1.2 (after immersion for 144 h)	
WE43	Hanks' Balanced Salt Solution	7.04 (as-build) 2.11 (grinded batches-SiC4000)	
WE43 [40]	0.1 M NaCl solution	5–7.2	
ZK30 [69]		3.70 ± 0.10	
ZK30–0.6GO [69]	Simulated body fluid	3.38 ± 0.07	
ZK60 [71]	Simulated body fluid	1.01	
ZK60 [59]	Hank's solution		0.006
Mg–2Mn [93]	Simulated body fluid		0.017
Mg1Zn [94]	C-simulated body fluid		0.17

Shuai et al. [71] found that the adding of zinc can alleviate the degradation rate of Mg–Sn alloy and improve the corrosion resistance by formation of $Zn(OH)_2$ protective layer during the degradation process. However, with the further increase of Zn content, the content of Mg_7Zn_3 in the second phase increases, and galvanic corrosion occurs between the second phase and the matrix, which will accelerate the degradation of magnesium alloy. The research of Zhou et al. [95] shows that the addition of a certain amount of Sn can alleviate the corrosion of magnesium alloys. With the increase of Sn element, the second phase Mg_2Sn will also increase and accelerate the degradation rate due to its galvanic corrosion. Therefore, adding a certain amount of Sn element will reduce the degradation rate of magnesium alloy. Similarly, ZK60–0.4Cu [71] alloy showed certain anti-degradation ability and strong antibacterial ability caused by alkaline environment and copper ions. He et al. [65] formed AZ61 alloy by SLM technology. The reduced the formation of β-$Mg_{17}Al_{12}$ phase can reduced the galvanic corrosion between α-Mg and β-$Mg_{17}Al_{12}$ phase and alleviated the corrosion of magnesium alloy to a certain extent. In Yin's study of ZK30 [96], the corrosion degradation rate of the composites decreased by adding bioactive amorphous bioactive glass. Furthermore, the evenly distributed Mg_7Zn_3 phase was refined and acts as a barrier to slow down the corrosion rate. By adding 2 wt% Mn during the SLM processing of Mg–Mn alloy, Yang et al. [93] observed that the corrosion resistance of pure magnesium can be improved. It is attributed to the increase of corrosion potential and grain refinement caused by manganese solid solution.

Besides the factors mentioned above, grain refinement and uniform structure can also improve the corrosion resistance of the material [59]. In the study of AZ61 alloy [65], the formation of uniform and fine equiaxed grains improves its corrosion resistance. Due to the existence of more grain boundary areas in fine grains, the corroded α-Mg grain boundary can act as a certain corrosion barrier. In addition, fine grains can form a denser surface oxide film to prevent the reaction between Cl^- and Mg^{2+} ions in body fluid. In the study of the corrosion resistance of Mg–Nb–Zn–Zr (JDBM) alloy, Zhang et al. [97] used reciprocating extrusion technology to process JDBM alloy and found that its grain size was reduced to 1 μm, and its corrosion resistance has also been improved.

5. Potential Applications

Magnesium alloys have elastic modulus similar to human bones (2–30 GPA) [98], and Mg (Mg^{2+}) is an essential element and regulatory ion in human body. Therefore, magnesium alloys have great application potential in biomedical materials [99,100]. Selective laser melting has a significant advantage in manufacturing medical porous metal structures, which can produce fine and porous structures while adapting to various shapes. This makes it the preferred technology for the production of metal stents and implants [101,102]. Many engineering materials have been used to produce complex porous/cellular structures through SLM technology. So, it is possible to produce porous structures with magnesium alloys. Li et al. [103] produced biodegradable porous magnesium, which is likely to meet all the functional requirements of an ideal bone substitute material. First of all, its mechanical performance is high enough to be used as a mechanical support. Secondly, the manufactured parts showed a fully interconnected porous structure, which can accurately control the topology. Finally, the biodegradation rate of it was slow, and the volume loss is about 20% after 4 weeks. However, the corrosion rate of magnesium alloy is relatively fast, accompanied by the concentrated release of a large amount of hydrogen, which makes the human body unable to diffuse and absorb it in time, which will form bubbles on the surface of human skin, affecting the physiological function of tissues around the implant and the restoration treatment of the implant site. Yuan's team developed a patented biomedical JDBM material, which has good biocompatibility, good strength and toughness, and low corrosion rate. The process parameters for manufacturing magnesium alloy biomaterials by SLM are currently being developed and are hoped to be applied in the near future [60,104–107].

In addition to the medical field, magnesium alloys are also widely used in automotive industries. By replacing vehicle components with magnesium-based materials, the vehicle weight can be reduced by 20–70%. Volkswagen began to formally apply magnesium alloy to the automotive industry in 1970. Nowadays, benefitting from the reduced fuel use and carbon dioxide emissions by weight reduction, automobile companies are going to use more magnesium alloys and composites in their products [108]. In the past two decades, the use of magnesium and its alloys in the automotive industry has generally shown an upward trend. Although the cost of parts made by Mg in cars is higher than that of aluminum, it is worthwhile to compensate Mg for its contribution to reducing fuel consumption and carbon dioxide emissions [109]. However, due to its poor ductility and easy corrosion property, magnesium and its alloys cannot fully replace aluminum and its alloys in the application of automotive parts. However, magnesium also has the characteristics of various automotive applications, such as better damping performance than aluminum. Additive manufacturing processes such as SLM are widely used in the lightweight process of automotive parts and the integral molding of complex parts such as support brackets for clutch and brakes, housing for transmission [110]. When the preparation process and other problems are solved, SLM magnesium alloy will also make a difference in the field of automotive weight reduction.

6. Opportunities and Challenges

Additive manufacturing technology shows promising potential for development of the future, and it will change the production mode and production site of the manufacturing industry. Selective Laser Melting (SLM) is one of the most attractive metal additive manufacturing (MAM) technologies that allows high precision metal products to be manufactured directly without the need for any molds. Thanks to its high energy input ($10–10^3$ J/mm^3)

and ultra-high cooling rate (10^4–10^6 K/s), tailored microstructures and tunable complex structures can be obtained from microscale to macroscopic scale combined. In addition, the advantages of reusable recyclable powders, high manufacturing accuracy, and little or no post-processing also expand the possibilities for SLM to manufacture magnesium alloy products.

However, due to the low boiling point (~1091 °C) and the good oxygen affinity of magnesium, there are still many challenges to achieve large-scale applications of SLM in magnesium alloys.

The first is the high manufacturing cost. SLM technology makes magnesium alloy raw materials from spherical metal powders with high purity, narrow particle size distribution and low oxygen content. Due to its easy oxidation and spontaneous combustion problems, magnesium alloy powders are difficult to prepare which leads to additional processing and costs, so the raw material cost of SLM process to manufacture magnesium alloy is much higher than that of traditional technology. At the same time, the current production speed of the SLM process is too slow, resulting in a high depreciation rate of equipment, which further increases the manufacturing cost of SLM. At present, magnesium alloys prepared by SLM are only suitable for high value-added industries such as aerospace, high-end automobiles and biomedicine.

Second, the size of magnesium alloy products manufactured by the SLM process is limited. Due to the limitations of equipment and interlayer resolution, the size of the products manufactured by the current SLM process is usually less than 1 metre, so it is not suitable for SLM technology to prepare large parts.

The third is the structural defects of the magnesium alloy manufactured by SLM. One is that due to the large coefficient of thermal expansion of magnesium alloys and the high cooling rate of the SLM manufacturing process, the manufactured products are prone to defects such as thermal cracks. The other one is that the products produced by additive manufacturing usually leads to anisotropy of the products due to the defect of inter-layer bonding.

7. Summary

This paper summarizes the research progress in the field of laser selective melting of magnesium alloys, mainly introduces the influence of magnesium alloy powder and process parameters on laser selective melting (SLM), then discusses the influence on the microstructure, mechanical properties and corrosion resistance of magnesium alloys, and finally prospects the potential application fields of SLM magnesium alloys. Summarized as follows:

(1) High quality powder is the key to improve the manufacturing of magnesium alloy additive. Raw materials are one of the other factors that affect the synthetic properties of SLMed magnesium. Reducing the distribution range of powder particle size, improving the quality of powder, and combining magnesium with other alloy elements can improve product performance. At present, there is no validated commercial magnesium alloy powder material for SLM, the development of new generation high performance magnesium alloy powder for SLM is the key to realize application of the SLMed Magnesium Alloys.

(2) SLM process parameters involve laser power, scanning speed, overlapping distance, layer thickness, scanning angle and others. Adjusting printing parameters can improve tensile strength, hardness and finer microstructure. Due to the low boiling point and the good oxygen affinity of magnesium, violent evaporation splash, large amounts of smoke and dust and recoil pressure on the molten pool will affect the stable melting of the molten pool. During the forming process, appropriate laser energy input and effective gas circulation system can suppress the negative impact of evaporation splashing, thus achieving stable forming quality. It is of great significance to control the defects and increase the window of high-density forming process by

adjusting the process parameters. Therefore, more research needs to be done to obtain the optimal printing parameters of various magnesium alloys.

(3) Compared with other materials, Mg has some advantages such as weight, which makes it a potential material to reduce the weight of components in aerospace and automotive industries. In addition, the superior performance of magnesium as a biocompatible and biodegradable material, especially through solid-state laser processing, has attracted more attention to the use of solid-state laser processing of magnesium in the pharmaceutical industry. However, due to the poor corrosion resistance of magnesium alloys, their applications are limited. How to improve the degradation rate of magnesium and its magnesium alloys is the key problem in the application of magnesium alloys.

Author Contributions: Conceptualization, Y.D. and Y.L.; methodology, Y.L.; validation, Y.Z. and Y.L.; investigation, K.Z.; resources, K.Z.; data curation, Y.Z.; writing—original draft preparation, Y.Z.; writing—review and editing, Y.L.; supervision, Y.D.; project administration, J.C.; funding acquisition, Y.D. and J.C. All authors have read and agreed to the published version of the manuscript.

Funding: This work was supported by Scientific research project of Hunan Education Department (20C1796), National Natural Science Foundation of China (51901193); Natural Science Foundation of Changsha, China (kq2202306); State Key Laboratory of Powder Metallurgy, Central South University, Changsha, China.

Institutional Review Board Statement: Not applicable.

Informed Consent Statement: Not applicable.

Data Availability Statement: Not applicable.

Conflicts of Interest: The authors declare no conflict of interest.

References

1. Zhang, C.; Lin, J.; Liu, H. Magnesium-based Biodegradable Materials for Biomedical Applications. *MRS Adv.* **2018**, *3*, 2359–2364. [CrossRef]
2. Zhang, W.-n.; Wang, L.-z.; Feng, Z.-x.; Chen, Y.-m. Research progress on selective laser melting (SLM) of magnesium alloys: A review. *Optik* **2020**, *207*, 163842. [CrossRef]
3. Ke, W.C.; Oliveira, J.P.; Ao, S.S.; Teshome, F.B.; Chen, L.; Peng, B.; Zeng, Z. Thermal process and material flow during dissimilar double-sided friction stir spot welding of AZ31/ZK60 magnesium alloys. *J. Mater. Res. Technol.* **2022**, *17*, 1942–1954. [CrossRef]
4. Song, J.; She, J.; Chen, D.; Pan, F. Latest research advances on magnesium and magnesium alloys worldwide. *J. Magnes. Alloy.* **2020**, *8*, 1–41. [CrossRef]
5. Yan, X.; Yin, S.; Chen, C.; Huang, C.; Bolot, R.; Lupoi, R.; Kuang, M.; Ma, W; Coddet, C.; Liao, H. Effect of heat treatment on the phase transformation and mechanical properties of Ti6Al4V fabricated by selective laser melting. *J. Alloys Compd.* **2018**, *764*, 1056–1071. [CrossRef]
6. Chang, C.; Yan, X.; Bolot, R.; Gardan, J.; Gao, S.; Liu, M.; Liao, H.; Chemkhi, M.; Deng, S. Influence of post-heat treatments on the mechanical properties of CX stainless steel fabricated by selective laser melting. *J. Mater. Sci.* **2020**, *55*, 8303–8316. [CrossRef]
7. Chen, C.; Xie, Y.; Yan, X.; Yin, S.; Fukanuma, H.; Huang, R.; Zhao, R.; Wang, J.; Ren, Z.; Liu, M.; et al. Effect of hot isostatic pressing (HIP) on microstructure and mechanical properties of Ti6Al4V alloy fabricated by cold spray additive manufacturing. *Addit. Manuf.* **2019**, *27*, 595–605. [CrossRef]
8. Yan, X.; Chen, C.; Zhao, R.; Ma, W.; Bolot, R.; Wang, J.; Ren, Z.; Liao, H.; Liu, M. Selective laser melting of WC reinforced maraging steel 300: Microstructure characterization and tribological performance. *Surf. Coat. Technol.* **2019**, *371*, 355–365. [CrossRef]
9. Olakanmi, E.O.; Cochrane, R.F.; Dalgarno, K.W. A review on selective laser sintering/melting (SLS/SLM) of aluminium alloy powders: Processing, microstructure, and properties. *Prog. Mater. Sci.* **2015**, *74*, 401–477. [CrossRef]
10. Yap, C.Y.; Chua, C.K.; Dong, Z.L.; Liu, Z.H.; Zhang, D.Q.; Loh, L.E.; Sing, S.L. Review of selective laser melting: Materials and applications. *Appl. Phys. Rev.* **2015**, *2*, 041101. [CrossRef]
11. Wang, X.; Gong, X.; Chou, K. Review on powder-bed laser additive manufacturing of Inconel 718 parts. *Proc. Inst. Mech. Eng. Part B J. Eng. Manuf.* **2016**, *231*, 1890–1903. [CrossRef]
12. Thijs, L.; Verhaeghe, F.; Craeghs, T.; Humbeeck, J.V.; Kruth, J.-P. A study of the microstructural evolution during selective laser melting of Ti–6Al–4V. *Acta Mater.* **2010**, *58*, 3303–3312. [CrossRef]
13. Zhang, L.-C.; Attar, H. Selective Laser Melting of Titanium Alloys and Titanium Matrix Composites for Biomedical Applications: A Review. *Adv. Eng. Mater.* **2016**, *18*, 463–475. [CrossRef]
14. Lu, Z.L.; Cao, J.W.; Jing, H.; Liu, T.; Lu, F.; Wang, D.X.; Li, D.C. Review of main manufacturing processes of complex hollow turbine blades. *Virtual Phys. Prototyp.* **2013**, *8*, 87–95. [CrossRef]

15. Shen, J.; Zeng, Z.; Nematollahi, M.; Schell, N.; Maawad, E.; Vasin, R.N.; Safaei, K.; Poorganji, B.; Elahinia, M.; Oliveira, J.P. In-situ synchrotron X-ray diffraction analysis of the elastic behaviour of martensite and H-phase in a NiTiHf high temperature shape memory alloy fabricated by laser powder bed fusion. *Addit. Manuf. Lett.* **2021**, *1*, 100003. [CrossRef]
16. Sercombe, T.B.; Li, X. Selective laser melting of aluminium and aluminium metal matrix composites. *Mater. Technol.* **2016**, *31*, 77–85. [CrossRef]
17. Oliveira, J.P.; LaLonde, A.D.; Ma, J. Processing parameters in laser powder bed fusion metal additive manufacturing. *Mater. Des.* **2020**, *193*, 108762. [CrossRef]
18. Zhan, X.; Chen, J.; Liu, J.; Wei, Y.; Zhou, J.; Meng, Y. Microstructure and magnesium burning loss behavior of AA6061 electron beam welding joints. *Mater. Des.* **2016**, *99*, 449–458. [CrossRef]
19. Liao, H.; Zhu, H.; Xue, G.; Zeng, X. Alumina loss mechanism of Al2O3-AlSi10 Mg composites during selective laser melting. *J. Alloy. Compd.* **2019**, *785*, 286–295. [CrossRef]
20. Czerwinski, F. Controlling the ignition and flammability of magnesium for aerospace applications. *Corros. Sci.* **2014**, *86*, 1–16. [CrossRef]
21. Tekumalla, S.; Gupta, M. An insight into ignition factors and mechanisms of magnesium based materials: A review. *Mater. Des.* **2017**, *113*, 84–98. [CrossRef]
22. Lee, S.-J.; Do, L.H.T. Effects of copper additive on micro-arc oxidation coating of LZ91 magnesium-lithium alloy. *Surf. Coat. Technol.* **2016**, *307*, 781–789. [CrossRef]
23. Fan, J.; Yang, C.; Xu, B. Effect of Ca and Y additions on oxidation behavior of magnesium alloys at high temperatures. *J. Rare Earths* **2012**, *30*, 497–502. [CrossRef]
24. Tan, Q.; Mo, N.; Lin, C.-L.; Zhao, Y.; Yin, Y.; Jiang, B.; Pan, F.; Atrens, A.; Huang, H.; Zhang, M.-X. Generalisation of the oxide reinforcement model for the high oxidation resistance of some Mg alloys micro-alloyed with Be. *Corros. Sci.* **2019**, *147*, 357–371. [CrossRef]
25. Tan, Q.; Mo, N.; Lin, C.-L.; Jiang, B.; Pan, F.; Huang, H.; Atrens, A.; Zhang, M.-X. Improved oxidation resistance of Mg-9Al-1Zn alloy microalloyed with 60 wt ppm Be attributed to the formation of a more protective (Mg, Be)O surface oxide. *Corros. Sci.* **2018**, *132*, 272–283. [CrossRef]
26. Gunduz, K.O.; Oter, Z.C.; Tarakci, M.; Gencer, Y. Plasma electrolytic oxidation of binary Mg-Al and Mg-Zn alloys. *Surf. Coat. Technol.* **2017**, *323*, 72–81. [CrossRef]
27. Tan, Q.; Mo, N.; Jiang, B.; Pan, F.; Atrens, A.; Zhang, M.-X. Oxidation resistance of Mg–9Al–1Zn alloys micro-alloyed with Be. *Scr. Mater.* **2016**, *115*, 38–41. [CrossRef]
28. Leleu, S.; Rives, B.; Bour, J.; Causse, N.; Pébère, N. On the stability of the oxides film formed on a magnesium alloy containing rare-earth elements. *Electrochim. Acta* **2018**, *290*, 586–594. [CrossRef]
29. Fan, J.F.; Yang, C.L.; Han, G.; Fang, S.; Yang, W.D.; Xu, B.S. Oxidation behavior of ignition-proof magnesium alloys with rare earth addition. *J. Alloys Compd.* **2011**, *509*, 2137–2142. [CrossRef]
30. Ahmadi, M.; Tabary, S.A.A.B.; Rahmatabadi, D.; Ebrahimi, M.S.; Abrinia, K.; Hashemi, R. Review of selective laser melting of magnesium alloys: Advantages, microstructure and mechanical characterizations, defects, challenges, and applications. *J. Mater. Res. Technol.* **2022**, *19*, 1537–1562. [CrossRef]
31. Spierings, A.B.; Voegtlin, M.; Bauer, T.; Wegener, K. Powder flowability characterisation methodology for powder-bed-based metal additive manufacturing. *Prog. Addit. Manuf.* **2015**, *1*, 9–20. [CrossRef]
32. Yuan, J.; Liu, J.; Zhou, Y.; Wang, J.; Xv, T. Aluminum agglomeration of AP/HTPB composite propellant. *Acta Astronaut.* **2019**, *156*, 14–22. [CrossRef]
33. Attar, H.; Prashanth, K.G.; Zhang, L.-C.; Calin, M.; Okulov, I.V.; Scudino, S.; Yang, C.; Eckert, J. Effect of Powder Particle Shape on the Properties of In Situ Ti–TiB Composite Materials Produced by Selective Laser Melting. *J. Mater. Sci. Technol.* **2015**, *31*, 1001–1005. [CrossRef]
34. Liu, J.; Wen, P. Metal vaporization and its influence during laser powder bed fusion process. *Mater. Des.* **2022**, *215*, 110505. [CrossRef]
35. Jinye, W.; Zhipeng, C.; Yanfang, Y.; Hongjie, C.; Guang, Y. Effect of partilce size distribution of AZ91D magneisum alloy powder on selective laser melting process. *Hebei J. Ind. Sci. Tech.* **2022**, *39*, 7.
36. Hu, D.; Wang, Y.; Zhang, D.; Hao, L.; Jiang, J.; Li, Z.; Chen, Y. Experimental Investigation on Selective Laser Melting of Bulk Net-Shape Pure Magnesium. *Mater. Manuf. Processes* **2015**, *30*, 1298–1304. [CrossRef]
37. Olatunde Olakanmi, E.; Dalgarno, K.W.; Cochrane, R.F. Laser sintering of blended Al-Si powders. *Rapid Prototyp. J.* **2012**, *18*, 109–119. [CrossRef]
38. Manakari, V.; Parande, G.; Gupta, M. Selective Laser Melting of Magnesium and Magnesium Alloy Powders: A Review. *Metals* **2017**, *7*, 2. [CrossRef]
39. Deng, Q.; Wang, X.; Lan, Q.; Su, N.; Wu, Y.; Peng, L. Limitations of Linear Energy Density for Laser Powder Bed Fusion of Mg-15Gd-1Zn-0.4Zr Alloy. *Mater. Charact.* **2022**, *190*, 112071. [CrossRef]
40. Esmaily, M.; Zeng, Z.; Mortazavi, A.N.; Gullino, A.; Choudhary, S.; Derra, T.; Benn, F.; D'Elia, F.; Müther, M.; Thomas, S.; et al. A detailed microstructural and corrosion analysis of magnesium alloy WE43 manufactured by selective laser melting. *Addit. Manuf.* **2020**, *35*, 101321. [CrossRef]

41. Olakanmi, E.O.; Cochrane, R.F.; Dalgarno, K.W. Densification mechanism and microstructural evolution in selective laser sintering of Al–12Si powders. *J. Mater. Process. Technol.* **2011**, *211*, 113–121. [CrossRef]
42. Zhou, X.; Liu, X.; Zhang, D.; Shen, Z.; Liu, W. Balling phenomena in selective laser melted tungsten. *J. Mater. Process. Technol.* **2015**, *222*, 33–42. [CrossRef]
43. Deng, Q.; Wu, Y.; Luo, Y.; Su, N.; Xue, X.; Chang, Z.; Wu, Q.; Xue, Y.; Peng, L. Fabrication of high-strength Mg-Gd-Zn-Zr alloy via selective laser melting. *Mater. Charact.* **2020**, *165*, 110377. [CrossRef]
44. Ross Cunningham, C.Z.; Niranjan, P.; Christopher, K.; Joseph, P.; Kamel, F.; Tao, S.; Anthony, D.R. Keyhole threshold and morphology in laser melting revealed by ultrahigh-speed X-ray imaging. *Science* **2019**, *363*, s849–s852. [CrossRef]
45. Wang, W.; Wang, D.; He, L.; Liu, W.; Yang, X. Thermal behavior and densification during selective laser melting of Mg-Y-Sm-Zn-Zr alloy: Simulation and experiments. *Mater. Res. Express* **2020**, *7*, 116519. [CrossRef]
46. Zhang, B.; Liao, H.; Coddet, C. Effects of processing parameters on properties of selective laser melting Mg–9%Al powder mixture. *Mater. Des.* **2012**, *34*, 753–758. [CrossRef]
47. Wei, K.; Gao, M.; Wang, Z.; Zeng, X. Effect of energy input on formability, microstructure and mechanical properties of selective laser melted AZ91D magnesium alloy. *Mater. Sci. Eng. A* **2014**, *611*, 212–222. [CrossRef]
48. Yang, Y.; Wu, P.; Lin, X.; Liu, Y.; Bian, H.; Zhou, Y.; Gao, C.; Shuai, C. System development, formability quality and microstructure evolution of selective laser-melted magnesium. *Virtual Phys. Prototyp.* **2016**, *11*, 173–181. [CrossRef]
49. Ng, C.C.; Savalani, M.M.; Lau, M.L.; Man, H.C. Microstructure and mechanical properties of selective laser melted magnesium. *Appl. Surf. Sci.* **2011**, *257*, 7447–7454. [CrossRef]
50. Krishna, B.V.; B, S.; Bandyopadhyay, A. Low stiffness porous Ti structures for load-bearing implants. *Acta Biomater.* **2007**, *3*, 997–1006. [CrossRef]
51. Wen, P.; Voshage, M.; Jauer, L.; Chen, Y.; Qin, Y.; Poprawe, R.; Schleifenbaum, J.H. Laser additive manufacturing of Zn metal parts for biodegradable applications: Processing, formation quality and mechanical properties. *Mater. Des.* **2018**, *155*, 36–45. [CrossRef]
52. Zheng, Y.; Xia, D.; Shen, Y.; Liu, Y.; Xu, Y.; Wen, P.; Tian, Y.; Lai, Y. Additively Manufactured Biodegradable Metal Implants. *Acta Metall. Sin.* **2021**, *57*, 1499–1520. [CrossRef]
53. Ng, C.C.; Savalani, M.M.; Man, H.C.; Gibson, I. Layer manufacturing of magnesium and its alloy structures for future applications. *Virtual Phys. Prototyp.* **2010**, *5*, 13–19. [CrossRef]
54. Wei, K.; Wang, Z.; Zeng, X. Influence of element vaporization on formability, composition, microstructure, and mechanical performance of the selective laser melted Mg–Zn–Zr components. *Mater. Lett.* **2015**, *156*, 187–190. [CrossRef]
55. Savalani, M.M.; Pizarro, J.M. Effect of preheat and layer thickness on selective laser melting (SLM) of magnesium. *Rapid Prototyp. J.* **2016**, *22*, 115–122. [CrossRef]
56. Shi, Q.; Gu, D.; Xia, M.; Cao, S.; Rong, T. Effects of laser processing parameters on thermal behavior and melting/solidification mechanism during selective laser melting of TiC/Inconel 718 composites. *Opt. Laser Technol.* **2016**, *84*, 9–22. [CrossRef]
57. Peel, M.; Steuwer, A.; Preuss, M.; Withers, P.J. Microstructure, mechanical properties and residual stresses as a function of welding speed in aluminium AA5083 friction stir welds. *Acta Mater.* **2003**, *51*, 4791–4801. [CrossRef]
58. Zumdick, N.A.; Jauer, L.; Kersting, L.C.; Kutz, T.N.; Schleifenbaum, J.H.; Zander, D. Additive manufactured WE43 magnesium: A comparative study of the microstructure and mechanical properties with those of powder extruded and as-cast WE43. *Mater. Charact.* **2019**, *147*, 384–397. [CrossRef]
59. Shuai, C.; Yang, Y.; Wu, P.; Lin, X.; Liu, Y.; Zhou, Y.; Feng, P.; Liu, X.; Peng, S. Laser rapid solidification improves corrosion behavior of Mg-Zn-Zr alloy. *J. Alloys Compd.* **2017**, *691*, 961–969. [CrossRef]
60. Zhang, X.; Yuan, G.; Mao, L.; Niu, J.; Fu, P.; Ding, W. Effects of extrusion and heat treatment on the mechanical properties and biocorrosion behaviors of a Mg-Nd-Zn-Zr alloy. *J. Mech. Behav. Biomed. Mater.* **2012**, *7*, 77–86. [CrossRef]
61. Wen, H.; Topping, T.D.; Isheim, D.; Seidman, D.N.; Lavernia, E.J. Strengthening mechanisms in a high-strength bulk nanostructured Cu–Zn–Al alloy processed via cryomilling and spark plasma sintering. *Acta Mater.* **2013**, *61*, 2769–2782. [CrossRef]
62. Cáceres, C.H.; Rovera, D.M. Solid solution strengthening in concentrated Mg–Al alloys. *J. Light Met.* **2001**, *1*, 151–156. [CrossRef]
63. Mercelis, P.; Kruth, J.P. Residual stresses in selective laser sintering and selective laser melting. *Rapid Prototyp. J.* **2006**, *12*, 254–265. [CrossRef]
64. Liu, C.; Zhang, M.; Chen, C. Effect of laser processing parameters on porosity, microstructure and mechanical properties of porous Mg-Ca alloys produced by laser additive manufacturing. *Mater. Sci. Eng. A* **2017**, *703*, 359–371. [CrossRef]
65. He, C.; Bin, S.; Wu, P.; Gao, C.; Feng, P.; Yang, Y.; Liu, L.; Zhou, Y.; Zhao, M.; Yang, S.; et al. Microstructure Evolution and Biodegradation Behavior of Laser Rapid Solidified Mg–Al–Zn Alloy. *Metals* **2017**, *7*, 105. [CrossRef]
66. Pawlak, A.; Szymczyk, P.E.; Kurzynowski, T.; Chlebus, E. Selective laser melting of magnesium AZ31B alloy powder. *Rapid Prototyp. J.* **2019**, *26*, 249–258. [CrossRef]
67. Wang, X.; Chen, C.; Zhang, M. Effect of laser power on formability, microstructure and mechanical properties of selective laser melted Mg-Al-Zn alloy. *Rapid Prototyp. J.* **2020**, *26*, 841–854. [CrossRef]
68. Liu, S.; Guo, H. Influence of hot isostatic pressing (HIP) on mechanical properties of magnesium alloy produced by selective laser melting (SLM). *Mater. Lett.* **2020**, *265*, 127463. [CrossRef]
69. Tao, J.-X.; Zhao, M.-C.; Zhao, Y.-C.; Yin, D.-F.; Liu, L.; Gao, C.; Shuai, C.; Atrens, A. Influence of graphene oxide (GO) on microstructure and biodegradation of ZK30-xGO composites prepared by selective laser melting. *J. Magnes. Alloy.* **2020**, *8*, 952–962. [CrossRef]

70. Wang, W.; He, L.; Yang, X.; Wang, D. Microstructure and microhardness mechanism of selective laser melting Mg-Y-Sm-Zn-Zr alloy. *J. Alloy. Compd.* **2021**, *868*, 159107. [CrossRef]
71. Shuai, C.; Liu, L.; Zhao, M.; Feng, P.; Yang, Y.; Guo, W.; Gao, C.; Yuan, F. Microstructure, biodegradation, antibacterial and mechanical properties of ZK60-Cu alloys prepared by selective laser melting technique. *J. Mater. Sci. Technol.* **2018**, *34*, 1944–1952. [CrossRef]
72. Gao, C.; Li, S.; Liu, L.; Bin, S.; Yang, Y.; Peng, S.; Shuai, C. Dual alloying improves the corrosion resistance of biodegradable Mg alloys prepared by selective laser melting. *J. Magnes. Alloy.* **2021**, *9*, 305–316. [CrossRef]
73. Ng, C.C.; Savalani, M.; Chung Man, H. Fabrication of magnesium using selective laser melting technique. *Rapid Prototyp. J.* **2011**, *17*, 479–490. [CrossRef]
74. Peng, L.; Deng, Q.; Wu, Y.; Fu, P.; Liu, Z.; Wu, Q.; Chen, K.; Ding, W. Additive Manufacturing of Magnesium Alloys by Selective Laser Melting Technology: A Review. *Acta Metall. Sin.* **2022**, 00166. [CrossRef]
75. Liu, S.; Yang, W.; Shi, X.; Li, B.; Duan, S.; Guo, H.; Guo, J. Influence of laser process parameters on the densification, microstructure, and mechanical properties of a selective laser melted AZ61 magnesium alloy. *J. Alloys Compd.* **2019**, *808*, 151160. [CrossRef]
76. Shamsaei, N.; Yadollahi, A.; Bian, L.; Thompson, S.M. An overview of Direct Laser Deposition for additive manufacturing; Part II: Mechanical behavior, process parameter optimization and control. *Addit. Manuf.* **2015**, *8*, 12–35. [CrossRef]
77. Kurth, M.; Graat, P.C.J.; Mittemeijer, E.J. The oxidation kinetics of magnesium at low temperatures and low oxygen partial pressures. *Thin Solid Film.* **2006**, *500*, 61–69. [CrossRef]
78. Jeurgens, L.P.H.; Vinodh, M.S.; Mittemeijer, E.J. Initial oxide-film growth on Mg-based MgAl alloys at room temperature. *Acta Mater.* **2008**, *56*, 4621–4634. [CrossRef]
79. Xia, M.; Gu, D.; Yu, G.; Dai, D.; Chen, H.; Shi, Q. Porosity evolution and its thermodynamic mechanism of randomly packed powder-bed during selective laser melting of Inconel 718 alloy. *Int. J. Mach. Tools Manuf.* **2017**, *116*, 96–106. [CrossRef]
80. King, W.E.; Barth, H.D.; Castillo, V.M.; Gallegos, G.F.; Gibbs, J.W.; Hahn, D.E.; Kamath, C.; Rubenchik, A.M. Observation of keyhole-mode laser melting in laser powder-bed fusion additive manufacturing. *J. Mater. Process. Technol.* **2014**, *214*, 2915–2925. [CrossRef]
81. Savalani, M.M.; Ng, C.C.; Man, H.C. Selective Laser Melting of Magnesium for Future Applications in Medicine. In Proceedings of the 2010 International Conference on Manufacturing Automation, Hong Kong, China, 13–15 December 2010; pp. 50–54.
82. Liu, Y.; Yang, Y.; Wang, D. A study on the residual stress during selective laser melting (SLM) of metallic powder. *Int. J. Adv. Manuf. Technol.* **2016**, *87*, 647–656. [CrossRef]
83. Liang, J.; Lei, Z.; Chen, Y.; Fu, W.; Wu, S.; Chen, X.; Yang, Y. Microstructure evolution of laser powder bed fusion ZK60 Mg alloy after different heat treatment. *J. Alloys Compd.* **2022**, *898*, 163046. [CrossRef]
84. Hyer, H.; Zhou, L.; Benson, G.; McWilliams, B.; Cho, K.; Sohn, Y. Additive manufacturing of dense WE43 Mg alloy by laser powder bed fusion. *Addit. Manuf.* **2020**, *33*, 101123. [CrossRef]
85. Deng, Q.; Wu, Y.; Zhu, W.; Chen, K.; Liu, D.; Peng, L.; Ding, W. Effect of heat treatment on microstructure evolution and mechanical properties of selective laser melted Mg-11Gd-2Zn-0.4Zr alloy. *Mater. Sci. Eng. A* **2022**, *829*, 142139. [CrossRef]
86. Fu, P.-h.; Wang, N.-q.; Liao, H.-g.; Xu, W.-y.; Peng, L.-m.; Chen, J.; Hu, G.-q.; Ding, W.-j. Microstructure and mechanical properties of high strength Mg−15Gd−1Zn−0.4Zr alloy additive-manufactured by selective laser melting process. *Trans. Nonferrous Met. Soc. China* **2021**, *31*, 1969–1978. [CrossRef]
87. Deng, Q.; Wu, Y.; Wu, Q.; Xue, Y.; Zhang, Y.; Peng, L.; Ding, W. Microstructure evolution and mechanical properties of a high-strength Mg-10Gd-3Y-1Zn-0.4Zr alloy fabricated by laser powder bed fusion. *Addit. Manuf.* **2022**, *49*, 102517. [CrossRef]
88. Rong, W.; Wu, Y.; Zhang, Y.; Sun, M.; Chen, J.; Peng, L.; Ding, W. Characterization and strengthening effects of γ′ precipitates in a high-strength casting Mg-15Gd-1Zn-0.4Zr (wt.%) alloy. *Mater. Charact.* **2017**, *126*, 1–9. [CrossRef]
89. Chang, C.; Yue, S.; Li, W.; Lu, L.; Yan, X. Study on microstructure and tribological behavior of the selective laser melted MgZnCa alloy. *Mater. Lett.* **2022**, *309*, 131439. [CrossRef]
90. Atrens, A.; Song, G.L.; Cao, F.; Shi, Z.; Bowen, P.K. Advances in Mg corrosion and research suggestions. *J. Magnes. Alloy.* **2013**, *1*, 24. [CrossRef]
91. Song, G.; Atrens, A. Understanding Magnesium Corrosion—A Framework for Improved Alloy Performance. *Adv. Eng. Mater.* **2003**, *5*, 837–858. [CrossRef]
92. Ghali, E. *Corrosion Resistance of Aluminum and Magnesium Alloys (Understanding, Performance, and Testing) Active and Passive Behaviors of Aluminum and Magnesium and Their Alloys*; John Wiley & Sons: Hoboken, NJ, USA, 2010; pp. 78–120.
93. Yang, Y.; Wu, P.; Wang, Q.; Wu, H.; Liu, Y.; Deng, Y.; Zhou, Y.; Shuai, C. The Enhancement of Mg Corrosion Resistance by Alloying Mn and Laser-Melting. *Materials* **2016**, *9*, 216. [CrossRef] [PubMed]
94. Benn, F.; D'Elia, F.; van Gaalen, K.; Li, M.; Malinov, S.; Kopp, A. Printability, mechanical and degradation properties of Mg-(x)Zn elemental powder mixes processed by laser powder bed fusion. *Addit. Manuf. Lett.* **2022**, *2*, 100025. [CrossRef]
95. Zhou, Y.; Wu, P.; Yang, Y.; Gao, D.; Feng, P.; Gao, C.; Wu, H.; Liu, Y.; Bian, H.; Shuai, C. The microstructure, mechanical properties and degradation behavior of laser-melted Mg-Sn alloys. *J. Alloys Compd.* **2016**, *687*, 109–114. [CrossRef]
96. Yin, Y.; Huang, Q.; Liang, L.; Hu, X.; Liu, T.; Weng, Y.; Long, T.; Liu, Y.; Li, Q.; Zhou, S.; et al. In vitro degradation behavior and cytocompatibility of ZK30/bioactive glass composites fabricated by selective laser melting for biomedical applications. *J. Alloy. Compd.* **2019**, *785*, 38–45. [CrossRef]

97. Zhang, X.; Yuan, G.; Wang, Z. Mechanical properties and biocorrosion resistance of Mg-Nd-Zn-Zr alloy improved by cyclic extrusion and compression. *Mater. Lett.* **2012**, *74*, 128–131. [CrossRef]
98. Staiger, M.P.; Pietak, A.M.; Huadmai, J.; Dias, G. Magnesium and its alloys as orthopedic biomaterials: A review. *Biomaterials* **2006**, *27*, 1728–1734. [CrossRef]
99. Gieseke, M.; Noelke, C.; Kaierle, S.; Wesling, V.; Haferkamp, H. Selective Laser Melting of Magnesium and Magnesium Alloys. In *Magnesium Technology 2013*; Hort, N., Mathaudhu, S.N., Neelameggham, N.R., Alderman, M., Eds.; Springer International Publishing: Cham, Switzerland, 2016; pp. 65–68. [CrossRef]
100. Rodrigues, T.A.; Duarte, V.; Miranda, R.M.; Santos, T.G.; Oliveira, J.P. Current Status and Perspectives on Wire and Arc Additive Manufacturing (WAAM). *Materials* **2019**, *12*, 1121. [CrossRef]
101. Sing, S.L.; An, J.; Yeong, W.Y.; Wiria, F.E. Laser and electron-beam powder-bed additive manufacturing of metallic implants: A review on processes, materials and designs. *J. Orthop. Res.* **2016**, *34*, 369–385. [CrossRef]
102. Xie, K.; Wang, N.; Guo, Y.; Zhao, S.; Tan, J.; Wang, L.; Li, G.; Wu, J.; Yang, Y.; Xu, W.; et al. Additively manufactured biodegradable porous magnesium implants for elimination of implant-related infections: An in vitro and in vivo study. *Bioact. Mater.* **2022**, *8*, 140–152. [CrossRef]
103. Li, Y.; Zhou, J.; Pavanram, P.; Leeflang, M.A.; Fockaert, L.I.; Pouran, B.; Tumer, N.; Schroder, K.U.; Mol, J.M.C.; Weinans, H.; et al. Additively manufactured biodegradable porous magnesium. *Acta Biomater.* **2018**, *67*, 378–392. [CrossRef]
104. Xiaobo, Z.; Lin, M.; Guangyin, Y.; Zhangzhong, W. Performances of biodegradable Mg-Nd-Zn-Zr magnesium alloy for cardiovascular stent. *Rare Met. Mater. Eng.* **2013**, *42*, 1300–1305.
105. Mao, L.; Shen, L.; Niu, J.; Zhang, J.; Ding, W.; Wu, Y.; Fan, R.; Yuan, G. Nanophasic biodegradation enhances the durability and biocompatibility of magnesium alloys for the next-generation vascular stents. *Nanoscale* **2013**, *5*, 9517–9522. [CrossRef] [PubMed]
106. Zong, Y.; Yuan, G.; Zhang, X.; Mao, L.; Niu, J.; Ding, W. Comparison of biodegradable behaviors of AZ31 and Mg–Nd–Zn–Zr alloys in Hank's physiological solution. *Mater. Sci. Eng. B* **2012**, *177*, 395–401. [CrossRef]
107. Zhang, X.; Yuan, G.; Niu, J.; Fu, P.; Ding, W. Microstructure, mechanical properties, biocorrosion behavior, and cytotoxicity of as-extruded Mg-Nd-Zn-Zr alloy with different extrusion ratios. *J. Mech. Behav. Biomed. Mater.* **2012**, *9*, 153–162. [CrossRef] [PubMed]
108. Kumar, D.S.; Suman, K.N.S. Selection of Magnesium Alloy by MADM Methods for Automobile Wheels. *Int. J. Eng. Manuf.* **2014**, *2*, 31–41. [CrossRef]
109. Kulekci, M.K. Magnesium and its alloys applications in automotive industry. *Int. J. Adv. Manuf. Technol.* **2008**, *39*, 851–865. [CrossRef]
110. Madhuri, N.; Jayakumar, V.; Sathishkumar, M. Recent developments and challenges accompanying with wire arc additive manufacturing of Mg alloys: A review. *Mater. Today Proc.* **2021**, *46*, 8573–8577. [CrossRef]

Article

Synthesis of Manganese Oxide Sorbent for the Extraction of Lithium from Hydromineral Raw Materials

Zaure Karshyga [1], Albina Yersaiynova [1,*], Azamat Yessengaziyev [1], Bauyrzhan Orynbayev [1], Marina Kvyatkovskaya [1] and Igor Silachyov [2]

[1] The Institute of Metallurgy and Ore Beneficiation, Satbayev University, Almaty 050013, Kazakhstan; z.karshyga@satbayev.university (Z.K.); a.yessengaziyev@satbayev.university (A.Y.); bauyrzhan.orynbayev@stud.satbayev.university (B.O.); kmn_55@mail.ru (M.K.)
[2] The Institute of Nuclear Physics, Almaty 050032, Kazakhstan; silachyov@inp.kz
* Correspondence: a.yersaiynova@stud.satbayev.university; Tel.: +7-778-9329565

Abstract: The article presents the research results for the synthesis of inorganic sorbents based on manganese oxide compounds. It shows the results of the lithium sorption from brines with the use of synthesized sorbents. The effect of temperature, the molar ratio of Li/Mn, and the duration for obtaining a lithium-manganese precursor and its acid treatment was studied. The sorption characteristics of the synthesized sorbents were studied. The effect of the ratio of the sorbent mass to the brine volume and the duration of the process on the sorption of lithium from brine were studied. In this case, the sorbent recovery of lithium was ~86%. A kinetic model of the lithium sorption from brine on a synthesized sorbent was determined. The kinetics of the lithium sorption was described by a pseudo-second-order model, which implies limiting the speed of the process due to a chemical reaction.

Keywords: brines; lithium; synthesis; calcination; precursor; sorbents

Citation: Karshyga, Z.; Yersaiynova, A.; Yessengaziyev, A.; Orynbayev, B.; Kvyatkovskaya, M.; Silachyov, I. Synthesis of Manganese Oxide Sorbent for the Extraction of Lithium from Hydromineral Raw Materials. Materials 2023, 16, 7548. https:// doi.org/10.3390/ma16247548

Academic Editors: Yilong Dai, Deqiao Xie and Youwen Yang

Received: 19 October 2023
Revised: 28 November 2023
Accepted: 2 December 2023
Published: 7 December 2023

Copyright: © 2023 by the authors. Licensee MDPI, Basel, Switzerland. This article is an open access article distributed under the terms and conditions of the Creative Commons Attribution (CC BY) license (https:// creativecommons.org/licenses/by/ 4.0/).

1. Introduction

Lithium is one of the most important energy materials and strategic resources of the 21st century. It is represented in high technologies covering many areas of human activity. Lithium has become extremely important in the production of rechargeable lithium-ion batteries (LIBs), which have revolutionized the market supply and demand of renewable energy due to their unique technical characteristics (specific energy density 100–265 Wh/kg, specific power 250–340 Wh/kg, service life 400–1200 cycles) [1,2]. LIBs are used in smartphones, computers, hybrid cars, and electric vehicles. Besides batteries, lithium has large areas of application in the production of glass and ceramics (30%), lubricants (11%), metallurgy (4%), as well as in the production of chemicals, pharmaceuticals, and rubbers [3].

The created high global demand for this metal contributes to research and the search for technological solutions involving the processing of lithium-containing hydromineral raw materials, including associated reservoir brines.

Currently, the use of natural mineral raw materials is proposed to recover valuable metals from various hydromineral sources and industrial solutions [4,5]. In [6], methods intended to modify natural aluminosilicate and carbon-mineral sorbents were used to increase their sorption capacity.

According to the literature, such methods as natural evaporation, deposition, electrolysis, and others are used to process lithium raw materials [7–11]. In [12], a method of evaporation and crystallization was proposed to process solutions. In [13], precipitation methods were used to recover lithium from brines. In [14–17], ion exchange and extraction methods along with a combination of these methods with precipitation [18] were used for brines containing high concentrations of calcium and magnesium. There are known methods for the sorption extraction of lithium from sea water and brines with the use of spinel-type manganese oxide; aluminum compounds have an extremely high selectivity for the extraction of lithium from sea water [19–22]. These materials have a high adsorption capacity; lithium was concentrated more than 400 times in alkaline media (pH ~8).

Sorbents obtained based on double compounds of aluminum and lithium LiCl·2Al(OH)$_3$·mH$_2$O (DHAL-Cl) have high performance and are stable in brines with low pH [23–26]. The interaction occurs during the crystallization of DGAL-Cl via an intercalation mechanism with the introduction of Li+ cations and Cl$^-$ anions into the interlayer space. In this case, an intermediate phase of aluminum hydroxide with a deformed structure is formed. It is preserved during complete or partial deintercalation of lithium chloride from DGAL-Cl and is characterized by increased reactivity. Defective DGAL-Cl with a lithium deficiency in its composition is a sorbent selective for lithium. However, the deposition of impurities and mechanical inclusions on it can result in a narrowing of transport channels and to undersaturation of the sorbent with lithium during repeated long-term operation of the sorbent in sorption–desorption cycles under dynamic conditions, and as a result, the lithium deficiency in the sorbent may increase above the required limits after desorption [27]. The lithium deficiency should not exceed 35% of its total content in it to avoid destruction of the sorbent. It determines, and thereby limits, the value of the total exchange capacity of 7 mg/g. Effective use of the sorbent requires strict adherence to technological regimes.

Recently, technology with the use of lithium-ion sieves (LIS) has become one of the most promising for the extraction of lithium from brines and seawater. LIS make it possible to recover lithium with high selectivity from complex solutions with a high content of accompanying components.

In general, LIS are divided into two types according to the chemical composition: type of oxide lithium and manganese (LMO) and type of oxide lithium and titanium (LTO).

Lithium-ion sieves based on titanium oxides are currently produced in two categories: H$_2$TiO$_3$ with a layered structure and H$_4$Ti$_5$O$_{12}$ with spinel structure. When sorbents based on titanium H$_2$TiO$_3$ synthesized with the sol–gel method by the interaction of CH$_3$COOLi and Ti(OC$_4$H$_9$)$_4$, are used, 31.2 mg/g of lithium can be adsorbed [28]. During sorption by a sorbent from a TiO$_2$ nanotube with a diameter of 50–70 nm and a length of 1–2 µm, synthesized with a soft hydrothermal method at 150 °C for 48 h, 39.4 mg/g of lithium can be adsorbed from a solution with a concentration of 120 mg/L at alkaline pH [29]. Studies [30] on the adsorption of lithium on various titanium oxides showed that the Li$_2$TiO$_3$ structure obtained from anatase was more suitable for lithium recovery than that obtained from rutile. However, titanium oxide-based LISs have limited application in recovery of lithium from aqueous solution by applying electrical potential that may hinder future industrial applications.

Lithium-ion oxides based on spinel-type manganese oxide are currently the most popular selective sorbents. The formation of a three-dimensional structure with lithium as the LiMn$_2$O$_4$ compound favors the sorption mechanism instead of the two-dimensional layered crystal structure of LiMnO$_2$. The smaller size of lithium ions compared to any other alkali metals contributes to the formation of a stable structure of LiMn$_2$O$_4$, while lithium in LiMnO$_2$ occupies the interlayer octahedral region [20,31]. Chitrakar et al. [32] synthesized low-crystalline orthorhombic LiMnO$_2$ by the interaction of γ-MnOOH or Mn$_2$O$_3$ with LiOH·H$_2$O in the solid phase in a steam atmosphere at 120 °C with subsequent heating of samples at 400 °C in an air atmosphere for 4 h to form the cubic structure pf Li$_{1.6}$Mn$_{1.6}$O$_4$. After acid treatment of the precursor, the lithium capacity of the resulting sorbent was 33 mg/g. In another study [33], manganese oxide adsorbent H$_{1.6}$Mn$_{1.6}$O$_4$ was obtained from Li$_{1.6}$Mn$_{1.6}$O$_4$ precursor prepared by calcination of LiMnO$_2$ at 400 °C. In this case, two different methods were used for the synthesis of LiMnO$_2$, hydrothermal and reflux; the lithium capacity of the resulting sorbents was 40.9 mg/g and 34.1 mg/g, respectively. In [34], Li$_{0.15}$H$_{0.76}$Mg$_{0.40}$–Mn$^{III}_{0.08}$Mn$^{IV}_{1.59}$O$_4$ adsorbent was studied. The adsorbent showed a maximum lithium adsorption capacity of 23 mg/g at pH 6.5. After adsorption, lithium can be desorbed by dilute HCl solution, and the adsorption efficiency of the sample does not decrease even after 10 cycles.

As the literature data show, the synthesis of LIS based on manganese oxide consists of several stages: preparation of the precursor, its calcination, and acid treatment of the precursor to obtain a sorbent. The decisive role may be played by the first stage of obtaining lithium-manganese oxide, which ensures good contact of the reacting substances with the use of the lithium reagent in the quantities required for the reaction. At the same time, the remaining stages of sorbent synthesis are also important. Therefore, to study all stages of sorbent preparation under various conditions, temperature, duration, etc. are of interest.

2. Materials and Methods

Materials: Lithium hydroxide single-mode LiOH·H$_2$O brand "puriss."; salt acid HCl qualification "puriss."; Mn$_2$O$_3$ "puriss. spec."; MnO "puriss.".

The object of this study is the formation brines of oil and gas fields of JSC "Mangystaumunaigas" (Kazakhstan) of the following composition, mg/L: 5.9–7.8 Li; 25,000–3,0000 Na; 4000–6000 Ca; 20–670 K; 1200–2000 Mg; 10–700 Fe; 160–320 Sr; 13–14 B; 42,000–60,000 Cl$^-$; 670–780 SO$_4^{2-}$.

Analysis methods: the quantitative content of basic elements in precursors and sorbents was determined on an atomic emission spectrometer with inductively coupled plasma Optima 8300DV (Perkin Elmer Inc., Waltham, MA, USA). X-ray phase analysis (XRD) was carried out on a diffractometer D8 ADVANCE "BRUKER AXS GmbH", (Karsruhe, Germany) radiation Cu-Kα, database PDF-2 International Center for Diffraction Data ICDD (Swarthmore, PA, USA).

Thermal analysis of the lithium-manganese oxide sample was performed using an STA 449 F3 Jupiter simultaneous (NETZSCH, Selb, Germany) thermal analysis device. Before heating, the furnace space was evacuated (the percentage of the evacuated volume was ~92%) and then purged with inert gas for 5 min. Heating was carried out at a speed of 10 °C/min. in an atmosphere of highly purified argon. The total volume of incoming gas was maintained within 120 mL/min. The results obtained with the STA 449 F3 Jupiter (NETZSCH, Selb, Germany) were processed with the use of the NETZSCH Proteus software, version 5.1.

Experimental procedure: Reagents were taken only in the required quantities, according to the given molar ratio; lithium hydroxide LiOH·H$_2$O was dissolved in 100–150 mL of hot distilled water, and then mixed in a porcelain cup with samples of manganese oxides Mn$_2$O$_3$ and MnO, taken in accordance with the stoichiometry of the reaction of the formation of lithium-manganese oxides and the specified molar ratios Li/Mn. The resulting mixture was kept in a drying cabinet with heating to a set temperature and kept for a set time while stirring and keeping wet, and then the sample was dried until moisture was completely removed. In this way, uniform mixing and contact of all components of the reaction mixture were achieved during interaction with a liquid solution of the lithium hydroxide reagent. After evaporation and drying, this mixture was calcined. After drying, the resulting lithium-manganese oxides (LMOs) were calcined in a muffle furnace with heating to a given temperature and held for a given time. The resulting LMO and calcined precursors were analyzed for lithium and manganese content, and the phase composition was determined.

Precursors were poured with the required amount of dilute hydrochloric acid solution according to the experimental procedure for acid treatment. The process was performed at a given temperature and contact time under stirring in a 3 dm^3 sealed thermostated cell equipped with a VELP Scientifica LS F201A0151 mechanical stirrer (Usmate Velate, Italy), providing a fixed speed. Constant temperature was maintained using an Aizkraukles TW 2.02 water bath thermostat (ELMI, Riga, Latvia).

The resulting sorbents were washed with distilled water to pH = 6–7 and dried in air at room temperature. The resulting sorbents were analyzed for lithium content, and the phase composition was determined. The filtrates were analyzed for lithium and manganese content.

Sorption was performed under static conditions on an orbital shaker with a rotation of 200 rpm. A given amount of sorbent was placed in 300 cm^3 dry flasks filled with a given volume of brine, set to a given temperature, and stirred for a certain time to perform sorption. The solution was separated from the sorbent by filtration after sorption. Sorption filtrates were analyzed for lithium, sodium, potassium, iron, calcium, and magnesium content.

The study of sorption kinetics was performed under static conditions on an orbital shaker at a rotation of 200 rpm. To carry out sorption, 0.2 g of the sorbent was placed in dry flasks with a volume of 300 cm^3, filled with a brine volume of 130 cm^3, a set temperature was established, and stirred for a certain time. After sorption, the solution was separated from the sorbent by filtration. Sorption filtrates were analyzed for lithium content.

Static exchange capacity, distribution coefficients K_d, and partition coefficients K_s were determined by the Formulas (1)–(3).

Static exchange capacity is calculated by the formula:

$$\text{SEC} = \frac{(C_0 - C_e) \cdot V}{m}, \tag{1}$$

where C_0 is metal concentration in initial solution, mg/dm^3; C_e is residual equilibrium concentration of metal in solution, mg/dm^3; V—volume of solution, dm^3; m—mass of dry sorbent, g.

The distribution coefficients K_d and partition coefficients K_s were determined by the following formulas:

$$K_d = \frac{(C_0 - C_e) \cdot V'}{C_e \cdot m}, \tag{2}$$

where C_0 is concentration of metal in initial brine, mg/dm^3; C_e is residual equilibrium concentration of metal in solution, mg/dm^3; V' is volume of solution, cm^3; m is mass of dry sorbent, g.

$$K_s = \frac{K_d^{Li}}{K_d^{Me}} \tag{3}$$

where Me is Ca, Mg, Na, K, Fe.

3. Results and Discussion

3.1. Study of Conditions for Production of Lithium-Manganese Precursors

3.1.1. Preparation of Lithium-Manganese Oxides

The main purpose of the research is to obtain lithium-manganese oxide with the main phase consisting more preferably of LiMnO$_2$, LiMn$_2$O$_4$ compounds.

The interaction of reacting substances can presumably take place in accordance with the following reactions:

$$Mn_2O_3 + MnO + 3\,LiOH \cdot H_2O + 0.25\,O_2 = 3\,LiMnO_2 + 4.5\,H_2O\uparrow \tag{4}$$

$$2\,Mn_2O_3 + 2\,MnO + 3\,LiOH \cdot H_2O + 1.25\,O_2 = 3\,LiMn_2O_4 + 4.5\,H_2O\uparrow \tag{5}$$

Study of the temperature effect. The experiments were carried out under the following conditions: temperature—125, 150, 175, and 200 °C, duration—13 h, the mass ratio of manganese oxides to lithium hydroxide monohydrate was taken from the calculation of Li/Mn molar ratio maintenance = 1.

The obtained LMOs were studied using the XRD analysis. The results of the XRD analysis are presented in Figure 1.

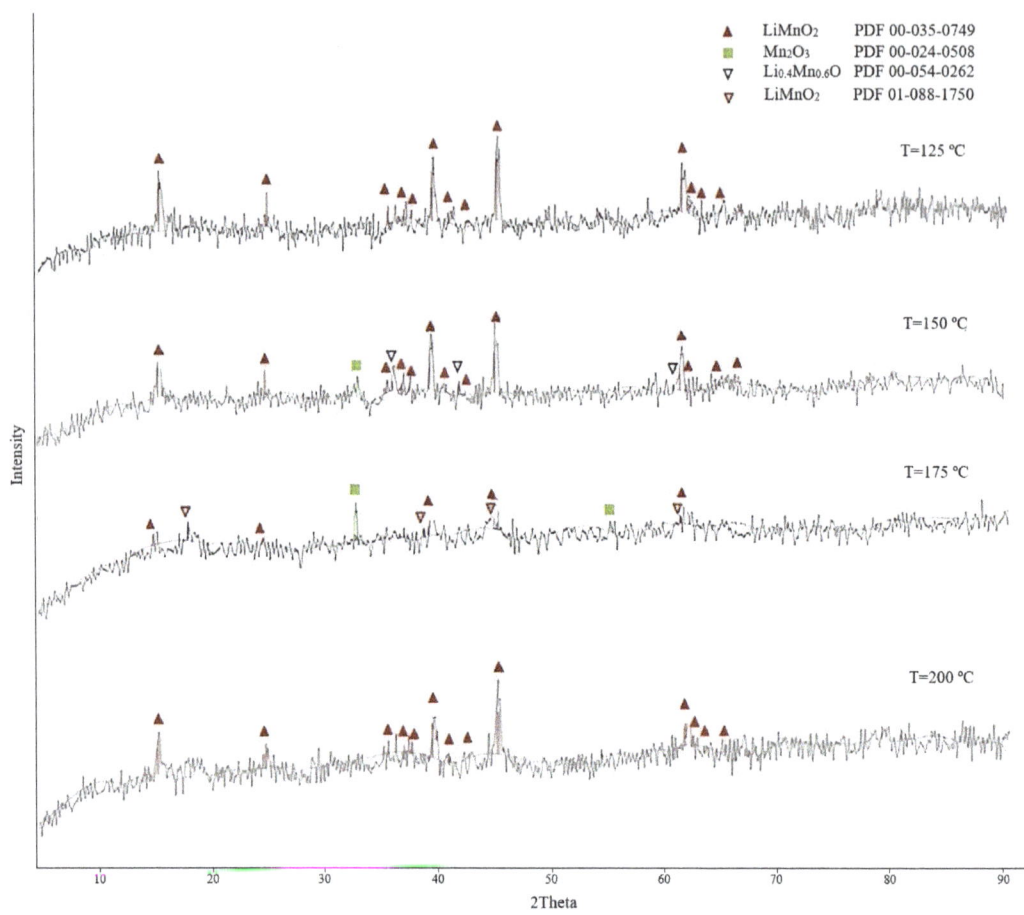

Figure 1. Diffractograms of LMOs obtained with exposure at different temperatures (LiMnO$_2$ triangle orthorhombic; LiMnO$_2$ rhombus—tetragonal).

As the results of studies show, at temperatures of 125 and 200 °C, the reaction of the interaction of manganese oxide with lithium hydroxide fully proceeded with the formation of lithium-manganese oxide LiMnO$_2$ with the orthorhombic structure of the crystal lattice (Figure 1). It should be noted that the XRD analysis of samples obtained at 150 and 175 °C showed the presence of the Mn$_2$O$_3$ phase in the samples; however, this phase was absent in the sample processed at 125 °C. Presumably, the manganese oxide phase was present in the sample obtained at 125 °C in an X-ray amorphous or amorphous state, and the XRD analysis could not identify it. In the sample at 150 °C, there was a Li$_{0.4}$Mn$_{0.6}$O phase with a spinel structure similar in composition to the LiMnO$_2$ phase, which presumably may indicate an intermediate stage of LMO formation. At a temperature of 200 °C, only the phase of orthorhombic LiMnO$_2$ was identified, as at a temperature of 125 °C. Research results show that a temperature of 125 °C was sufficient to form the lithium-manganese oxide phase LiMnO$_2$.

Study of the effect of Li/Mn molar ratio. The experiments were performed under the following conditions: the ratio of the mass of manganese oxides to lithium hydroxide monohydrate was taken based on the calculation to maintain the molar ratios Li/Mn = 0.5, 0.9, 1, and 1.5; temperature 125 °C, duration—13 h. The phase composition of the obtained

LMOs was studied using X-ray phase analysis. The XRD patterns of lithium-manganese oxides are presented in Figure 2.

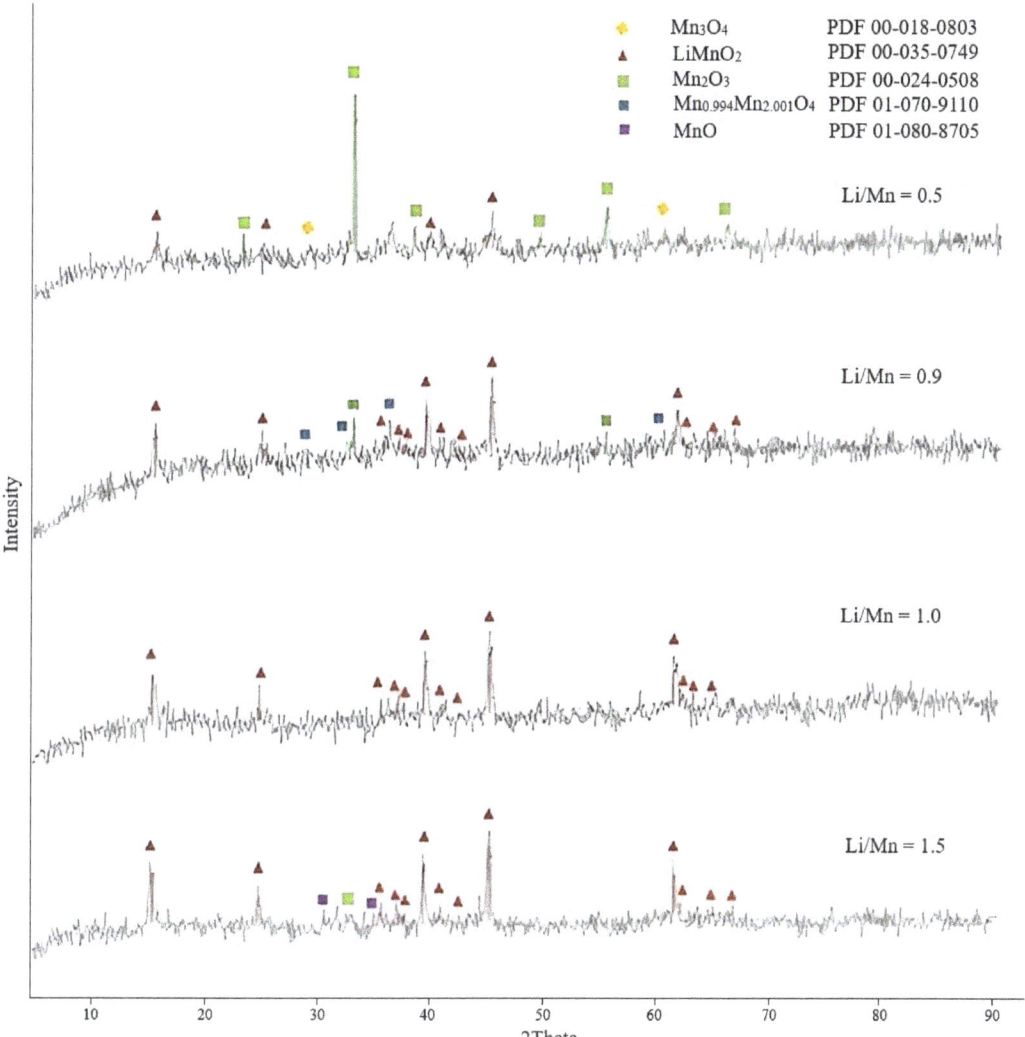

Figure 2. Diffractograms of LMOs obtained at different Li/Mn molar ratios.

The diffractogram of the sample with a molar ratio of Li/Mn = 0.5 showed that the while the process of LiMnO$_2$ formation was at an initial stage, the initial manganese oxides were largely present in the sample. With a molar ratio of Li/Mn = 0.9, the process of LMO formation was more active, as indicated by the presence of peaks on the diffractogram corresponding to the LiMnO$_2$ phase with a higher intensity. However, the sample also contained phases of the initial manganese oxides that did not react with lithium hydroxide. The XRD diffractogram of the sample substance with a molar ratio Li/Mn = 1 showed that it was represented by the orthorhombic LiMnO$_2$ phase, indicating the most complete passage of the process. With a molar ratio of Li/Mn equal to 1.5, lithium-manganese oxide was also actively formed. However, despite the increase in the intensity of LMO peaks, the diffractogram indicateed the presence of a small amount of initial manganese oxides (II), (III). At the same time, in [35], when calcining a mixture of the initial lithium carbonate or hydroxide with manganese dioxide or carbonate at molar ratios Li/Mn 0.75 and 1, the presence of the initial tetragonal MnO_2 was observed, and when increased to 1.5, the intensity of the spinel peak gradually decreased to an amorphous phase with low crystallinity. In [36], a sorbent prepared from a Li_2MnO_3 precursor with a monoclinic structure at a molar ratio of Li/Mn = 2 showed an inability to sorb lithium from a solution (−21.1 mg/g), while a sorbent prepared at a molar ratio of Li/Mn = 1 showed the highest capacity, which was 6.6 mg/g of lithium.

According to the research results, the optimal molar ratio of Li/Mn in the reaction mixture was equal to 1, which was characterized by a more complete interaction of manganese oxides with lithium hydroxide and the formation of $LiMnO_2$.

Study of the effect of the process duration. The experiments were performed under the following conditions: the ratio of the mass of manganese oxides to lithium hydroxide monohydrate was taken based on the calculation to maintain the molar ratio Li/Mn = 1; temperature 125 °C, duration 8, 13, 16, and 20 h. The XRD results of the obtained LMOs is presented in Figure 3.

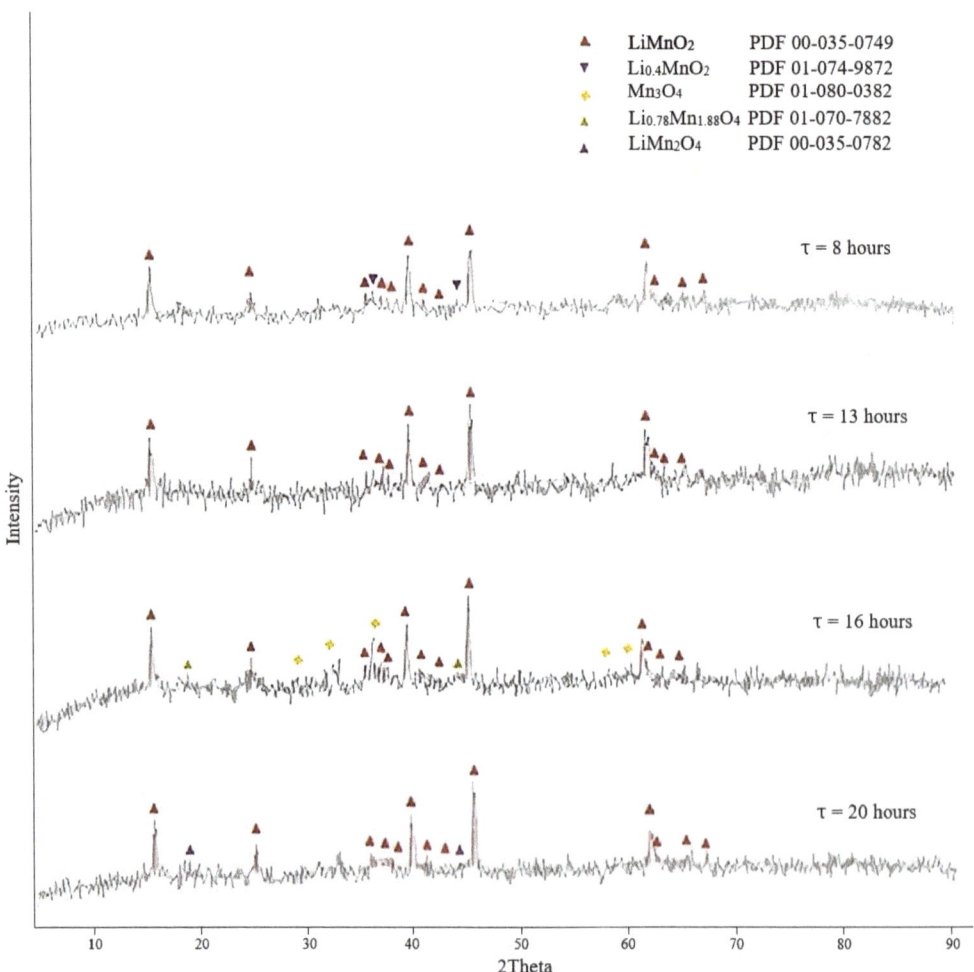

Figure 3. Diffractograms of LMOs obtained at different dwell times.

As the XRD results show (Figure 3), the reactions of the interaction of manganese oxide with lithium hydroxide took place at all durations. The XRD diffractogram of the sample obtained after exposure for 8 h showed the formation of the orthorhombic $LiMnO_2$ phase, as well as the presence of a small amount of $Li_{0.4}MnO_2$, most likely reflecting the intermediate process of the formation of the main $LiMnO_2$ phase. However, there was a phase of manganese oxide Mn_3O_4 that did not completely react with lithium hydroxide with a duration of 16 h. In addition, the spinel phase $Li_{0.78}Mn_{1.88}O_4$ appeared, indicating the beginning of the manganese oxidation process. The diffractogram of the product of the process with a duration of 20 h also showed, along with the $LiMnO_2$ phase, a lithium-manganese precursor with the spinel phase $LiMn_2O_4$, which can be represented as $Li[Mn(III)Mn(IV)]O_4$.

All samples contain the LMO phases of various compositions, except for the sample obtained with a 13 h process. It was characterized by a monophase $LiMnO_2$ that is most preferable for further synthesis of the sorbent. Therefore, the duration of 13 h was sufficient to pass the reactions of LMO formation.

Thus, the following conditions may be acceptable for the preparation of lithium-manganese oxides: temperature 125 °C, duration 13 h, the mass ratio of manganese oxides to lithium hydroxide monohydrate being based on the calculation to maintain the molar ratio Li/Mn = 1.

3.1.2. Obtaining Precursors

The diffractograms of Figures 1–3 are characterized by a very high background, indicating the presence of an amorphous component or insufficiently crystallized phase in the sample.

Various types of lithium-manganese spinels are promising precursors for obtaining sorbents for lithium extraction. Currently, there are only a few precursors for the production of sorbents or lithium-ion sieves (LIS) characterized by high lithium capacity, such as $LiMn_2O_4$, $Li_4Mn_5O_{12}$, and $Li_{1.6}Mn_{1.6}O_4$ [37]. LIS with one of the highest capacities was obtained from the $Li_{1.6}Mn_{1.6}O_4$ precursor with a cubic structure by calcination from LMO with an orthorhombic structure of the $LiMnO_2$ composition. In this case, the oxidation process for manganese from trivalent in the composition of $LiMnO_2$ to tetravalent was required to obtain a precursor with the composition $Li_{1.6}Mn_{1.6}O_4$.

Therefore, in order to obtain a lithium-manganese precursor with a sufficiently stable crystal structure, the next stage was calcination of the first-stage LMO. A batch of lithium-manganese oxide of the first stage was previously developed under the above selected conditions.

Before studying the calcination temperature, a sample of the produced batch of lithium-manganese oxide of the first stage was researched with the use of a thermal analysis method. The thermal analysis results for the sample are presented in Figure 4.

As it can be seen from Figure 4, the DTA curve showed endothermic effects of varying intensity with maximum development at 161.4, 199.9, and 726.1 °C. Additional effects were recorded on the dDTA curve. Their extremes were at 111.8, 130.1, 394, 680, and 731 °C. Additionally, exothermic peaks at 221.1 and 416.9 °C can be noted on the dDTA curve. All endothermic effects were developed against the background of a permanent decrease in the mass of the sample demonstrated by the TG curve course. The DTG curve formed a not very obvious maximum at 422.1 °C in the area of development of the exothermic effect (416.9 °C), and then a slight rise was observed. It indicated the occurrence of an oxidative process, i.e., on the oxidation of manganese (III) to manganese (IV), which was part of the lithium-manganese oxide $LiMnO_2$. Effects in the temperature range of 100–200 °C were associated with the dehydration process. Adsorbed moisture was removed. The endothermic effect with an extremum at 726.1 °C on the DTA curve presumably reflected the decomposition of MnO_2 with the release of oxygen. According to standards, the reaction occurred in the range of 600–700 °C. Perhaps the presence of lithium in the oxide effected the shift of the extremum toward higher temperatures. The combination of an endothermic effect with an extremum at 189.3 °C and an exothermic peak at 221.1 °C on the dDTA curve can be interpreted as a manifestation of an admixture of manganese dioxide gel.

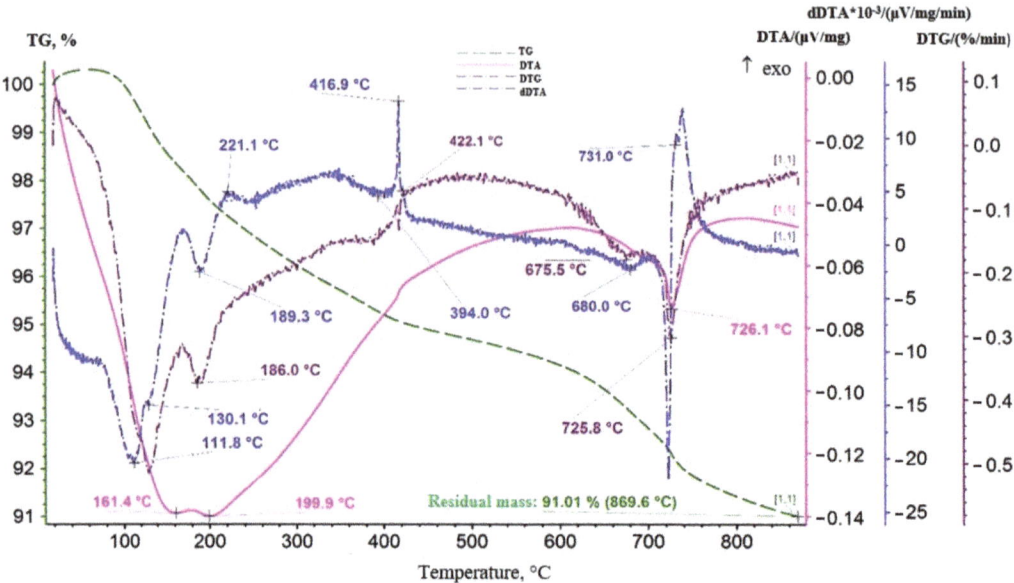

Figure 4. Thermogram of lithium-manganese oxide sample of the first stage of processing.

A repeat measurement was performed to obtain additional information. The sample was increased to 0.4 g, and the heating interval was extended. The results of the study are presented in Figure 5.

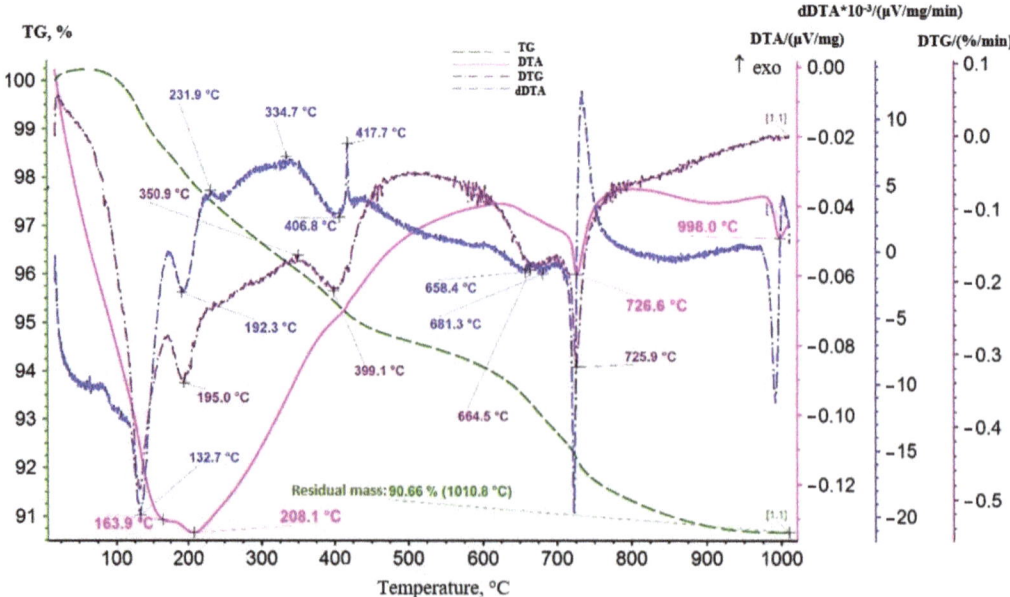

Figure 5. Thermogram obtained by repeated measurement of the lithium-manganese oxide sample.

As it can be seen from the thermogram in Figure 5, an additional endothermic effect appeared on the DTA curve in this measurement with the maximum development at 998 °C. This effect was not accompanied by a change in mass, and on the DTA curve obtained during sample cooling (Figure 6) it corresponded to an exothermic peak at 927.1 °C.

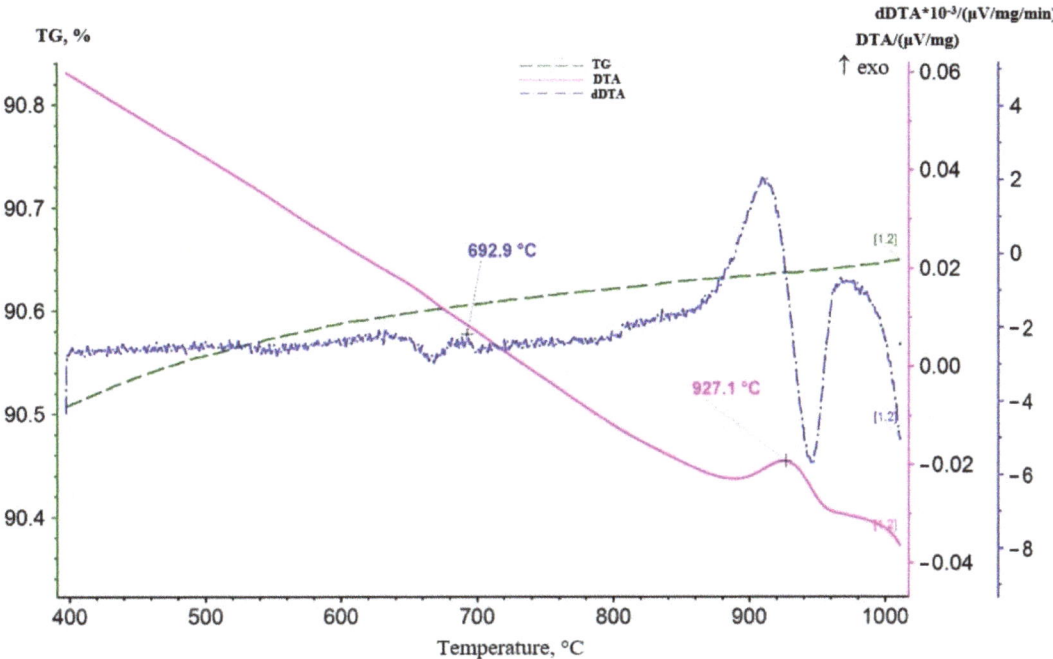

Figure 6. Plot of thermogram obtained during cooling of lithium-manganese oxide sample.

In general, presumably, it is an enantiotropic polymorphic transformation of hausmanite—α-Mn_3O_4 ($Mn^{4+}Mn_2^{2+}O_4$) → β-Mn_3O_4. The dDTA curve showed a more clearly exothermic peak at 334.7 °C, accompanied by an increase in the mass of the sample, as indicated by the maximum at 417.7 °C in the DTG curve. For example, oxidative processes also occurred in the area of development of these effects. The combination of an endothermic effect with an extremum at 394 °C on the dDTA curve and an endothermic effect with maximum development at 998 °C on the DTA curve can presumably be interpreted as a manifestation of manganite—MnOOH.

It is possible that the effect of lithium affected the shift in effect temperatures toward lower values. Thus, it can be assumed that the endothermic effect with extremum at 658.4 °C or 681.3 °C on the dDTA curve reflected the decomposition of lithium-manganese oxide—$LiMnO_2$. Probably β-$LiMn_2O_3$ was formed.

According to the standards, the decomposition of β-kurnakite occurred in the temperature range 900–1050 °C. In our case, this decomposition may reflect an endothermic effect with maximum development at 726.6 °C. As a result, β-Mn_3O_4 (hausmanite) was formed.

The last endothermic effect with maximum development at 998 °C reflected the enantiotropic polymorphic transformation of hausmanite—β-Mn_3O_4 ($Mn^{4+}Mn_2^{2+}O_4$) → γ-Mn_3O_4. It was also impossible to exclude the possibility of transformation in the area of development of this effect α-Mn_3O_4 ($Mn^{4+}Mn_2^{2+}O_4$) → β-Mn_3O_4.

As the results of thermal analysis show, for the calcination process of lithium-manganese oxide $LiMnO_2$ with an orthorhombic crystal lattice structure to form the cubic form $Li_{1.6}Mn_{1.6}O_4$, it was necessary to study the calcination process of lithium-manganese oxide obtained at stage

1 in the temperature range from 350 to 600 °C, within which the oxidation processes of manganese present in the LMO should take place, from degree +3 to degree +4.

The effect of temperature and duration of calcination of lithium-manganese oxide was studied.

Effect of calcination temperature. The experiments were conducted under the following conditions: temperature 350, 400, 450, 500, 550, 600 °C; duration 5 h. Samples of the obtained precursors were investigated using X-ray phase analysis. The results of XRD are presented in Figure 7. The diffractogram of the sample obtained at a temperature of 350 °C indicated that the process was at the initial stage, since there was mainly a completely unformed phase of the composition $Li_{0.27}Mn_2O_4$; there was also a residual phase of orthorhombic $LiMnO_2$. At a temperature of 400 °C, the diffractogram was characterized by the formation of a phase of a lithium-manganese precursor of the composition $Li_{1.27}Mn_{1.73}O_4$. It is clear from Figure 7 that the $Li_{1.6}Mn_{1.6}O_4$ precursor phase was formed when LMO was calcined at temperatures from 450 to 600 °C. The authors of [38] came to the conclusion that, as the temperature decreased from 550 to 450 °C, the adsorption of lithium increased and the higher the temperature of precursor calcination, the worse the extractability of lithium from the precursor during sorbent preparation.

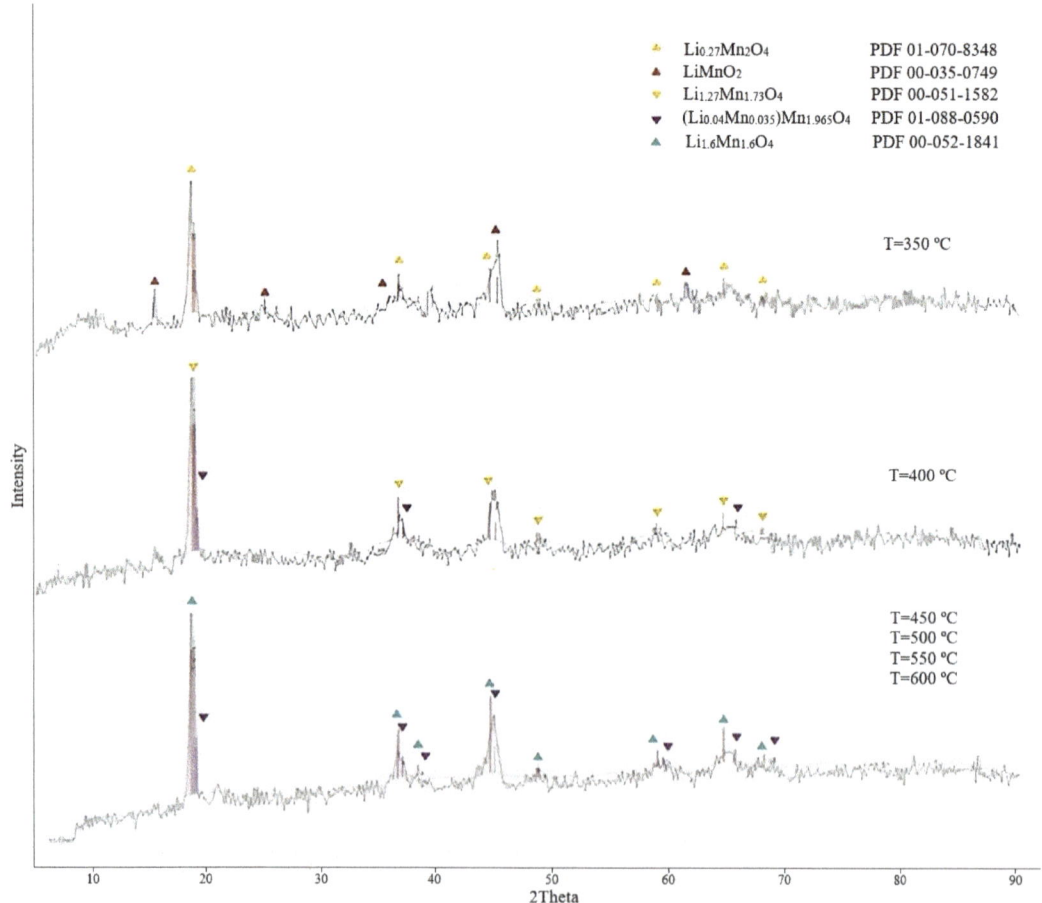

Figure 7. Diffractograms of lithium-manganese precursors obtained at different temperature exposures.

The most preferable calcination temperature is 450 °C based on the obtained research results.

The effect of the duration of calcination was carried out under the following conditions: temperature 450 °C; duration 4, 5, 6, 7, and 8 h. The XRD patterns of the obtained precursor samples are presented in Figure 8.

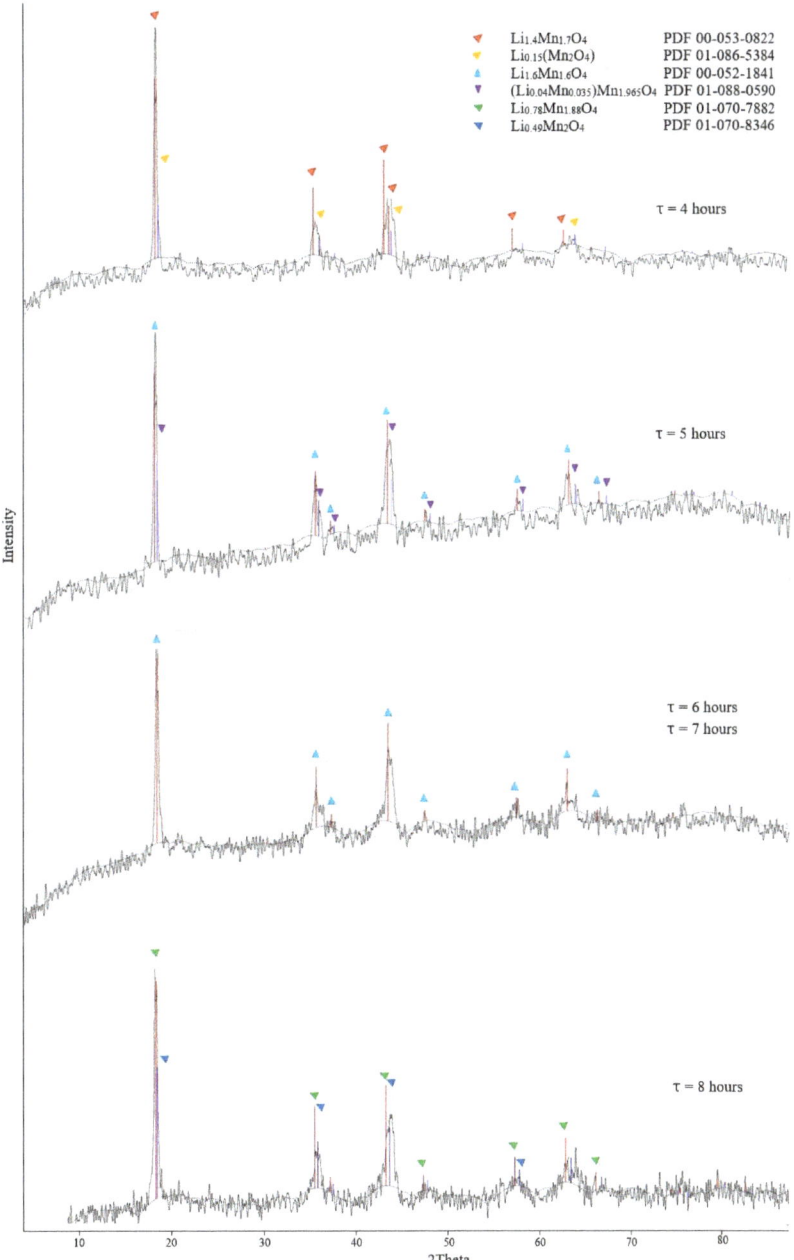

Figure 8. Diffractograms of lithium-manganese precursors obtained at different dwell times.

The XRD data of the precursors presented in Figure 8 showed that lithium-manganese oxides $Li_{1.4}Mn_{1.7}O_4$ were already formed at 4 h. However, the presence of phase $Li_{0.15}(Mn_2O_4)$ indicated that the process was not completed and additional time was required for the formation of precursor. The formation of the $Li_{1.6}Mn_{1.6}O_4$ phase occurred upon exposure for 5 h. X-ray phase analysis of the precursor obtained after calcination at 450 °C for 6 h identified the $Li_{1.6}Mn_{1.6}O_4$ monophase with a cubic structure in the sample; it can also be noted that the background of the diffractogram was significantly lower. It indicated good crystallization of the sample substance and a decrease in the amorphous component.

The most preferable calcination conditions are a temperature of 450 °C and a duration of 6 h, according to the results of the studies.

3.2. Study of Acid Treatment Conditions for Lithium-Manganese Precursors

Acid Treatment of Lithium-Manganese Precursors

Acid treatment was performed to remove lithium from the lithium-manganese precursor and obtain a sorbent. Free vacant cells must remain during the removal of lithium from the precursor and, at the same time, in the structure of the resulting sorbent. They are very small in size and can only be occupied by lithium during sorption, or the replacement of lithium with a hydrogen atom that can be exchanged for a lithium atom during sorption.

The effect of temperature, the ratio of precursor mass to acid volume, and duration on the acid treatment was studied.

The effect of the process temperature was studied under the following conditions: temperature 30, 40, 50, 60 °C; HCl concentration 0.5 M; duration 12 h; the ratio of the sorbent mass to the acid solution volume of acid solution (S:L) = 1:800. The research results are shown in Table 1.

Table 1. The effect of temperature on the acid treatment of the precursor.

Temperature, °C	Lithium Content in the Sorbent, %	Lithium Extraction into Solution, %	Loss of Manganese in Solution, %
30	1.698	83.77	12.51
40	1.041	91.02	12.95
50	1.003	97.49	13.38
60	0.692	98.00	13.94

The research results obtained show that the extraction of lithium into the solution increased with an increase in the process temperature; and the extraction of lithium reached above 91% at 40 °C. Manganese losses over the entire temperature range studied were ~12.5–14%. The most preferable temperature is 40 °C, at which lithium extraction is 91%, manganese losses are 12.95%.

Study of the effect of the ratio of the precursor mass to the volume of the acid. The studies were conducted under the following conditions: temperature 40 °C; HCl concentration 0.5 M; duration 12 h; ratio of sorbent mass to acid solution volume (S:L) = 1:600; 1:700; 1:800; 1:900. The research results are shown in Table 2.

Table 2. The effect of the S:L ratio on the acid treatment of the precursor.

Ratio S:L	Lithium Content in the Sorbent, %	Lithium Extraction into Solution, %	Loss of Manganese in Solution, %
1:600	1.117	88.90	11.34
1:700	1.033	90.38	12.28
1:800	1.041	91.02	12.95
1:900	1.150	88.33	12.72

The experimental results show that lithium extraction was the maximum and amounted to ~90–91% while manganese losses were in the range from 12.28 to 12.95% at S:L ratios of S:L 1:700 and 1:800.

The effect of acid treatment duration was studied under the following conditions: temperature 40 °C; HCl concentration 0.5 M; S:L ratio = 1:800; duration 2, 6, 12, 18, and 24 h. The obtained results are shown in Table 3.

Table 3. Effect of process duration on acid treatment of precursor.

Duration, h	Lithium Content in the Sorbent, %	Lithium Extraction into Solution, %	Loss of Manganese in Solution, %
2	2.095	75.58	12.35
6	1.693	81.55	12.31
12	1.041	91.02	12.95
18	0.332	89.13	12.51
24	0.277	93.26	12.91

An increase in the acid treatment duration resulted in an increase in the transition degree of lithium into solution. The extraction reached ~90% or more at 12 h or more, while the loss of manganese remained practically unchanged throughout the studied duration of the process. The most preferable duration is 24 h, which makes it possible to achieve lithium recovery above 93%, under the data obtained.

X-ray phase analysis of the obtained sorbent presented in Figure 9 shows that it consisted of manganese dioxide monophase with cubic crystal lattice structure.

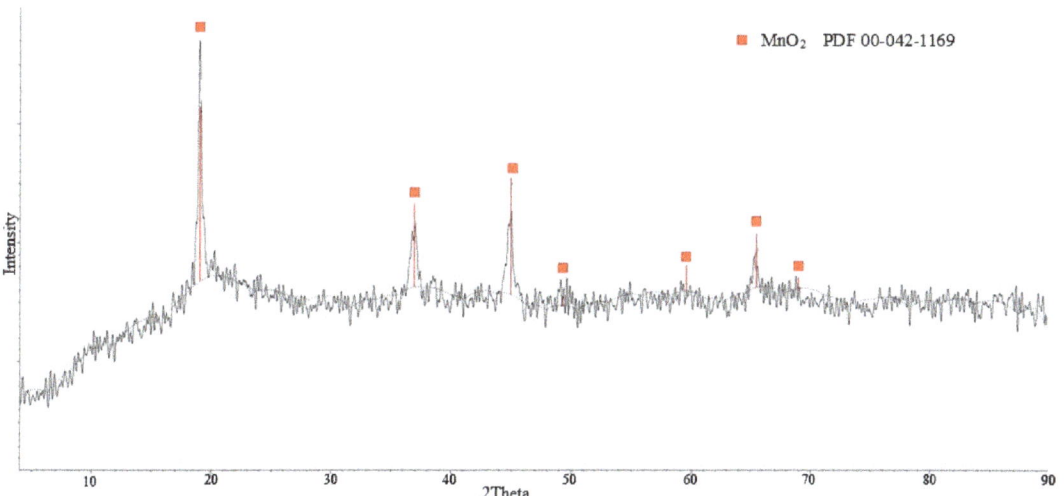

Figure 9. Diffractogram of the obtained sorbent.

A thermal analysis of the sample was conducted to clarify the composition of the resulting sorbent (Figure 10). The DTA curve showed endothermic effects of varying intensity with maximum development at 155.9, 554.2 and 620 °C. The most intense endothermic effect at 155.9 °C reflected the removal of chemically bound water, the protons of which can participate in the sorption process. The following two endothermic effects at 554.2 and 620 °C were possibly a manifestation of the decomposition of ß-MnO_2 with the formation of ß-Mn_2O_3. The water content in the samples was determined by the weight loss during heating the sorbent sample at 450 °C. The H_2O/Mn molar ratio was close to 0.5. The composition of the resulting sorbent apparently corresponded to the formula $MnO_2 \cdot 0.5H_2O$.

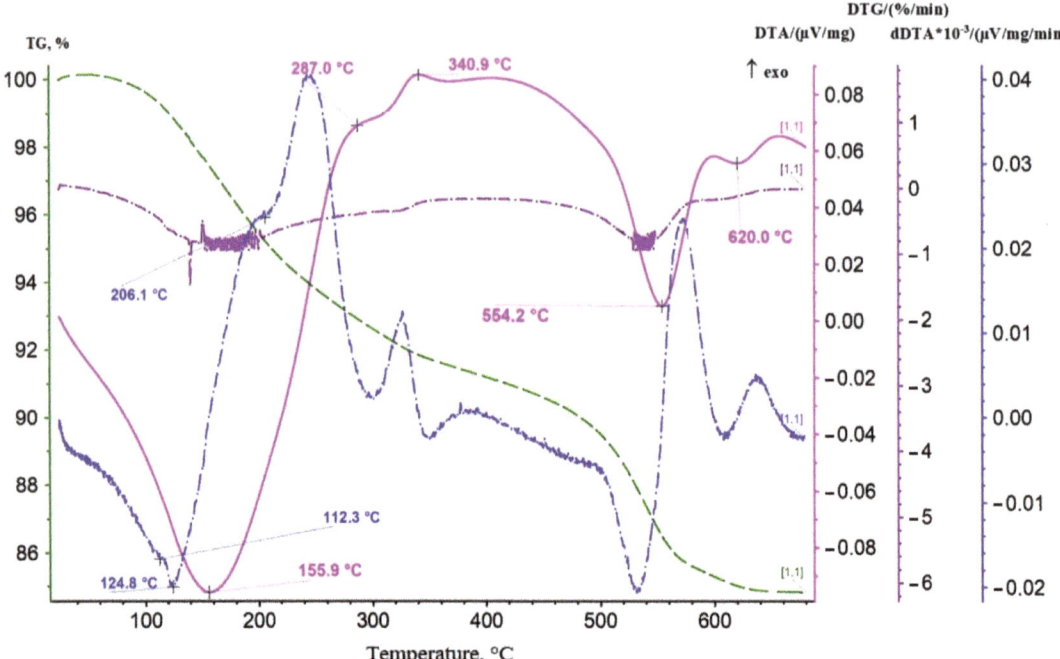

Figure 10. Thermogram of a sample of the obtained sorbent manganese dioxide.

Thus, the study results of acid treatment showed that the most acceptable conditions for the process are temperature 40 °C, HCl concentration 0.5 M; S:L ratio = 1:700 and 1:800 and duration 24 h. In this case, the lithium extraction into the solution from the precursor can reach ~93%, and the lithium content in the sorbent is 0.277%.

3.3. Study of the Sorption Characteristics of the Obtained Sorbents
Study of the Process Conditions on the Lithium Sorption Recovery Characteristics

The sorption capacity of the sorbent increased with the increase in the pH value of the initial brine under the literature data [10]. Therefore, it was of interest to study the sorption capacity of the obtained sorbent at different pH of the initial brine.

Effect of the initial brine pH. The studies were at the following sorption conditions: temperature 35 °C, duration 24 h, ratio of sorbent mass to brine volume—1:6000; brine pH—7.32; 8.08; 9.08; 10.04; 11.08, and 12.06. The brine with the appropriate pH was prepared by adding a concentrated NaOH solution to the original brine with a pH of 7.32. Precipitates formed from the brine were filtered off. The initial brines and solutions after sorption were analyzed for the content of the studied components. The research results are presented in Tables 4 and 5.

Table 4. Compositions of initial brines and sorbent capacity for lithium.

pH of Initial Brine	Concentration in Initial Brine, g/L					SEC, mg/g
	Li, mg/L	Ca	Mg	Na	K	
7.32	6.389	2.329	0.759	19.74	0.376	10.00
8.08	8.094	2.204	0.721	17.88	0.348	9.39
9.08	8.392	2.317	0.756	18.83	0.449	12.39
10.04	8.129	2.298	0.690	18.59	0.408	18.948
11.08	7.838	2.185	0.032	21.45	0.387	12.408
12.06	7.661	2.047	0.001	21.86	0.373	21.204

Table 5. Distribution and partition coefficients for the lithium sorption from brine depending on the pH.

pH of Initial Brine	Distribution Coefficient, K_d					Partition Coefficient, K_s			
	Li	Ca	Mg	Na	K	Li/Ca	Li/Mg	Li/Na	Li/K
7.32	2116	224	221	0	0	9.4	9.6	-	-
8.08	1438	71	67	0	0	20	21	-	-
9.08	1958	73	0	0	207	27	-	-	9.4
10.04	3811	0	0	0	0	-	-	-	-
11.08	2150	72	0	318	0	30	-	6.8	-
12.06	5138	110	0	398	0	46	-	13	-

As can be seen from Table 4, the capacity of the sorbent for lithium increased, reaching a maximum capacity of 21.204 mg of lithium per 1 g of sorbent, at pH 12.06 with an increase in the pH of the medium and with the exception of pH 11.08. The capacity indicators of the synthesized sorbents based on manganese oxide obtained by various researchers have different values. In [36], a synthesized sorbent based on manganese oxide during the lithium sorption from geothermal fluid of Lumpur Sidoarjo (Lusi) showed a maximum capacity of 6.6 mg/g. In another work [10], the capacity of the synthesized sorbent $H_{1.6}Mn_{1.6}O_4$ at the lithium sorption from brine was 22–27 mg/g, and a higher capacity of 34–40 mg/g was shown by the sorbent $H_{1.6}Mn_{1.6}O_4$ at the lithium sorption from seawater [33].

Along with the capacity, the values of the distribution and separation coefficients during the sorption of lithium from brines by the synthesized sorbents are of great interest. As the calculated values of the distribution and partition coefficients presented in Table 5 show, lithium has the highest sorbent distribution values, and calcium—to a much lesser extent. In most cases, magnesium, sodium, and potassium are practically not sorbed on the manganese dioxide sorbent; and accordingly, the degree of separation of these metals from lithium is the maximum in all these cases.

For the initial brine with pH 7.32, the distribution coefficient for lithium was quite good, the separation of lithium from impurity macrocomponents also occurred at a fairly acceptable level.

Study of the effect of the ratio of sorbent mass to brine volume. The studies were performed at the following sorption conditions: temperature 35 °C, duration 24 h, pH of the initial brine 7.32. Varying the sorbent mass to the volume of the brine was performed at the following ratios—1:650, 1:1000, 1:2000 and 1:3000.

The sorption equilibrium characteristics during the extraction of lithium from brines with the use of a manganese dioxide sorbent were studied. The research results are presented in Table 6.

Table 6. Effect of the ratio of sorbent mass to brine volume on the characteristics of sorption extraction of lithium from brine.

Ratio $m_{sorbent}$ to $V_{solvent}$	Lithium Recovery onto Sorbent, %	SEC, mg/g	Distribution Coefficient, K_d
1:650	85.9	3.66	3952
1:1000	66.7	4.15	2003
1:2000	53.2	6.62	2273
1:3000	39.2	7.33	1937

As can be seen from Table 6, the highest lithium extraction rates onto the sorbent occurred at a ratio of sorbent mass to brine volume of 1:650 and amounted to 85.9%.

The process duration effect was studied at the following conditions: temperature 35 °C, duration 6, 12, 24, 48 h, ratio of sorbent mass to brine volume—1:650. The research results are presented in Table 7.

Table 7. Effect of process duration on the characteristics of sorption extraction of lithium from brine.

Duration, h	Extraction of Lithium onto Sorbent, %	SEC, mg/g	Distribution Coefficient, K_d
8	78.3	3.33	2340
16	83.7	3.57	3348
24	85.9	3.66	3952
48	86.1	3.67	4038

The research results show that the sorbent recovery of lithium increased from 78.3 to 86.1% with an increase in the sorption duration from 8 to 48 h. As it can be seen from Table 7, the sorption process reached ~86% of lithium extraction onto the sorbent at the studied process conditions and a duration of 24 h.

3.4. Determination of the Kinetic Model of the Lithium Sorption Process
Study of the Lithium Sorption Kinetics on a Synthesized Manganese Oxide Inorganic Sorbent

The study of the duration effect on the characteristics of the lithium sorption on synthesized manganese dioxide showed that the process takes a sufficiently long time to ensure acceptable extraction of the target metal from the brine onto the sorbent.

It is required to study the kinetics of the process in order to more thoroughly complete the sorption extraction of lithium from brines and optimize the process. Kinetic parameters can be useful to predict sorption rates and can also provide important information for the design and modeling of the sorption processes. Sorption is a complex and multistage process, and it is necessary to evaluate the adequacy of several kinetic models to identify the limiting stage.

Four kinetic models were used in the studies to analyze the kinetics of lithium sorption: pseudo-first and pseudo-second order models, the Elovich model, and the intraparticle diffusion model.

Brine with a Li concentration of 6.32 mg/dm^3 with pH 7.32 at temperatures of 25 and 35 °C were used during kinetic studies for lithium sorption.

The linear form of the pseudo-first order model (Lagergren model) can be represented by the following equation [39]:

$$\log(q_e - q_t) = \log q_e - \left(\frac{k_1}{2.303}\right) t \qquad (6)$$

where q_t and q_e (mg/g) are the number of lithium ions sorbed at time t (min) and at equilibrium with one g of sorbent, respectively, and k_1 is the adsorption rate constant (1/min).

The Lagergren equation describes the patterns of sorption at the initial stages of the sorption process when the phenomenon of film diffusion has a significant effect on the process [40].

The linear dependence presented in the $\log(q_e - q_t)$ coordinates on t is shown in Figure 11, which describes the lithium sorption process from brine by a synthesized inorganic sorbent of manganese dioxide in accordance with Lagergren's pseudo-first order model at temperatures of 25 and 35 °C.

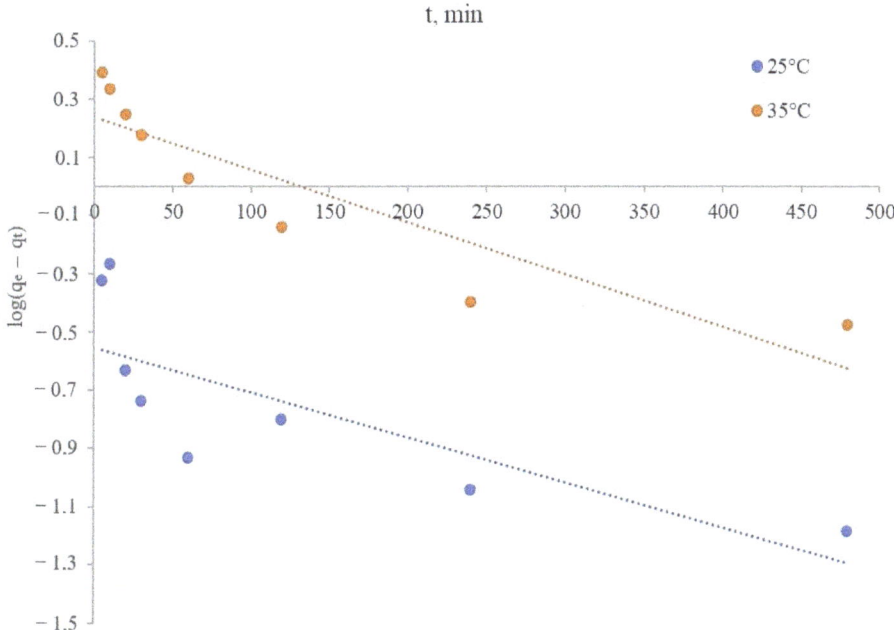

Figure 11. Description of the lithium sorption kinetics from brine in the coordinates of the pseudo-first order model.

The rate constant k_1 can be determined experimentally by plotting $\log(q_e - q_t)$ versus t.
Table 8 presents the kinetic parameters of the sorption process calculated from the data of the linear dependence $\log(q_e - q_t) - t$ and Equation (6).

Table 8. Kinetic parameters of the lithium sorption by manganese dioxide sorbent.

Kinetic Model	Parameters	Temperature of the Sorption Process, °C	
		35	25
Pseudo-first order	q_e, mg/g k_1, 1/min R^2	0.28 3.454×10^{-3} 0.6175	1.73 4.145×10^{-3} 0.8123
Pseudo-second order	q_e, mg/g k_2, g/(mg min) h, g/(mg min) R^2	1.31 0.170 0.292 0.9997	3.44 0.019 0.224 0.9995
Elovich model	α, mg/(g min) β_t, g/mg R^2	203.626 10.111 0.7604	1.162 1.982 0.9818
Intraparticle diffusion model	k_{id}, mg/(g min$^{1/2}$) C, mg/g R^2	0.0194 0.9625 0.5516	0.1068 1.4125 0.8290
Experimental sorbent capacity	$q_{exp.}$, mg/g	1.37	3.67

The values of sorption capacities q_e found from the graphical plots in Figure 11 for the process temperatures of 25 and 35 °C were 0.28 and 1.73 mg/g, and the correlation coefficients (R^2) were 0.617 and 0.812, respectively. The data obtained on the lithium sorbent sorption, both at temperatures of 25 and 35 °C, did not fit into the pseudo-first order model.

As it can be seen from the calculated data (Table 8), the sorbent sorption capacity and the extraction, respectively, increased significantly with an increase in temperature. The dependence of the process on the temperature may indicate the chemical nature of the process rate limitations.

The integral form of the classical pseudo-second-order velocity equation of Ho and Mackay has the following form [41,42]:

$$q_t = \frac{t}{\frac{1}{k_2 \cdot q_e^2} + \frac{t}{q_e}} \qquad (7)$$

where k_2—sorption rate constant of the pseudo-second order model, g/(mg·min); t—time, min.

The equation was used in the following transformed form to process the experimental data:

$$\frac{t}{q_t} = \frac{1}{k_2 q_e^2} + \left(\frac{1}{q_e}\right) t \qquad (8)$$

The straight lines of the t plots $\frac{t}{q_t}$ shown in Figure 12 give the slope $\frac{1}{q_e}$ and intercept $\frac{1}{k_2 q_e^2}$.

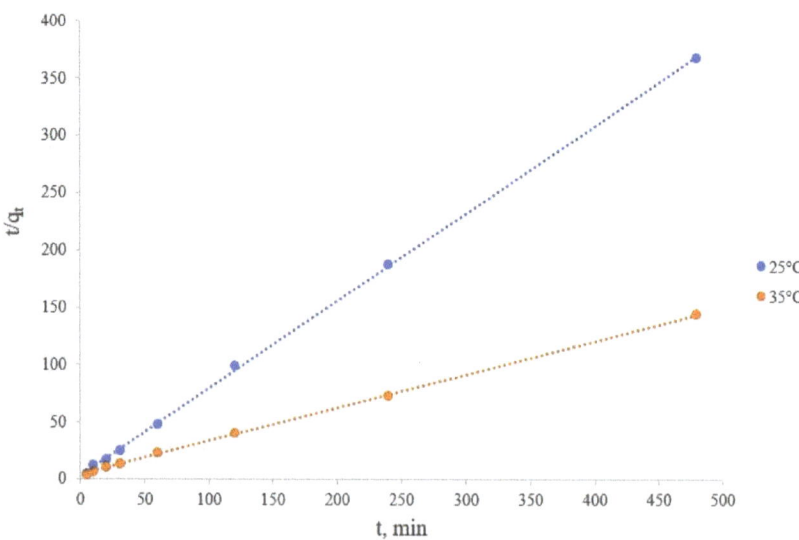

Figure 12. Description of the lithium sorption kinetics from brine in the coordinates of the pseudo-second order model.

The results of graphical constructions and calculations presented in Table 8 show good comparability with the experimental data and suggest the applicability of the pseudo-second order kinetic model.

The calculated values of q_e were found to be 1.31 and 3.44 mg/g with the correlation coefficients (R^2) of 0.9997 and 0.9995 at the process temperatures of 25 and 35 °C, respectively. The obtained data were approximated by a pseudo-second order model and were ideally close to the experimental results.

In accordance with the results obtained, the pseudo-second-order model is the most suitable for the description of the kinetic process of the lithium sorption on the synthesized manganese sorbent. It suggested that the kinetic process is mainly controlled by chemical sorption or chemisorption with the participation of valence forces due to the exchange of electrons between the sorbent and the sorbate [43]. Data from physical research methods of the sorbent indicate that it has a composition of $MnO_2 \cdot 0.5H_2O$.

Apparently, the sorption process occurred due to ion exchange between the lithium ion from the brine and the hydrogen proton that is part of the sorbent water molecule. A sorbent of a similar composition was obtained from its predecessor in accordance with studies [44], i.e., from the lithium-manganese precursor $Li_{1.6}Mn_{1.6}O_4$. During the synthesis of manganese oxide sorbent under selected optimal conditions, its lithium-manganese precursor had a similar formula (Figure 8).

The Elovich kinetic model describes cases of heteroganic chemisorption on solid surfaces [45], i.e., applicable to the process of chemisorption between lithium ions and active proton-containing sites of the sorbent. The Elovich equation takes the contribution to the kinetics of adsorption and desorption processes into account. The linear form of the Elovich model can be represented by the following equation:

$$q_t = \frac{1}{\beta}\ln(\alpha\beta) + \frac{1}{\beta}\ln(t) \qquad (9)$$

where α is the initial sorption rate (mg/(g·min)), and β is the desorption constant (g/mg).

The values of the Elovich parameters are calculated from the slope and intersection of linear graphs of qt versus ln(t) (Figure 13 and Table 8).

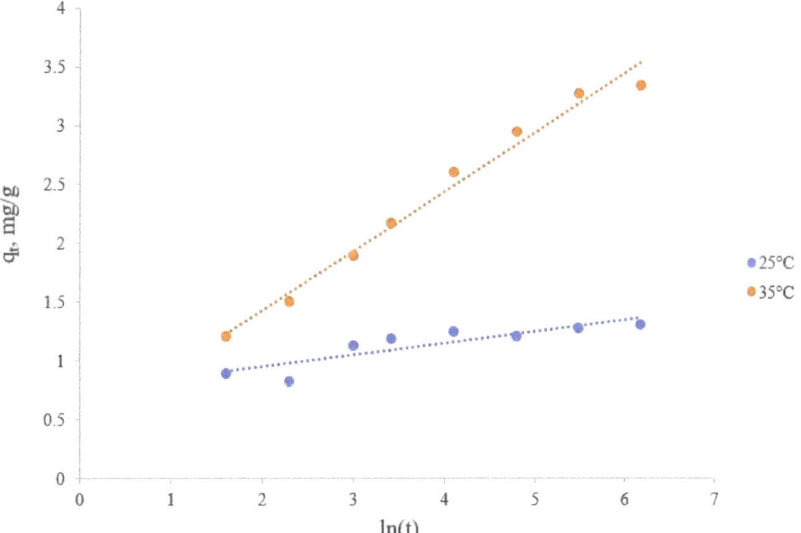

Figure 13. Description of the lithium sorption kinetics from brine in the coordinates of the Elovich model.

The α values for the sorption of lithium ions on the manganese sorbent decreased from 203.625 to 1.163 mg/(g·min), and the β values also decreased from 10.111 to 1.982 g/mg at 25 and 35 °C of the sorption process, and respectively, the values of the correlation coefficient (R^2) were 0.7604 and 0.9818, respectively. The data obtained confirmed that the Elovich model does not agree with the experimental data, which primarily concerns the process at 25 °C. According to the calculated values, the process with a lower temperature of 25 °C was characterized by a very high initial rate of sorption and desorption, while with an increase by 10 °C the initial rate of sorption and desorption decreased significantly.

On the other hand, the sorption of lithium ions from brines on a manganese dioxide sorbent can be represented as a multi-stage process. The first stage includes the transport of lithium ions from the bulk of the brine to the solid surface of inorganic sorbent particles, characterized by volumetric diffusion. Then, the second stage occurs by the diffusion of lithium ions into the boundary layer of solid manganese dioxide sorbent particles, considered as film diffusion. This is followed by a third stage where lithium ions are transported from the surface to the internal pores (pore diffusion or intraparticle diffusion). The last stage will likely be a slow process. The intraparticle diffusion model can be applied using the Weber and Morris equation [46]:

$$q_t = K_{id} t^{0.5} + C \qquad (10)$$

where q_t (mg/g) is the amount of lithium sorbed at time t, K_{id} (mg/(g min 0.5)) is the rate constant of intraparticle diffusion, C is the thickness of the boundary layer.

The linear dependence, presented in the q_t coordinates from $t^{0.5}$, describing the lithium sorption process from brine by a synthesized inorganic sorbent of manganese dioxide under the Weber and Morris model of intraparticle diffusion at temperatures of 25 and 35 °C, is shown in Figure 14. Intraparticle diffusion parameters were calculated from the slope and intercept of line graphs, as shown in Figure 14. Graphs of q_t versus $t^{0.5}$ show that the resulting straight lines did not pass through the origin (C > 0). It was found according to the data obtained as shown in Table 8 that the values of the correlation coefficient R^2 were 0.5516 and 0.8290, the intraparticle diffusion rate constants K_{id} were 0.0194 and 0.1068 mg/(g·min$^{0.5}$), and the thickness of the boundary layer C was 0.9625 and 1.4125 mg/g at sorption process temperatures of 25 and 35 °C, respectively. The data obtained confirmed the unsuitability of this model for description of the sorption kinetics.

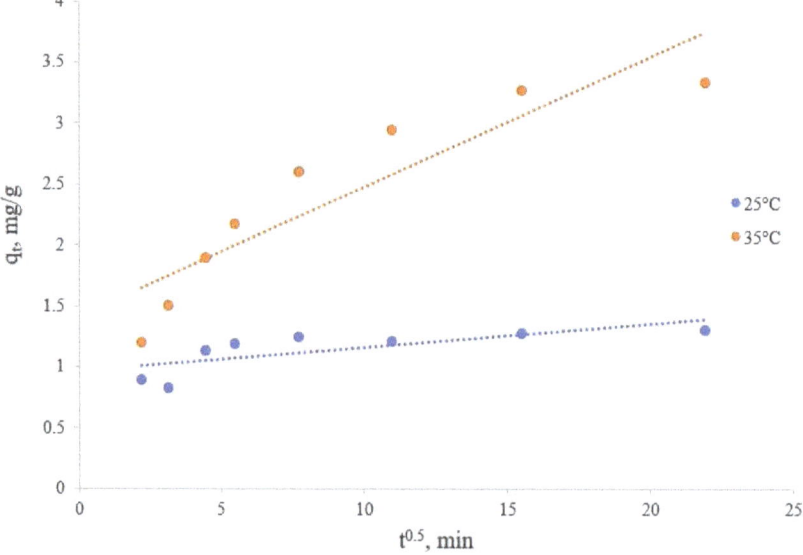

Figure 14. Description of the lithium sorption kinetics from brine using the intraparticle diffusion model.

Based on the kinetic parameters of the four kinetic models, as seen in Table 8, the sorption kinetics were estimated and are in good agreement with the pseudo-second order kinetic model.

Thus, the data obtained show that the process of the lithium sorption from brines on a synthesized inorganic sorbent of manganese dioxide could be described by the Ho and Mackay pseudo-second order equation, and that chemical kinetics occurred. The kinetic parameters were significantly affected by the sorption conditions, in particular temperature and duration.

4. Conclusions

The obtained research results show that, at the first stage of sorbent preparation at the exposure of the mixture of lithium hydroxide and manganese oxides taken from the calculation of Li/Mn molar ratio maintenance = 1, at temperature 125 °C, duration of 13 h, the lithium-manganese oxide of $LiMnO_2$ composition with orthorhombic structure is formed.

Thermal analysis showed that, in order to undergo the calcination process of lithium-manganese oxide $LiMnO_2$ with an orthorhombic crystal lattice structure until the formation of a cubic form $Li_{1.6}Mn_{1.6}O_4$, calcination should preferably be carried out at least at a temperature of 450 °C, but not higher than 600 °C, within which the processes of oxidation of manganese present in the composition of LMO should take place, from degree +3 to degree +4. At temperatures above 600 °C, the decomposition reaction of manganese dioxide can begin with the formation of oxides of lower valencies.

Acid treatment of the precursor 0.5 M HCl is preferably carried out under the following conditions: temperature 40–50 °C, HCl concentration 0.5 M; ratio S:L = 1:700 and 1:800, and duration 24 h. At the same time, the extraction of lithium into solution from the precursor can reach ~93–97%.

It is possible to recover lithium by ~86% from brine with a low content of the target component—5.9–7.8 mg/L lithium with the use of resulting sorbent.

A kinetic model of the lithium sorption process was determined. The adequacy of several kinetic models was assessed to identify the rate-limiting stage.

Four kinetic models were used in the studies to analyze the kinetics of lithium sorption—pseudo-first and pseudo-second order models, the Elovich model, and the intraparticle diffusion model. According to the research results, the pseudo-second-order model is the most suitable for description of the lithium sorption process kinetics on a synthesized manganese sorbent and assumes that the chemical exchange reaction limits the process.

Author Contributions: Conceptualization, Z.K. and A.Y. (Albina Yersaiynova); methodology, A.Y. (Albina Yersaiynova), M.K. and I.S.; software, B.O., M.K. and I.S.; validation, Z.K. and A.Y. (Albina Yersaiynova); formal analysis, A.Y. (Albina Yersaiynova); investigation, A.Y. (Albina Yersaiynova), A.Y. (Azamat Yessengaziyev) and B.O.; resources, A.Y. (Azamat Yessengaziyev), B.O. and I.S.; data curation, Z.K. and A.Y. (Albina Yersaiynova); writing—original draft preparation, Z.K.; writing—review and editing, Z.K. and A.Y. (Albina Yersaiynova); visualization, A.Y. (Azamat Yessengaziyev) and M.K.; supervision, Z.K.; project administration, Z.K.; funding acquisition, Z.K. All authors have read and agreed to the published version of the manuscript.

Funding: This research is funded by the Science Committee of the Ministry of Science and Higher Education of the Republic of Kazakhstan, Programme Targeted Funding BR18574018.

Institutional Review Board Statement: Not applicable.

Informed Consent Statement: Not applicable.

Data Availability Statement: The data and results presented in this study are available in the article.

Acknowledgments: The authors would like to express their deepest gratitude to the head of the BR18574018 program Abdulvaleyev Rinat for his participation and contribution to the scientific work.

Conflicts of Interest: The authors declare that there are no conflict of interest regarding the publication of this manuscript.

References

1. Choubey, P.K.; Chung, K.S.; Kim, M.S.; Lee, J.C.; Srivastava, R.R. Advance review on the exploitation of the prominent energy-storage element Lithium. Part II: From sea water and spent lithium ion batteries (LIBs). *Miner. Eng.* **2017**, *110*, 104–121. [CrossRef]
2. Choubey, P.K.; Kim, M.S.; Srivastava, R.R.; Lee, J.C.; Lee, J.Y. Advance review on the exploitation of the prominent energy-storage element: Lithium. Part I: From mineral and brine resources. *Miner. Eng.* **2016**, *89*, 119–137. [CrossRef]

3. Tedjar, F. Challenges for recycling advanced Li-ion batteries. In Proceedings of the International Battery Association (IBA2013), Barcelona, Spain, 11–15 March 2013. Available online: http://congresses.icmab.es/iba2013/images/Oral_26FEB.pdf (accessed on 11 March 2013).
4. Samoilova, V.I.; Zelenin, V.I.; Saduakasova, A.T.; Kulenova, N.A. Uranium sorption from lake water by natural sorbents and products of their modification. *Kompleks. Ispolz. Miner. Syra Complex Use Miner. Resour.* **2016**, *3*, 91–96.
5. Kenzhaliyev, B. Innovative technologies providing enhancement of non-ferrous, precious, rare and rare earth metals extraction. *Kompleks. Ispolz. Miner. Syra Complex Use Miner. Resour.* **2019**, *310*, 64–75. [CrossRef]
6. Abdikerim, B.E.; Kenzhaliyev, B.K.; Surkova, T.Y.; Didik, N.; Berkinbayeva, A.N.; Dosymbayeva, Z.D.; Umirbekova, N.S. Uranium extraction with modified sorbents. *Kompleks. Ispolz. Miner. Syra Complex Use Miner. Resour.* **2020**, *3*, 84–90. [CrossRef]
7. Guo, T.; Wang, S.; Ye, X.; Liu, H.; Gao, X.; Li, Q.; Guo, M.; Wu, Z. Competitive adsorption of Li, K, Rb, and Cs ions onto three ion-exchange resins. *Desalination Water Treat.* **2013**, *51*, 3954–3959. [CrossRef]
8. Guijosa, A.N.; Casas, R.N.; Calahorro, C.V.; Gonzalez, J.D.L.; Rodríguez, A.G. Lithium adsorption by acid and sodium amberlite. *J. Colloid Interface Sci.* **2003**, *264*, 60–66. [CrossRef]
9. Poluektov, N.S.; Meshkova, S.B.; Poluektova, E.N. *Analytical Chemistry of Lithium*; Nauka: Moscow, Russia, 1975; p. 204.
10. Shi, X.; Zhou, D.; Zhang, Z.; Yu, L.; Xu, H.; Chen, B.; Yang, X. Synthesis and properties of $Li_{1.6}Mn_{1.6}O_4$ and its adsorption application. *Hydrometallurgy* **2011**, *110*, 99–106. [CrossRef]
11. Kan, S.M.; Berstenev, S.V. To the technology extraction of lithium from the formation waters of oil and gas fields of southern Mangyshlak. *News Acad. Sci. Repub. Kazakhstan Ser. Geol. Tech. Sci.* **2017**, *5*, 149–155.
12. Shaoju, B.; Dongdong, L.; Dandan, G.; Jiaoyu, P.; Yaping, D.; Wu, L. Hydrometallurgical processing of lithium, potassium, and boron for the comprehensive utilization of Da Qaidam lake brine via natural evaporation and freezing. *Hydrometallurgy* **2017**, *173*, 80–83.
13. Yoshinaga, T.; Kawano, K.; Imoto, H. Basic study on lithium recovery from lithium containing solution. *Bull. Chem. Soc. Jpn.* **1986**, *59*, 1207–1213. [CrossRef]
14. Sun, Y.; Wang, Q.; Wang, Y.; Yun, R.; Xiang, X. Recent Advances in Magnesium/Lithium Separation and Lithium Extraction Technologies from Salt Lake Brine. *Sep. Purif. Technol.* **2020**, *256*, 117807. [CrossRef]
15. Amer, A.M. The hydrometallurgical extraction of lithium from Egyptian montmorillonite-type clay. *J. Met.* **2008**, *60*, 55–57. [CrossRef]
16. An, J.W.; Kang, D.J.; Tran, K.T.; Kim, M.J.; Lim, T.; Tran, T. Recovery of lithium from Uyuni salar brine. *Hydrometallurgy* **2012**, *117–118*, 64–70. [CrossRef]
17. Ho, P.C.; Nelson, F.; Kraus, K.A. Adsorption on inorganic materials: VII. Hydrous tin oxide and SnO_2 filled carbon. *J. Chromatogr. A* **1978**, *147*, 263–269. [CrossRef]
18. Bukowsky, H.; Uhlemann, E.; Steinborn, D. The recovery of pure lithium chloride from "brines" containing higher contents of calcium chloride and magnesium chloride. *Hydrometallurgy* **1991**, *27*, 317–325. [CrossRef]
19. Kitajo, A.; Suzuki, T.; Nishihama, S.; Yoshizuka, K. Selective recovery of lithium from seawater using a novel MnO_2 type adsorbent II—Enhancement of lithium ion selectivity of the adsorbent. *Ars Separatoria Acta* **2003**, *2*, 97–106.
20. Ooi, K.; Miyai, Y.; Sakakihara, J. Mechanism of Li^+ insertion in spinel-type manganese oxide. Redox and ion-exchange reactions. *Langmuir* **1991**, *7*, 1167–1171. [CrossRef]
21. Umeno, A.; Miyai, Y.; Takagi, N.; Chitrakar, R.; Sakane, K.; Ooi, K. Preparation and adsorptive properties of membrane-type adsorbents for lithium recovery from seawater. *Ind. Eng. Chem. Res.* **2002**, *41*, 4281–4287. [CrossRef]
22. Yoshizuka, K.; Fukui, K.; Inoue, K. Selective recovery of lithium from seawater using a novel MnO_2 type adsorbent. *Ars Separatoria Acta* **2002**, *1*, 79–86.
23. Menzheres, L.T.; Ryabcev, A.D.; Mamylova, E.V.; Kocupalo, N.P. Method of Producing a Sorbent for the Extraction of Lithium from Brines. Patent RF 2223142, IPC B01J 20/02, C01D 15/00, 10 February 2004.
24. Menzherez, L.T.; Kotsupalo, N.P. Granular sorbents based on $LiCl·2Al(OH)_3·mH_2O$. *Appl. Chem. J.* **1999**, *10*, 1623–1627.
25. Ryabtsev, A.D.; Menzheres, L.T.; Kotsupalo, N.P.; Serikova, L.A. Obtainment of a granular sorbent based on $LiCl·2Al(OH)_3·mH_2O$ by the non-waste method. *Chem. Sustain. Dev.* **1999**, *7*, 343–349.
26. Menzheres, L.T.; Kocupalo, N.P.; Orlova, L.B. Method of Obtaining Granular Sorbent. Patent RF 2050184, IPC B01J 20/00, 20/30, 20 December 1995.
27. Kotsupalo, N.P.; Ryabcev, A.D. *Chemistry and Technology for Producing Lithium Compounds from Lithium-Bearing Hydromineral Raw Materials*; Geo: Novosibirsk, Russia, 2008; p. 291.
28. Zhang, L.; Zhou, D.; Yao, Q.; Zhou, J. Preparation of H_2TiO_3-lithium adsorbent by the sol–gel process and its adsorption performance. *Appl. Surf. Sci.* **2016**, *368*, 82–87. [CrossRef]
29. Moazeni, M.; Hajipour, H.; Askari, M.; Nusheh, M. Hydrothermal synthesis and characterization of titanium dioxide nanotubes as novel lithium adsorbents. *Mater. Res. Bull.* **2014**, *61*, 70–75. [CrossRef]
30. Zhang, L.; Zhou, D.; He, G.; Wang, F.; Zhou, J. Effect of crystal phases of titanium dioxide on adsorption performance of H_2TiO_3-lithium adsorbent. *Mater. Lett.* **2014**, *135*, 206–209. [CrossRef]
31. Nazri, G.A.; Pistoia, G. *Lithium Batteries-Science and Technology*; Kluwer Academic Publishers: Boston, MA, USA, 2003; p. 708.
32. Chitrakar, R.; Sakane, K.; Umeno, A.; Kasaishi, S.; Takagi, N.; Ooi, K. Synthesis of orthorhombic $LiMnO_2$ by solid-phase reaction under steam atmosphere and a study of its heat and acid-treated phases. *J. Solid State Chem.* **2002**, *169*, 66–74. [CrossRef]

33. Chitrakar, R.; Kahon, H.; Miyai, Y.; Ooi, K. Recovery of Lithium from Seawater Using Manganese Oxide Adsorbent ($H_{1.6}Mn_{1.6}O_4$) Derived from $Li_{1.6}Mn_{1.6}O_4$. *Ind. Eng. Chem. Res.* **2001**, *40*, 2054–2058. [CrossRef]
34. Chitrakar, R.; Makita, Y.; Ooi, K.; Sonoda, A. Magnesium-doped manganese oxide with lithium ion-sieve property: Lithium adsorption from salt lake brine. *Bull. Chem. Soc. Jpn.* **2013**, *86*, 850–855. [CrossRef]
35. Sorour, M.H.; Hani, H.A.; El-Sayed, M.M.H.; Mostafa, A.A.; Shaalan, H.F. Synthesis, Characterization and Performance Evaluation of Lithium Manganese Oxide Spinels for Lithium Adsorption. *Egypt. J. Chem.* **2017**, *60*, 697–710.
36. Lukman, N.; Gita, A.S.; Diah, S.; Amien, W. Synthesis and Characterization of Lithium Manganese Oxide with Different Ratio of Mole on Lithium Recovery Process from Geothermal Fluid of Lumpur Sidoarjo. *J. Mater. Sci. Chem. Eng.* **2015**, *3*, 56–62.
37. Xu, X.; Chen, Y.; Wan, P.; Gasem, K.; Wang, K.; He, T.; Adidharma, H.; Fan, M. Extraction of lithium with functionalized lithium ion-sieves. *Prog. Mater. Sci.* **2016**, *84*, 276–313. [CrossRef]
38. Zandevakili, S.; Ranjbar, M. Synthesis of Lithium Ion Sieve Nanoparticles and Optimizing Uptake Capacity by Taguchi Method. *Iran. J. Chem. Chem. Eng.* **2014**, *33*, 15–24.
39. Lagergren, S. About the theory of so called adsorption of soluble substance. *K. Sven. Vetenskapsakademiens Handl.* **1898**, *24*, 1–39.
40. Ho, Y.S.; Ng, J.C.Y.; McKay, G. Kinetics of pollutant sorption by biosorbents: Review. *Sep. Purif. Methods* **2000**, *2*, 189–232. [CrossRef]
41. Ho, Y.S.; McKay, G. Kinetic models for the sorption of dye from aqueous solution by wood. *Process Saf. Environ. Prot.* **1998**, *76*, 183–191. [CrossRef]
42. Ho, Y.S.; McKay, G. A comparison of chemisorption kinetic models applied to pollutant removal on various sorbents. *Trans. IChemE* **1998**, *76 Pt B*, 332–340. [CrossRef]
43. Ho, Y.S.; McKay, G. Pseudo-second order model for sorption processes. *Process Biochem.* **1999**, *34*, 451–465. [CrossRef]
44. Chitrakar, R.; Kanoh, H.; Miyai, Y.; Ooi, K. A New Type of Manganese Oxide ($MnO_2 \cdot 0.5H_2O$) Derived from $Li_{1.6}Mn_{1.6}O_4$ and Its Lithium Ion-Sieve Properties. *Chem. Mater.* **2000**, *12*, 3151–3157. [CrossRef]
45. Javadian, H. Application of kinetic, isotherm and thermodynamic models for the adsorption of Co(II) ions on polyamidine/polypyrrole copolymer nanofibers from aqueous solution. *J. Ind. Eng. Chem.* **2014**, *6*, 4233–4241. [CrossRef]
46. Weber, W.J.; Morris, J.C. Kinetics of adsorption on carbon from solution. *J. Sanit. Eng. Div. Proc. Am. Soc. Civ. Eng.* **1963**, *89*, 31–59. [CrossRef]

Disclaimer/Publisher's Note: The statements, opinions and data contained in all publications are solely those of the individual author(s) and contributor(s) and not of MDPI and/or the editor(s). MDPI and/or the editor(s) disclaim responsibility for any injury to people or property resulting from any ideas, methods, instructions or products referred to in the content.

Article

Properties of Padding Welds Made of CuAl2 Multiwire and CuAl7 Wire in TIG Process

Jarosław Kalabis [1,*], Aleksander Kowalski [1] and Santina Topolska [2]

[1] Center of Advanced Materials Technologies, Łukasiewicz Research Network—Institute of Non-Ferrous Metals, 44-100 Gliwice, Poland
[2] Department of Welding, Silesian University of Technology, 44-100 Gliwice, Poland
* Correspondence: jaroslaw.kalabis@imn.lukasiewicz.gov.pl

Abstract: This paper presents the influence of the Hot Isostatic Pressing (HIP) process on the structure, mechanical properties and corrosion resistance of padding welds made using the TIG method from aluminium bronzes—CuAl7 and CuAl2 (a composite bundled wire). The tested CuAl7 material was a commercial welding wire, while the CuAl2 composite was an experimental one (a prototype of the material produced in multiwire technology). The wire contains a bundle of component materials—in this case, copper in the form of a tube and aluminium in the form of rods. The padding welds were manufactured for both the CuAl7 wire and the CuAl2 multiwire. The prepared samples were subjected to the Hot Isostatic Pressing (HIP) process, chemical composition tests were performed, and then the samples were subjected to observations using light microscopy, Vickers hardness testing, electrical conductivity tests, and apparent density determination using Archimedes' Principle. Tribological tests (the 'pin on disc' method) and neutral salt spray corrosion tests were conducted. The padding weld made of CuAl2 multifiber material subjected to the HIP process is characterized by an improvement in density of 0.01 g/cm^3; a homogenization of the hardness results across the sample was also observed. The average hardness of the sample after the HIP process decreased by about 15HV, however, the standard deviation also decreased by about 8HV. The electrical conductivity of the CuAl2 welded sample increased from 16.35 MS/m to 17.49 MS/m for the CuAl2 sample after the HIP process. As a result of this process, a visible increase in electrical conductivity was observed in the case of the wall made of the CuAl2 multiwire—an increase of 1.14 MS/m.

Keywords: welding wires; CuAl7; multiwire; CuAl2; HIP; corrosion

Citation: Kalabis, J.; Kowalski, A.; Topolska, S. Properties of Padding Welds Made of CuAl2 Multiwire and CuAl7 Wire in TIG Process. *Materials* 2023, 16, 6199. https://doi.org/10.3390/ma16186199

Academic Editors: Yilong Dai and Murali Mohan Cheepu

Received: 11 July 2023
Revised: 6 September 2023
Accepted: 11 September 2023
Published: 13 September 2023

Copyright: © 2023 by the authors. Licensee MDPI, Basel, Switzerland. This article is an open access article distributed under the terms and conditions of the Creative Commons Attribution (CC BY) license (https://creativecommons.org/licenses/by/4.0/).

1. Introduction

Copper and aluminum alloys are the materials most commonly used in the marine industry for components operating in the corrosive seawater environment. CuAl alloys, due to their good mechanical properties, high corrosion resistance and wide availability, are most often used for the production of ship propellers, steering systems, machine parts and shafts, and in the aviation industry for engine parts. New technologies enable the production of modified construction materials, which facilitates the longer use of machines and significantly increases their reliability. The multifiber material suggested in the work is an innovative approach to the production of new welding materials, which allows for omitting the expensive melting and continuous casting processes. It is possible to make multifiber materials with a modified chemical composition based on plastic working. Such solutions are not known. After welding, the produced multi-filament material was subjected to a series of mechanical and structural tests, and pressing was performed using the HIP press. Hot isostatic pressing is a technological process used to remove the porosity of cast materials and objects produced, for example, using 3D printing techniques [1]. Additionally, the author of reference [1] notes the positive aspect of the HIP process application, which is revealed by the increase in mechanical properties and the elimination of gas porosities, e.g., in copper alloys manufactured in an unconventional way. Tungsten Inert Gas (TIG) is a welding method that is used very often due to the stability of the arc, good padding welds formation and the use of various welding materials [2,3]. This method uses a welding wire in the form of sections of about 1 m in length, which makes it easier to test new welding materials that can be manufactured in smaller series. The TIG method uses a shielding gas, which is usually compressed argon 5.0. Its role is to isolate the welded surface from the access of atmospheric air during the process. This protection guarantees the cleanliness of the padding weld, especially when welding with active metals such as copper, aluminum, and their alloys. Corrosion-resistant materials are desirable construction materials for use in many branches of industry. The group of bronzes with the addition of aluminum, which improves mechanical properties and corrosion resistance, is developing strongly, hence new applications for these types of materials are possible. So far, research on modified CuAl alloys has focused largely on improving the strength of the material by optimizing the crystal structure with the use of microalloying additives such as Ni, Zr, Cr and Ag [4,5]. The author of reference [6] describes the properties of a clad wire consisting of an aluminum core with a copper coating; this material has not been subjected to welding tests. The authors of references [7,8] study aluminum bundled in a copper pipe and describe the effect of heat treatment on the properties of copper and aluminum. The authors describe the positive aspects of heat treatment on the change in the yield strength and the influence of intermetallic compounds on the bonding of metals. Studies on new composite materials bundled in various configurations have been published; for example, the authors of references [8–10] describe their research on materials characterized by exceptional mechanical or electrical properties. The metals described in these publications are Cu-Nb, Cu-Ag, Cu-Fe and Cu-Al material configurations. The latter was tested by Chen et al. [11] as an alloy with low Al content in terms of tribological properties. The author describes the influence of aluminum content on the abrasion resistance of aluminum bronzes. These metals are successfully used in welding processes, and the author of reference [12] describes aluminum bronzes with additives welded, using the TIG method, onto steel. In reference [13], the author examines construction materials welded using the TIG method. The advantage of this welding method is the possibility of welding in positions convenient for the welder. Welded materials are characterized by high mechanical properties, which is demonstrated in the publication [12]. The described welding methods do not cause changes in the composition of the welded layer in relation to the electrode material (no effect of burning elements). The author of reference [14] describes the mechanisms occurring between the steel substrate and tin bronze welded using the TIG method. Bronzes are one of the most commonly used materials for welding using this method. There is a manual welding method using fluxes (A-TIG), a method described by Rana et al. in reference [15]. This method is characterized by the increased penetration of the weld material into the base material. In references [16–22], the authors discuss the use of welding methods in 3D printing. The presented WAAM (Wire Arc Additive Manufacturing) method is characterized by the high accuracy of the manufactured elements [17] compared to classic castings; moreover, it significantly affects the modification of the macro and microstructure [20].

2. Novelty of Idea

In this paper, the new concept of manufacturing materials for TIG applications was presented. There are many works which consider the conventional CuAl materials in the form of solid wires, but there is a lack of information in broadly accessible works that could be strictly connected to research on this kind of multiwire. The manufacturing of the CuAl2 multiwires is the first stage of the research on the manufacturing of the new groups of materials for TIG from the CuAl bronze group. The next step will be the production of CuAl multiwires with an increased content of aluminium, above the content currently available in the commercial welding wires—the manufacturing of multiwires provides this possibility. The idea of the work is to produce a welding wire from the broadly available semi-products in the form of pipes and rods/wires omitting the expensive casting process. It is assumed that the proposed manufacturing method of high-aluminium bronzes will be characterized by a low price, and the obtained materials will be characterized by high functional properties.

3. Experiment

The aim of the work was to determine the impact of the HIP process on the properties of padding welds manufactured using the TIG method using wires: Cu with the addition of Al 7% by weight (CuAl7) and 2% by weight (multiwire bundled three times). The CuAl7 welding wire is a material available on the welding consumables market. The multiwire with 2% aluminum is a new type of wire produced during plastic processing by drawing a bundle of materials: the matrix, which is a copper tube, and the core, which is a composite-clad wire with an aluminum core and a copper matrix. Figure 1 presents the input material for the drawing process (the copper pipe and aluminum wire), the SKET (Magdeburg, Germany) drum drawing machine (for the first stage of plastic processing) and the wet drawing machine (for drawing wires below 3 mm).

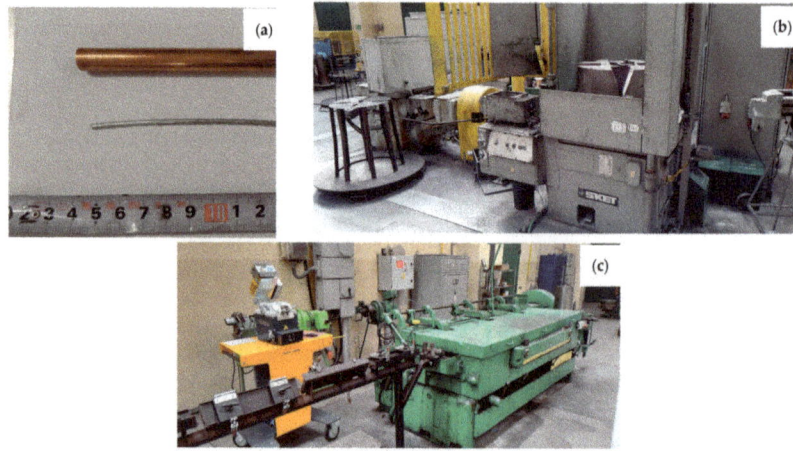

Figure 1. Photos of the input materials (**a**) and facilities for drawing process: SKET drum drawing machine (**b**) and wet drawing machine for drawing wires of small diameters (**c**).

The multiwire concept is shown in Figure 2. Figure 2a presents the first bundle of wires (7^1—7 aluminum fibers) which is used for the next stage, Figure 2b—the second stage of bundling (7^2—49 aluminum fibers) and Figure 2c—the third and last stage of bundling (7^3—343 aluminum fibers). The exponent means the number of bundling processes to which the multiwire was subjected.

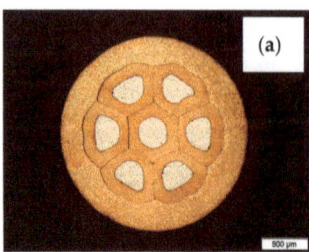 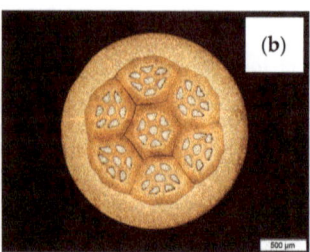 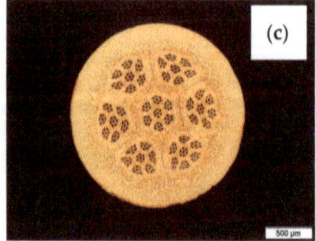

Figure 2. Photos of CuAl multiwires after subsequent stages of bundling: (**a**) bundling process I, (**b**) bundling process II and (**c**) bundling process III.

The trials included the manufacturing of the CuAl multifiber wires using aluminum (Al) wires by preparing composite clad wires and their double and triple bundling into copper (Cu) pipes. The final form of the multiwires was obtained by drawing the prepared bundles to a diameter of ⌀2.15 mm. Stage I of the work on the production of multimetal wires was the production of clad wires. A section of an aluminum wire with a diameter of ⌀3.15 mm was inserted into a copper pipe; its diameter was ⌀6.90 mm. Next, the drawing operation was performed using a chain drawbench to obtain a diameter of ⌀2.15 mm. Stage II involved cutting the obtained the CuAl clad wire into seven equal sections and bundling them into a copper tube with a diameter of ⌀9.00 mm. The bundles prepared this way were drawn to obtain a diameter of ⌀2.15 mm. The cutting and bundling operations were performed once more. The effect of the subsequent stages of bundling is shown in Figure 1. After the third stage of bundling, the multiwire was tested for fiber content, applying an automatic grain measurement method using a digital optical microscope (Keyence VHX7000, Birmingham, AL, USA) and the ratio of the surface area of aluminum fibers to the surface area of the multiwire was determined. The resulting diameter of ⌀2.15 mm is convenient for use in the TIG method. Compared to the automatic MIG/MAG method, the larger diameter of the multiwire facilitates the manual feeding of the material into the weld pool. The CuAl2 multiwires were subjected to TIG welding tests in an argon 5.0 shield. The TIG welding tests allowed for checking the quality of the produced multiwire and its functional properties. The multiwire was cut into approx. 400 mm sections. The padding welds were made on a 4 mm thick, X12Cr13 stainless steel baseplate, which enabled the controlling of the temperature of the process, so as not to allow the baseplate to bend after the welding process. As part of the work, welding tests were performed using two types of materials: the CuAl2 multiwire and the commercially available CuAl7 wire. The welding tests were conducted using a Fronius TransTig 2200 (Pettenbach, Austria) welding machine. The welding parameters are presented in Table 1. The differences in used welding parameters result from the susceptibility to welding of each particular wire and the observation of welding process stability; the multi-fiber material smoked significantly when increasing the current parameters.

Table 1. TIG welding process parameters.

Welding Process Parameters	CuAl2	CuAl7
Welding machine operating mode	2-cycle	2-cycle
Power type	AC	AC
Welding current, [A]	88	132
Arc voltage (observed range), [V]	13.0–16.5	15.8–18.0
Shielding gas, purity	Argon, 5.0	Argon, 5.0
Gas flow rate [L/min]	10	10

Two samples of each material were cut out of the manufactured padding welds and one of them was subjected to the HIP process (AIP8-30H, Columbus, OH, USA) at a temperature of 850 °C and a pressure of 150 MPa. The content of elements in the produced padding welds was tested using a spark spectrometer (spectromaxx LMA09, Kleve, Germany). Macroscopic and microscopic observations were made using light microscopy (Olympus GX71F, Tokyo, Japan) to determine the grain size and the quality of the padding weld (cavities, microshrinkage, gasification). Vickers HV1 hardness measurements (FutureTech FM700, Kawasaki, Japan) and electrical conductivity measurements were conducted (contact device Sigmatest Foerster 2.069, Baden-Württemberg, Germany). The samples were tested for changes in density. The Archimedes Principle test was carried out using a density measuring balance by immersing the samples in ethyl alcohol. Tribological tests (Anton Paar THT, Graz, Austria) were also performed using the 'pin on disc' method. The tribological tests were carried out on the side wall of the

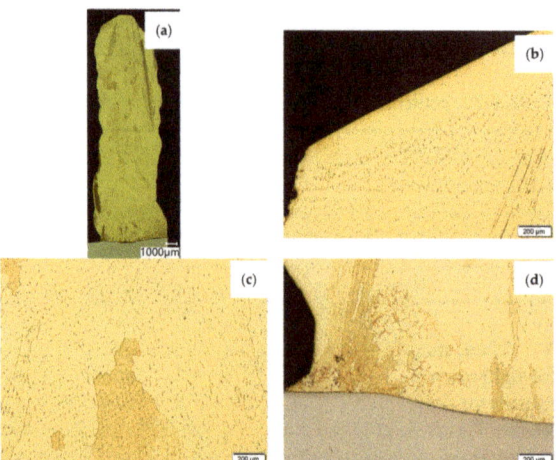

Figure 6. Structure of the wall made with CuAl7 wire: (**a**) macrostructure, (**b**) padding weld face, (**c**) padding weld middle area, (**d**) fusion area of CuAl7 with steel.

Then, the samples were subjected to the Hot Isostatic Pressing process. At a temperature of 850 °C and a pressure of 150 MPa, the process of material densification applying compressed Argon 5.0 was performed. The results of the observations of the CuAl2 sample subjected to the HIP process are shown in Figure 7, and Figure 8 shows the results of the observations of the CuAl7 sample subjected to the HIP process. The microstructure of the manufactured walls in the padding weld face area after the HIP process (Figures 7 and 8) is characterized by the presence of grains with clear boundaries. Nevertheless, there are visible longitudinal grains in the direction from the weld face to the area of fusion with the steel baseplate. The areas of large grains at the padding weld face were refined (Figures 7b and 8b). Observations of the microstructure after the HIP process indicate the absence of defects, discontinuities or voids in the middle area of the sample.

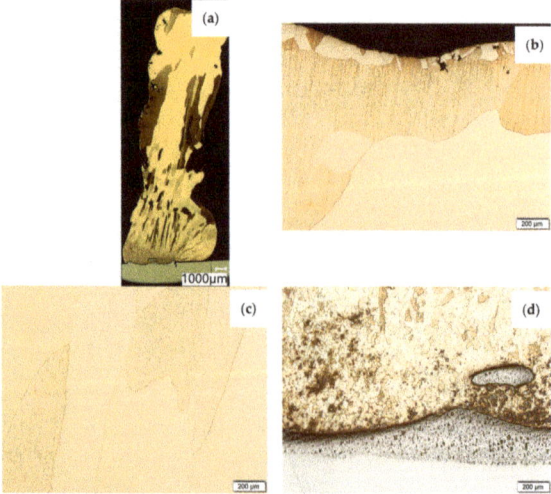

Figure 7. The structure of the wall produced using CuAl2 multiwire subjected to the HIP process: (**a**) macrostructure, (**b**) padding weld face, (**c**) middle area, (**d**) fusion area of CuAl2 with steel.

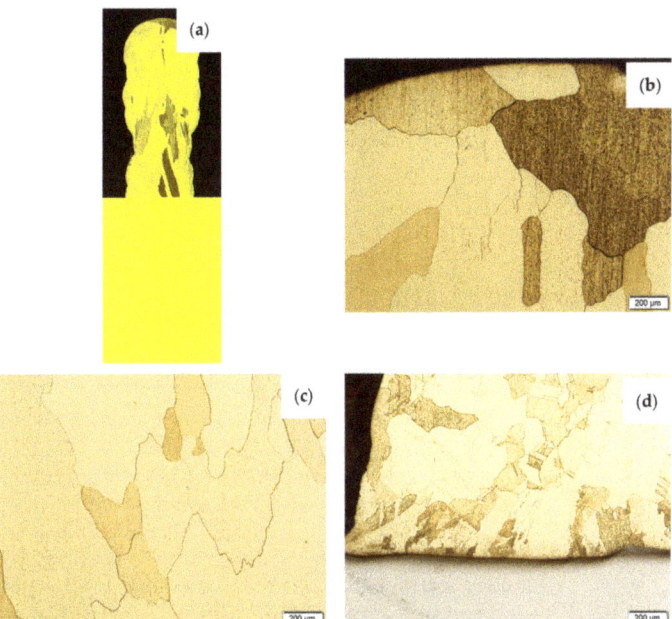

Figure 8. The structure of the wall made of CuAl7 wire subjected to the HIP process: (**a**) macrostructure, (**b**) padding weld face, (**c**) middle area, (**d**) fusion area of CuAl7 with steel.

In Table 4, the results of Vickers hardness measurements at a load of 1 kgf are presented. The measurements were made in accordance with the diagram shown in Figure 9 (Points from +1 to +6 representing the locations where the measurement was taken).

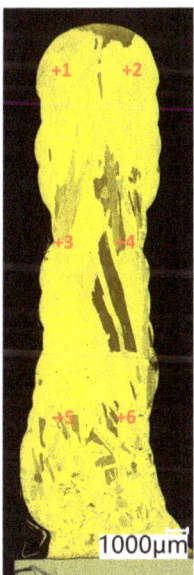

Figure 9. Test points of Vickers hardness test.

Table 4. HV1 hardness test results of the walls produced using the CuAl2 multiwire and the CuAl7 wire.

	No of Dents	CuAl2	CuAl2 after HIP	CuAl7	CuAl7 after HIP
face	1	85.5	56.4	115.5	83.8
face	2	86.5	56.6	117.2	82.6
middle	3	75.8	59.7	102.1	84.4
middle	4	76.1	60.8	105.7	70.1
bottom—fusion with the baseplate	5	64.2	61.2	95.5	65.1
bottom—fusion with the baseplate	6	62.0	63.2	102.7	69.0
MEAN VALUE		75.0	59.7	106.5	75.8
STANDARD DEV.		10.29	2.69	8.37	8.68

The mean hardness value of the CuAl2 wall is 75.0HV1 with a standard deviation of 10.29HV1, while the average hardness of the CuAl2 sample after HIP processing is 59.7HV1 with a standard deviation of 2.69HV1. The hardness after the HIP process decreased, and the standard deviation also decreased significantly. The average hardness of the CuAl7 sample was 106.7HV1 with a standard deviation of 8.37HV1, while the average hardness of the CuAl7 sample after the HIP process was 75.8HV1 with a standard deviation of 8.68HV1. The HIP process contributed to the unification of the grain structure. The conductivity of the walls made using the CuAl2 multiwire is 16.35 MS/m, while after HIP it is 17.49 MS/m. Electrical conductivity tests for the walls made with the CuAl7 wire before the HIP process showed a conductivity of 7.52 MS/m, while after HIP it was 7.55 MS/m. The obtaining of clearly higher values of electrical conductivity, in the case of the wall made with the CuAl2 multiwire compared to that made with the CuAl7 wire, is due to the presence of a higher copper content in the material. As a result of the HIP process, a visible increase in electrical conductivity was observed in the case of the wall made using the CuAl2 multiwire—an increase of 1.14 MS/m.

The results of the density measurement, using Archimedes' Principle, of both materials, before and after the Hot Isostatic Pressing process, are presented in Table 5.

Table 5. Results of density measurements using the Archimedes method.

Sample	Density [g/cm^3]
CuAl2	8.67
CuAl2 after HIP	8.68
CuAl7	7.80
CuAl7 after HIP	7.82

The increase in density caused by the HIP application is small and, in the case of the CuAl7 sample, it was 0.02 g/cm^3 while, for the CuAl2 sample after the HIP process, an increase of 0.01 g/cm^3 was noted. Table 5 presents the parameters of the 'pin-on-disc' tribological test. In addition, before and after the process, the weights of the samples and counter-samples involved in the study were measured. The list of mass values is presented in Table 6.

Tribological tests were conducted to determine the mean value of the coefficient of friction for the friction pair: a sample made of the tested materials and a counter-sample made of 100Cr6 steel.

Figure 10 shows the change in the weight of the samples subjected to the HIP process and, before and after the pin-on-disc process, a reduced weight loss in the CuAl2 sample after HIP can be seen while the mass of the counter-samples in the tested variants of materials does not change; therefore, these values are not included at Figure 10.

Table 6. Tribological tests parameters.

Sampling rate, Hz	60
Load, N	10
Radius, mm	8
Distance, m	2000
Linear speed, cm/s	50
Counter-sample material	100Cr6
Temperature, °C	23.6
Humidity, %	23.1

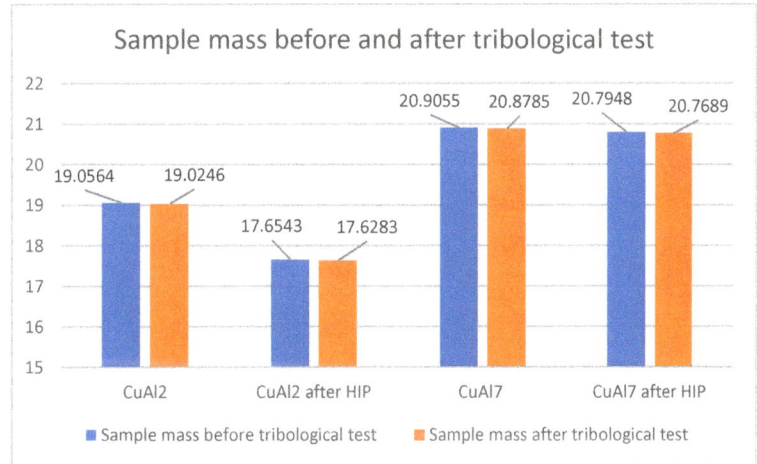

Figure 10. Sample weight before and after the tribological process.

The mean value of the coefficient of friction for the tested friction pairs was, for the samples from the CuAl2 multiwire, $\bar{\mu} = 0.57$, while for the samples from the CuAl7 wire the coefficient was $\bar{\mu} = 0.41$. Figure 11 shows the graph of the coefficient of friction as a function of the distance for the CuAl2 material, and Figure 12 for the CuAl7 material.

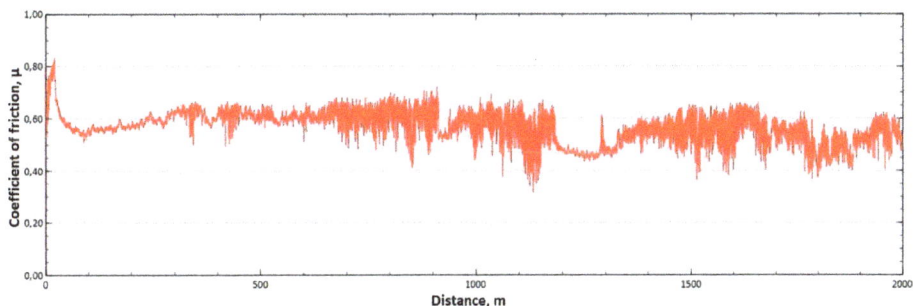

Figure 11. Graph of the coefficient of friction as a function of distance—CuAl2 sample.

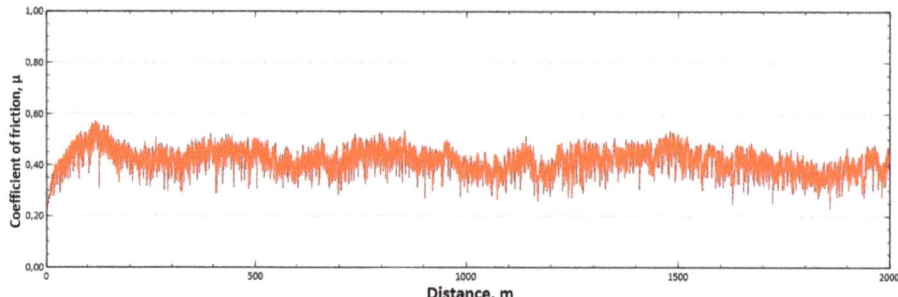

Figure 12. Graph of the coefficient of friction as a function of distance—CuAl7 sample.

The analysis of the course of changes in the coefficient of friction as a function of the sliding distance for both samples after tribological tests at room temperature showed that, at the initial distance of the sliding distance, the value of the coefficient of friction increases (the so-called material grinding-in), and then, after reaching approx. 50 m for the CuAl2 sample and about 100 m for the CuAl7 sample, this value reaches a maximum, then decreases and stabilizes. Further testing in the case of the CuAl7 sample proceeded in a stable manner, with a mean value of the coefficient of friction of approx. 0.41. In the case of the CuAl2 sample, the test after the grinding-in stage was less stable and, after covering a distance of approx. 700 m, the value of the coefficient of friction fluctuates with increasing amplitude. Tribological tests on the samples after the HIP process were also performed. The aim of the study was to check the effect of the process on the value of the coefficient of dry friction. The test was conducted in accordance with the parameters in Table 6.

The mean value of the coefficient of friction for the tested friction pairs was, for the samples from the CuAl2 multiwire after HIP, $\bar{\mu} = 0.64$, while for the samples from the CuAl7 wire after HIP the mean value of the coefficient $\bar{\mu} = 0.50$. In Figure 13, a graph of the coefficient of friction as a function of distance for the CuAl2 material after HIP is shown, while in Figure 14 the same graph for the CuAl7 material after HIP is presented. The analysis of the course of changes in the coefficient of friction as a function of sliding distance, for both samples after HIP after tribological tests at room temperature, showed that the overall trend of changes is similar to the test conducted for samples before HIP. At the initial distance of the sliding distance, the value of the coefficient of friction, and then, after reaching approx. 300 m for the CuAl2 sample after HIP and about 100 m for the CuAl7 sample after HIP, this value reaches a maximum, then decreases and stabilizes. Further testing in the case of the CuAl7 sample after HIP proceeded in a stable manner with a mean value of the coefficient of friction of about 0.50. In the case of the CuAl2 sample, after the grinding-in stage, the test was less stable and, after covering a distance of approx. 500 m, the value of the coefficient of friction fluctuates with increasing amplitude.

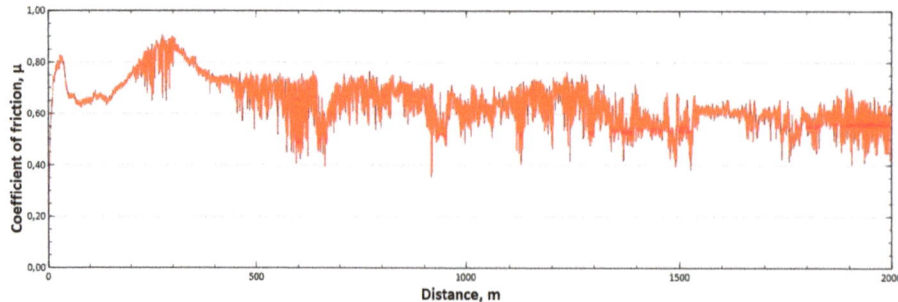

Figure 13. Graph of the coefficient of friction as a function of distance—CuAl2 sample after HIP.

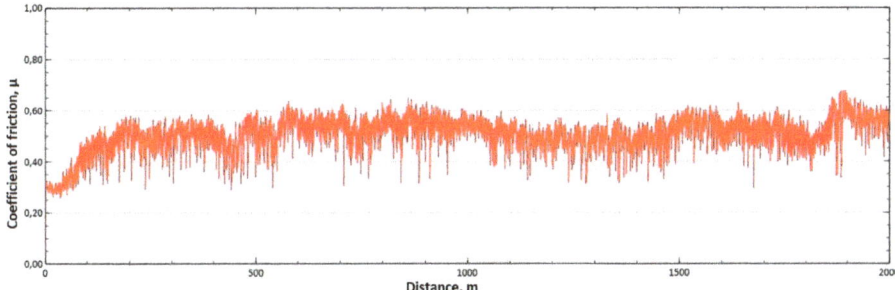

Figure 14. Graph of the coefficient of friction as a function of distance—CuAl7 sample after HIP.

The neutral salt spray corrosion test (NSS) of samples made of the CuAl2 multiwire and the CuAl7 wire was conducted in accordance with the PN-EN ISO 9227:2017-06 standard [23]. The test parameters are listed in Table 7.

Table 7. NSS test parameters.

Test time, h	48.0
Time interval between checks, h	24
Chamber temperature, °C	35.0
Humidifier temperature, °C	48.0
NaCl solution, g/L	50.0
Mass loss of steel reference samples, g/m^2	72.95

The results of the corrosion test describing the state of the solution and the process parameters for the conducted test are shown in Table 8.

Table 8. Corrosion tests results of CuAl2 and CuAl7 samples.

pH of the solution	6.41
pH of the condensate	6.96
Pluviometric constant, mL/h	1.38
Condensate density, g/cm^3	1.034
Volume of the collected solution, mL	130
Ionic conductivity of the water used to prepare the solution, μS/cm	3.68

After completion of the corrosion tests, macroscopic photos of the samples were taken showing the effect of neutral salt fog on the surface of the tested materials. In Figure 15, samples made from the CuAl2 multiwire subjected to the corrosion test are shown, including (a and b) samples directly after welding and (c and d) additionally subjected to the HIP process. The sample before the HIP process (Figure 15a,b) is covered with a visible layer of corrosion products, most likely copper oxide (a typical green tint). The oxide occurs both on the surface perpendicular to the direction of welding and on the side surface of the padding weld. In the case of the samples after the HIP process (Figure 15c,d), copper oxide is present in a smaller amount on the surface perpendicular to the direction of welding, and it is concentrated in the area close to the padding weld face. In Figure 15d, the characteristic greenish tint is not visible on the side surface, which may suggest the absence of corrosion. In Figure 16, samples made from the CuAl7 multiwire subjected to the corrosion test are shown, including (a and b) samples directly after welding and (c and d) additionally subjected to the HIP process. Corrosion products were observed on the entire surface of the samples. The sample not subjected to the HIP process has less visible traces of corrosion on both tested surfaces (Figure 16a,b). The sample after the HIP process has a clear trace of green tarnish on the surface perpendicular to the direction of welding, and a green layer of corrosion products can also be seen on the side surface (Figure 16c,d).

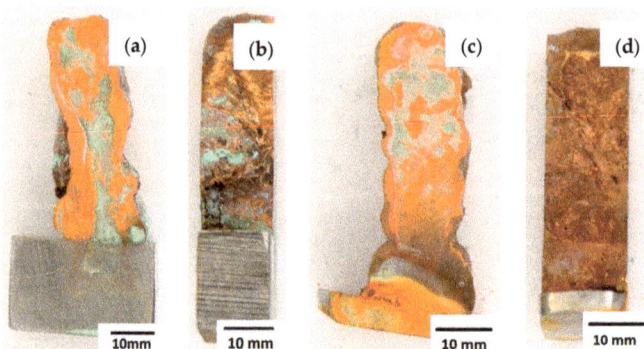

Figure 15. Surface of samples made with CuAl2 multiwire before the HIP process: (**a**) surface perpendicular to the direction of welding, (**b**) side surface of the wall; and after the HIP process: (**c**) surface perpendicular to the direction of welding, (**d**) side surface of the wall.

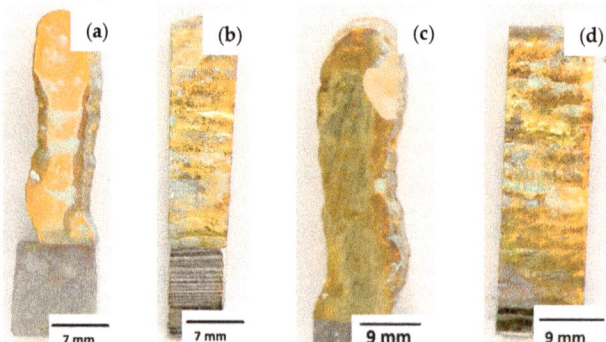

Figure 16. Surface of samples made using CuAl7 multiwire before the HIP process: (**a**) surface perpendicular to the direction of welding, (**b**) side surface of the wall; and after the HIP process: (**c**) surface perpendicular to the direction of welding, (**d**) side surface of the wall.

Samples before and after the HIP process, made using the CuAl2 multiwire and the CuAl7 wire, were subjected to metallographic observations after exposure to salt fog (Figures 17 and 18). The area at the padding weld face and the middle area of the manufactured wall were analysed. Figure 17 shows $Na_2Cr_2O_7$-etched CuAl2 samples, while, in Figure 18, CuAl7 samples are presented. The wall made with the CuAl2 multiwire in the area of the padding weld face (Figure 17a) is characterized by a structure in which the effect of salt fog is clearly visible. Despite the polishing and etching necessary to make the structure visible, there are areas on the micro-section where the metal has been removed by the salt spray. The surface reveals existing pits in the area perpendicular to the direction of welding. In the middle area (Figure 17c), clear large grains of regular shape are visible; no negative effects of salt fog were observed. The wall made using CuAl2 multiwire in the area of the padding weld face after HIP (Figure 17b) is characterized by a fine-grained structure with no traces of salt fog influence on the structure. The micro-section of the

CuAl2 padding weld after HIP in the middle zone also shows no traces of salt fog on the sample.

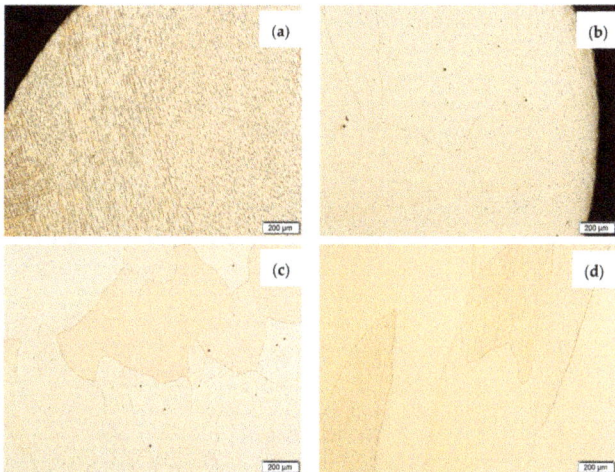

Figure 17. The microstructure of the wall made using CuAl2 multiwire after the NSS corrosion test before HIP: (**a**) padding weld face, (**c**) middle area of the wall; and after HIP: (**b**) padding weld face, (**d**) middle area of the wall.

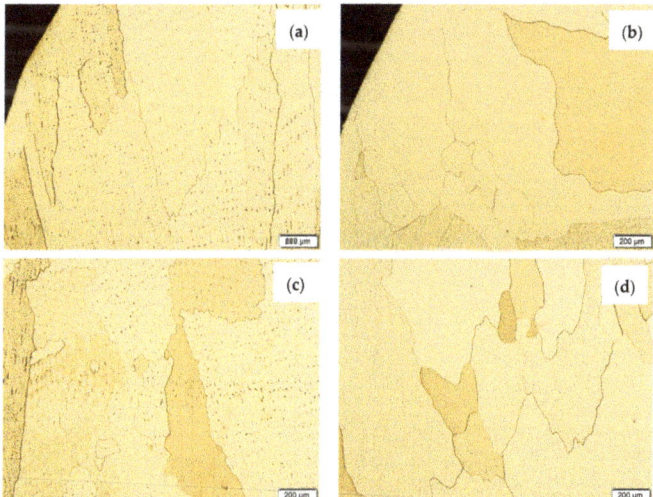

Figure 18. The microstructure of the wall made with CuAl7 wire after the NSS corrosion test before HIP: (**a**) padding weld face, (**c**) middle area of the wall; and after HIP: (**b**) padding weld face, (**d**) middle area of the wall.

The wall made using the CuAl7 wire in the area of the padding weld face (Figure 18a) is characterized by the presence of irregularly shaped grains; there are no visible areas where the effect of salt fog was observed. In the central area of the sample not subjected to the HIP process (Figure 18c), slight losses were observed due to the action of salt fog on the surface of the sample. In the middle area of the sample subjected to the HIP process (Figure 18d), no traces of salt fog influence were observed.

5. Conclusions

As a result of this work, objects in the form of walls were manufactured using both the commercially available CuAl7 wire and the CuAl2 multiwire, in laboratory conditions. Based on the conducted research, the following conclusions were drawn:

1. The HIP process had a significant impact on the structure of the tested padding welds. The structure has been homogenized, and the effect of the heat-affected area in the welding process has been eliminated (homogenization of the hardness).
2. The mass loss during the tribological tests was slightly smaller for the samples subjected to the HIP process in the case of both materials.
3. The density tests using Archimedes' Principle revealed that the CuAl2 sample showed an increase in density of 0.01g/cm^3, while the density of the CuAl7 sample changed by 0.02g/cm^3 after the HIP process.
4. It was found that the corrosion products were observed only on the walls' surfaces after the test in salt fog. The samples subjected to the HIP process were less affected by the corrosion process. The microscopic images of the samples subjected to the HIP process do not show the effects of the salt fog; there are no clear traces of corrosion.

Author Contributions: Conceptualization, J.K.; Methodology, J.K.; Writing—original draft, J.K.; Writing—review & editing, A.K. and S.T.; Supervision, A.K. and S.T. All authors have read and agreed to the published version of the manuscript.

Funding: This research was funded by Implementation Doctorate program financed by the Ministry of Education and Science of the Republic of Poland.

Informed Consent Statement: Not applicable.

Data Availability Statement: Not applicable.

Conflicts of Interest: The authors declare that they have no competing financial interest or personal relationships that could have seemed to influence the study reported in this paper.

References

1. Atkinson, H.V.; Davies, S. Fundamental aspects of hot isostatic pressing: An overview. *Metall. Mater. Trans. A* **2000**, *31*, 2981–3000. [CrossRef]
2. Chen, Q.; Lin, S.; Yang, C.; Fan, C.; Ge, H. Grain fragmentation in ultrasonic-assisted TIG weld of pure aluminium. *Ultrason. Sonochem.* **2017**, *39*, 403–413. [CrossRef] [PubMed]
3. Zhang, D.; Wang, G.; Wu, A.; Zhao, Y.; Li, Q.; Liu, X.; Meng, D.; Song, J.; Zhang, Z. Study on the inconsistency in mechanical properties of 2219 aluminium alloy TIG-welded joints. *J. Alloys Compd.* **2019**, *777*, 1044–1053. [CrossRef]
4. Lin, Y.T.; Wang, M.C.; Zhang, Y.; He, Y.Z.; Wang, D.P. Investigation of microstructure evolution after post-weld heat treatment and cryogenic fracture toughness of the weld metal of AA2219 VPTIG joints. *Mater. Des.* **2017**, *113*, 54–59. [CrossRef]
5. Dong, B.; Pan, Z.; Shen, C.; Ma, Y.; Li, H. Fabrication of Copper-Rich Cu-Al Alloy Using the Wire-Arc Additive Manufacturing Process. *Metall. Mater. Trans. B* **2017**, *48*, 3143–3151. [CrossRef]
6. Amiri, F.S.; Hosseinipour, S.J.; Aval, H.J.; Jamaati, R. Fabrication of a novel high-strength and high-conductivity copper-clad aluminum composite wire. *CIRP J. Manuf. Sci. Technol.* **2023**, *41*, 144–159. [CrossRef]
7. Keller, C.; Moisy, F.; Nguyen, N.; Eve, S.; Dashti, A.; Vieille, B.; Guillet, A.; Sauvage, X.; Hug, E. Microstructure and mechanical properties characterization of architectured copper aluminum composites manufactured by cold-drawing. *Mater. Charact.* **2021**, *172*, 110824. [CrossRef]
8. Głuchowski, W.; Rdzawski, Z.; Domagała-Dubiel, J.; Sobota, J. Microstructure and Properties of Multifibre Composites. *Arch. Metall. Mater.* **2016**, *61*, 911–916. [CrossRef]
9. Rdzawski, Z.; Głuchowski, W.; Stobrawa, J.; Kempiński, W.; Andrzejewski, B. Microstructure and properties of Cu–Nb and Cu–Ag nanofiber composites. *Arch. Civ. Mech. Eng.* **2015**, *15*, 689–697. [CrossRef]
10. Dupouy, F.; Snoeck, E.; Casanove, M.J.; Roucau, C.; Peyrade, J.P.; Askenazy, S. Microstructural characterization of high strength and high conductivity nanocomposite wires. *Scr. Mater.* **1996**, *34*, 1067–1073. [CrossRef]
11. Chen, X.; Han, Z.; Lu, K. Wear mechanism transition dominated by subsurface recrystallization structure in Cu–Al alloys. *Wear* **2014**, *320*, 41–50. [CrossRef]
12. Maurya, A.K.; Kumar, N.; Chhibber, R.; Pandey, C. Study on microstructure—Mechanical integrity of the dissimilar gas tungsten arc weld joint ofsDSS 2507/X-70 steels for marine applications. *Met. Corros.* **2023**, *58*, 11392–11423.
13. Białas, K.; Białucki, P.; Kozerski, S.; Ziemba, H. Brązy aluminiowe napawane na stalach metodami TIG I plazmą. *Rudy Metale Nieżelazne* **2002**, *6*, 281–283.

14. Lv, S.X.; Xu, Z.W.; Wang, H.T.; Yang, S.Q. Investigation on TIG cladding of copper alloy on steel plate. *Sci. Technol. Weld. Join.* **2013**, *13*, 10–15. [CrossRef]
15. Rana, H.; Badheka, V.; Patel, P.; Patel, V.; Li, W.; Andersson, J. Augmentation of weld penetration by flux assisted TIG welding and its distinct variants for oxygen free copper. *J. Mater. Res. Technol.* **2020**, *10*, 138–151. [CrossRef]
16. Wang, Y.; Chen, X.; Konovalov, S.; Su, C.; Siddiquee, A.N.; Gangil, N. In-situ wire-feed additive manufacturing of Cu-Al alloy by addition of silicon. *Appl. Surf. Sci.* **2019**, *487*, 1366–1375. [CrossRef]
17. Kim, J.; Kim, J.; Pyo, C. Comparison of mechanical properties of Ni-Al bronze alloy fabricated through WAAM with Ni-Al bronze fabricated through casting. *Metals* **2020**, *10*, 1164. [CrossRef]
18. DebRoy, T.; Wei, H.L.; Zuback, J.S.; Mukharjee, T.; Elmer, J.W.; Milewski, J.O.; Beese, A.M.; Wilson-Heid, A.; De, A.; Zhanh, W. Additive manufacturing of metallic components—Process, stucture and properties. *Prog. Mater. Sci.* **2017**, *92*, 112–224. [CrossRef]
19. Ding, D.; Pan, Z.; Cuiuri, D.; Li, H. Wire-feed additive manufacturing of metal components: Technologies, developments and future interests. *Int. J. Adv. Manuf. Technol.* **2015**, *81*, 456–481. [CrossRef]
20. Zhang, Q.-K.; Yang, J.; Sun, W.-S.; Song, Z.-L. Evolution in microstructure and mechanical properties of Cu alloy during wire and arc additive manufacturing. *J. Cent. South Univ.* **2023**, *30*, 400–411. [CrossRef]
21. Rodrigues, T.A.; Duarte, V.R.; Miranda, R.M.; Santos, T.G.; Oliveira, J.P. Ultracold-wire and arc additive manufacturing (UC-WAAM). *J. Mater. Process. Technol.* **2021**, *296*, 117196. [CrossRef]
22. Shiran, M.K.G.; Khalaj, G.; Pouraliakbar, H.; Jandaghi, M.R.; Denhavi, A.S.; Bakhatiari, H. Multilayer Cu/Al/Cu explosive welded joints: Characterizing heat treatment offect on the interface microstructure and mechanical properties. *J. Manuf. Process.* **2018**, *35*, 657–663. [CrossRef]
23. *ISO 9227:2017*; Corrosion Tests in Artificial Atmospheres—Salt Spray Tests. ISO: Geneva, Switzerland, 2017.

Disclaimer/Publisher's Note: The statements, opinions and data contained in all publications are solely those of the individual author(s) and contributor(s) and not of MDPI and/or the editor(s). MDPI and/or the editor(s) disclaim responsibility for any injury to people or property resulting from any ideas, methods, instructions or products referred to in the content.

Article
Effect of Grain Structure and Quenching Rate on the Susceptibility to Exfoliation Corrosion in 7085 Alloy

Puli Cao [1,2], Chengbo Li [1,2,3,*], Daibo Zhu [1,2,*], Cai Zhao [1,2], Bo Xiao [1,2] and Guilan Xie [1,2]

[1] School of Mechanical Engineering and Mechanics, Xiangtan University, Xiangtan 411105, China; xtucpl@126.com (P.C.); m17369283823@163.com (B.X.); xieguilan@xtu.edu.cn (G.X.)
[2] Engineering Research Center of the Ministry of Education for Complex Trajectory Processing Technology and Equipment, Xiangtan University, Xiangtan 411105, China
[3] Guangdong Xingfa Aluminium Co., Ltd., Foshan 528137, China
* Correspondence: csulicb@163.com (C.L.); daibozhu@xtu.edu.cn (D.Z.); Tel.: +86-731-58292209 (C.L.); +86-731-52898553 (D.Z.)

Citation: Cao, P.; Li, C.; Zhu, D.; Zhao, C.; Xiao, B.; Xie, G. Effect of Grain Structure and Quenching Rate on the Susceptibility to Exfoliation Corrosion in 7085 Alloy. *Materials* **2023**, *16*, 5934. https://doi.org/10.3390/ma16175934

Academic Editor: Francesco Iacoviello

Received: 6 August 2023
Revised: 19 August 2023
Accepted: 26 August 2023
Published: 30 August 2023

Copyright: © 2023 by the authors. Licensee MDPI, Basel, Switzerland. This article is an open access article distributed under the terms and conditions of the Creative Commons Attribution (CC BY) license (https://creativecommons.org/licenses/by/4.0/).

Abstract: The influence of grain structure and quenching rates on the exfoliation corrosion (EXCO) susceptibility of 7085 alloy was studied using immersion tests, optical microscopy (OM), scanning electron microscopy (SEM), electron backscatter diffraction (EBSD), and scanning transmission electron microscopy (STEM). The results show that as the cooling rate decreases from 1048 °C/min to 129 °C/min; the size of grain boundary precipitates (GBPs); the width of precipitate-free zones (PFZ); and the content of Zn, Mg, and Cu in GBPs rise, leading to an increase in EXCO depth and consequently higher EXCO susceptibility. Meanwhile, there is a linear relationship between the average corrosion depth and the logarithm of the cooling rate. Corrosion cracks initiate at the grain boundaries (GBs) and primarily propagate along the HAGBs. In the bar grain (BG) sample at lower cooling rates, crack propagation along the sub-grain boundaries (SGBs) was observed. Compared to equiaxed grain (EG) samples, the elongated grain samples exhibit larger GBPs, a wider PFZ, and minor compositional differences in the GBPs, resulting in higher EXCO susceptibility.

Keywords: 7085 alloy; grain structure; cooling rate; quenching-induced phase; exfoliation corrosion

1. Introduction

7XXX-series aluminum alloys are characterized by high strength and low density, making them widely utilized in industries such as automotive manufacturing and aerospace. However, this series of alloys exhibit a tendency for localized corrosion in corrosive environments, which encompasses phenomena such as pitting, intergranular corrosion (IGC), EXCO, and stress corrosion cracking (SCC) [1–4]. Corrosion issues lead to a decline in material performance [5], shortened service life, and restrict the wide application of this alloy series, making it a pressing problem to be addressed. Extensive research has been conducted by researchers to enhance the corrosion resistance of this alloy series.

The EXCO susceptibility of Al-Zn-Mg-Cu alloys is significantly affected by grain size and shape. Masoud et al. [6] demonstrated that high-angle grain boundaries (HAGBs) in Al-Zn-Mg-Cu alloys serve as relatively strong hydrogen trapping sites, possessing higher hydrogen desorption energies. This propensity results in the easy aggregation and retention of hydrogen atoms within HAGBs. Simultaneously, statistically stored dislocations exhibit moderate hydrogen desorption energies, continuously supplying hydrogen atoms to HAGBs, consequently leading to a sustained increase in hydrogen content within HAGBs. As a consequence, HAGBs tend to become the initiation points for corrosion and the propagation paths for crack propagation, thereby facilitating intergranular fracture. Wloka et al. [7] investigated the influence of recrystallized coarse grains on the EXCO susceptibility of AA7010 and AA7349 alloys. The findings indicated that the coarse grain layer increased resistance to EXCO. Huang et al. [8] discovered that in AA7075 and AA7178 alloys, the grain size and aspect ratio of grains are crucial factors in the kinetics of localized corrosion growth. For Al-Zn-Mg-Cu alloys, high aspect ratio grains generally exhibit higher EXCO susceptibility [9], as the accumulation of corrosion products during immersion in corrosive solutions leads to wedging stresses at GBs, resulting in elongated grain bubbling and a layered morphology. Equiaxed grains show intergranular corrosion but have low EXCO susceptibility [10–12]. For thick plates of Al-Zn-Mg-Cu alloys, during the solution and quenching processes, the grain structure often exhibits non-uniformity [13,14], with the cooling rate in the central layer being slower compared to that in the surface layer. As a result, quenching precipitates occur in the central layer and tend to reduce the mechanical and localized corrosion resistance properties [15,16]. For 7XXX alloys, the location of quenching precipitates varies with different grain structures, leading to varying losses in corrosion resistance and mechanical properties after subsequent aging [16,17].

Numerous studies indicate that the quenching rate has a significant influence on the EXCO resistance of 7XXX alloys. Sánchez-Amaya [18] found that slow quenching significantly enhanced the IGC susceptibility of the AA7075 alloy. Marlaud et al. [19] found that the resistance to EXCO of the 7449-T7651 alloy decreased with a decline in quenching rate. Li et al. [20] found that the EXCO grade of an Al-5Zn-3Mg-Cu alloy sheet gradually changed from P grade to ED grade as the quenching rate reduced from 2160 °C/min to 100 °C/min. Liu et al. [21] conducted exfoliation corrosion tests on 7055 alloy using end-quenching and found that the maximum corrosion depth of EXCO increased as the quenching rate decreased. Chen et al. [22] performed natural aging and artificial aging (T6) on a 7XXX-series aluminum alloy after end-quenching and compared their IGC results. They observed that the maximum corrosion depth of both aging samples increased with a decline in quenching rate, with the naturally aged sample exhibiting greater corrosion depth than the artificially aged sample. For 7XXX alloys, the characteristics of the quenched precipitates and the width of the PFZ are determining factors for the corrosion resistance of 7XXX alloys. The quenching rate affects the EXCO resistance of the 7XXX alloy by influencing the quantity, size, composition, and distribution of quenched precipitates, as well as the width of the PFZ [23–26].

The 7085 alloy holds a crucial position in the aerospace and defense industries, and its resistance to EXCO is a decisive factor in ensuring the long-term and safe utilization of the alloy. However, there has been limited research on the corrosion resistance of the 7085 alloy, and the influence of quenching-induced precipitation on the sensitivity to EXCO in the 7085 alloy with different grain structures is not yet clear; thus, its mechanism requires further investigation. This study simultaneously investigates the effects and mechanisms of grain structure and quenching rates on the EXCO sensitivity of the 7085 alloy, as well as the mechanisms of corrosion crack propagation.

2. Experimental
2.1. Materials

The experimental materials were homogenized ingots and hot-rolled thick plates of 7085 alloy. The hot-rolled deformation of the plates was 85%. The chemical composition is shown in Table 1. The end-quenching samples with dimensions of 25×25 mm and a length of 125 mm were cut from the homogenized ingots and hot-rolled plates. At one end of the samples, a circular groove with dimensions of 22 mm in diameter and 10 mm in depth was processed to serve as the water spray end. At the other end, a threaded hole with dimensions of 5 mm in diameter and 15 mm in depth was created to secure the samples on the sample holder for end-quenching. The specimens underwent solution treatment by heating in an SX-4-10 resistance furnace to 470 °C, followed by a holding time of 1 h, and then were rapidly removed for water quenching at the groove end (transfer time less than 15 s). The water temperature was approximately 20 °C. Following complete cooling to room temperature, the samples underwent artificial aging by being immersed in an oil bath set at 121 °C for a period of 24 h. Additionally, samples of the same dimensions were drilled and embedded with thermocouples at different distances of 3 mm, 13 mm, 23 mm,

53 mm, 78 mm, and 98 mm from the water spray end. The cooling curves at these positions during the end-quenching process were recorded, and the following average cooling rates were obtained (in the 420–230 °C range) [2]: 1048 °C/min, 782 °C/min, 526 °C/min, 152 °C/min, 132 °C/min, and 129 °C/min, as shown in Figure 1.

Table 1. Chemical compositions of the studied 7085 alloy (mass fraction, wt%).

Element	Zn	Mg	Cu	Zr	Fe	Si	Al
Content	7.5	1.6	1.7	0.11	<0.08	<0.06	Bal.

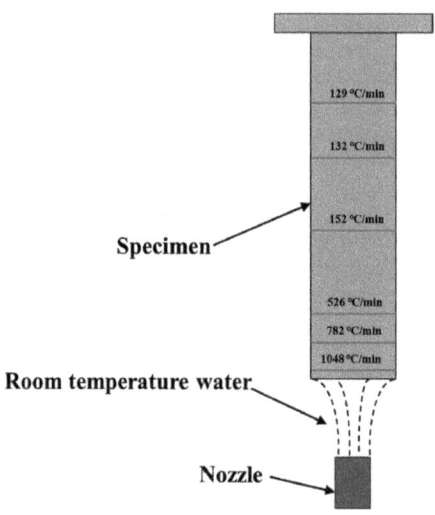

Figure 1. Schematic of end-quenching of the samples after solution heat treatment.

2.2. Immersion Tests

Cut end-quenched and aged specimens (2 mm thick) were used for peel corrosion experiments following the GB/T 22639-2008 standard [27]. The solution had an area to volume ratio of 25 cm^2/L, and the experimental temperature was maintained at 25 ± 1 °C. After 48 h of corrosion, the specimens were rated according to the standard. The solution system used was an EXCO solution (0.5 mol/L KNO_3 + 0.1 mol/L HNO_3 + 4 mol/L NaCl).

2.3. Microstructure Examination

Samples were taken from the corresponding positions of the specimens for microstructural analyses. The metallographic samples were subjected to coarse grinding, fine grinding, and polishing, followed by etching with the corrosion reagent (Graff Sargent). The composition of the reagent was 1% HF, 16% HNO_3, 83% H_2O (by volume), and 3 g of CrO_3, which effectively distinguished between the unrecrystallized and recrystallized regions in the alloy. The microstructure was observed using an XJP-6A (T-Bota, Nanjing, China) metallographic microscope, and further microstructural observations and energy-dispersive spectroscopy (EDS) analysis were conducted using the Quanta-200 SEM (FEI, San Jose, CA, America). The grain structure analyses were performed using a EVOMA10 SEM (Carl Zeiss, Jena, Germany) equipped with an OXFORD EBSD detector. Tecnai G2 F20 transmission electron microscopy (TEM) was employed for observing quenching precipitation phases. Both the EBSD and TEM specimens were prepared using the dual-jet electropolishing method, with electrolytic thinning carried out in a mixture of 20% nitric acid solution and 80% methanol. The electrolyte temperature was controlled at approximately −25 °C using liquid nitrogen.

3. Results
3.1. Grain Structure

Figure 2 displays the EBSD images of the homogenized cast ingot and hot-rolled samples of the 7085 alloy. In Figure 2a, it can be observed that almost all the grains exhibit an equiaxed shape with non-uniform grain sizes, and the GBs are predominantly HAGBs. The as-cast samples with this grain structure are subsequently referred to as equiaxed grain (EG). Figure 2b reveals that the grains after hot rolling exhibit a distinct elongated shape aligned along the rolling direction. In Figure 2c,f, the recrystallized structure is depicted by the blue region, the sub-grain structure is depicted by the yellow region, and the deformed structure is depicted by the red region. The deformed sample contains both recrystallized grains and sub-grains. The black lines indicate HAGBs, while the white lines represent low-angle grain boundaries (LAGBs). Along the deformation direction, there are numerous LAGBs in the sub-grain and deformed regions, which are typical grain structures of 7XXX aluminum alloys after hot deformation and solid solution treatment. Hot-rolled samples with this grain structure are subsequently labeled bar grain (BG). Furthermore, compared to the EG sample, the BG sample obtained through rolling exhibits a significant reduction in grain size. The average size has decreased from the original 148 µm to 70 µm, resulting in BG samples having a greater abundance of GBs, as shown in Figure 2d,e.

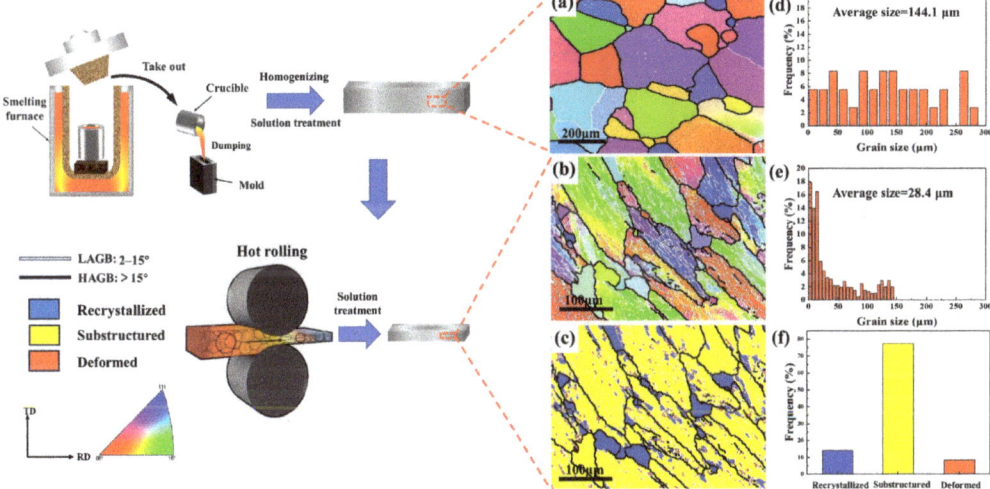

Figure 2. EBSD images of EG (a,d) and BG (b,c,e,f) samples: (a,b) IPF, (d,e) grain size, (c,f) DRX.

3.2. Grain Boundary Quenching Precipitated Phase

The SEM images of EG and BG specimens at different cooling rates are shown in Figure 3. In EG and BG samples, coarse white second-phase particles were observed. EDS analysis revealed that the chemical composition of these particles (wt%) was as follows: Al: 55.24~71.26, Fe: 8.58~14.20, and Cu: 18.44~28.32, with traces of Zn. These particles are likely to be Al_7Cu_2Fe phases [28]. The formation of this iron-containing phase occurs during the solidification process and exhibits good thermal stability. Subsequent processes such as solution treatment and quenching have a small influence on the distribution and size of this phase. In Figure 3a, it can be observed that in the EG sample at a cooling rate of 1048 °C/min, only the presence of white primary phases is observed, and no quenching precipitates are observed. Figure 3b reveals that at a cooling rate of 129 °C/min, fine quenching precipitates are observed at HAGBs, and some precipitates can also be seen within the grains. Based on the EDX analysis results, the main constituents of this phase are Zn and Mg, along with a small amount of Cu, and it corresponds to the η phase [23].

From Figure 3c, it is evident that in the BG sample at a cooling rate of 1048 °C/min, only white primary phases are observed, and no η phases are observed. Figure 3d shows that at a cooling rate of 129 °C/min, there are evident η phases at the HAGBs, and η phases can also be observed at the LAGBs, although the size of the precipitates is smaller. During the slow quenching process, due to the higher interfacial energy of both HAGBs and SGBs, favorable conditions are created for the nucleation and growth of the η phase. As a result, the η phase primarily forms at HAGBs and SGBs [29,30]. Additionally, SGBs have lower energy compared to HAGBs, resulting in slower solute atom diffusion along SGBs. As a result, the subgrain boundary phases (SGBPs) exhibit smaller sizes.

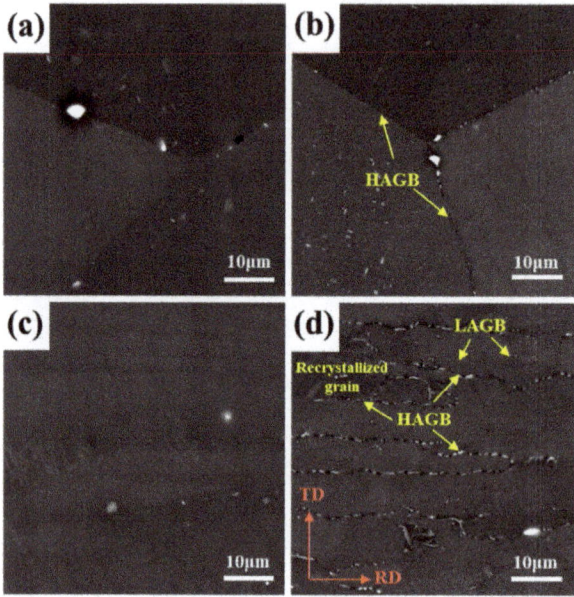

Figure 3. SEM images of EG (**a,b**) and BG (**c,d**) samples at different cooling rates: (**a,c**) 1048 °C/min, (**b,d**) 129 °C/min.

Figure 4 represents the TEM images of EG and BG samples at different cooling rates. From Figure 4a, it can be seen that in the EG sample at a cooling rate of 1048 °C/min, the size of the η precipitates at HAGBs is small and uniform, approximately 24 nm, and the PFZ is narrow, approximately 26 nm. As the cooling rate declines to 129 °C/min, the GBPs' size becomes larger, the distribution becomes discontinuous, and the PFZ becomes more pronounced. The η phase size and the PFZ width increase to approximately 64 nm and 86 nm, respectively, as shown in Figure 4b. This is because during the slow quenching process, vacancies diffuse and solute atoms to HAGBs, resulting in the formation of larger GBPs and a wider PFZ [31,32]. In Figure 4c, it can be observed that in the BG sample at a cooling rate of 1048 °C/min, which corresponds to a higher cooling rate, the precipitates are age precipitates that exhibit a continuous distribution with a size of approximately 14 nm, and the PFZ width is approximately 28 nm. However, the precipitation at the SGBs is not significant, as shown in Figure 4d. When the cooling rate reduces to 129 °C/min, the size of the GBPs increases, and the distance between the precipitates becomes wider. The PFZ also becomes very pronounced. The GBPs' size and the PFZ width are approximately 114 nm and 110 nm, respectively, as shown in Figure 4e. Additionally, quench precipitates can also be observed at the SGB (Figure 4f), with smaller sizes compared to the precipitates at the HAGB.

Figure 5 shows the size of the η phase at GBs and the width of the PFZ at different cooling rates. As shown in Figure 5a, the GBPs' size significantly augments as the cooling rate reduces, and this increase is more pronounced in the BG samples. At a cooling rate of 1048 °C/min, the GBPs' (η phase) average size in the BG samples is approximately 10 nm smaller than in the EG samples. However, when the cooling rate drops to 129 °C/min, the GBPs in the BG samples are 50 nm larger than in the EG samples. This indicates that the hot deformation grain structure promotes an increase in the size of the GBPs at lower cooling rates. In Figure 5b, it is evident that the width of the PFZ at GBs increases as the cooling rate declines, and this rise is more apparent in the BG samples. At a cooling rate of 1048 °C/min, the PFZ width at GBs in the BG samples is 2 nm larger than in the EG samples. However, when the cooling rate drops to 129 °C/min, the PFZ width at GBs in the BG samples is 24 nm larger than in the EG samples. The hot-deformed grain structure promotes an increase in the PFZ width at GBs at lower cooling rates. This indicates that hot deformation facilitates the nucleation and growth of the η phase at GBs and widens the PFZ.

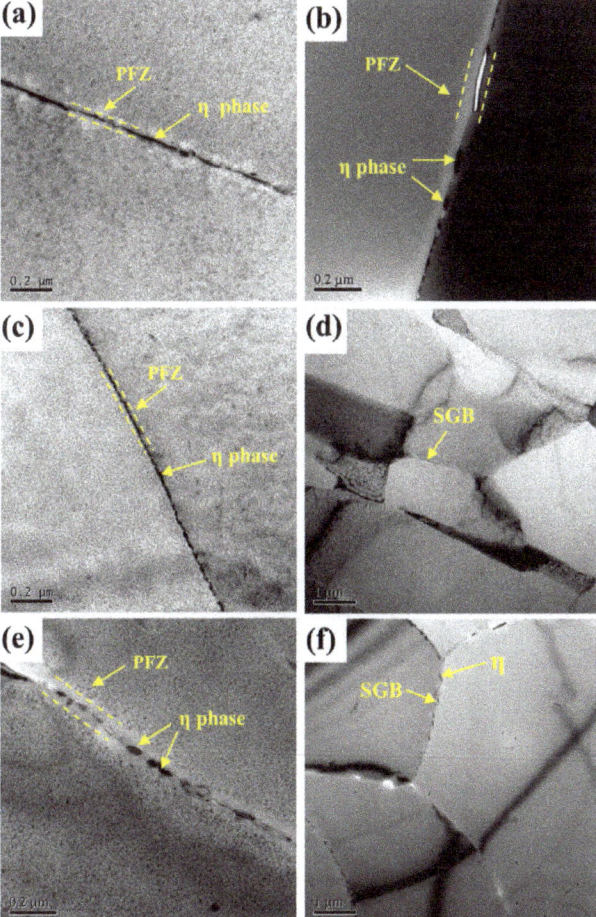

Figure 4. TEM images of EG (**a**,**b**) and BG (**c**–**f**) samples at different cooling rates: (**a**,**c**,**d**) 1048 °C/min, (**b**,**e**,**f**) 129 °C/min.

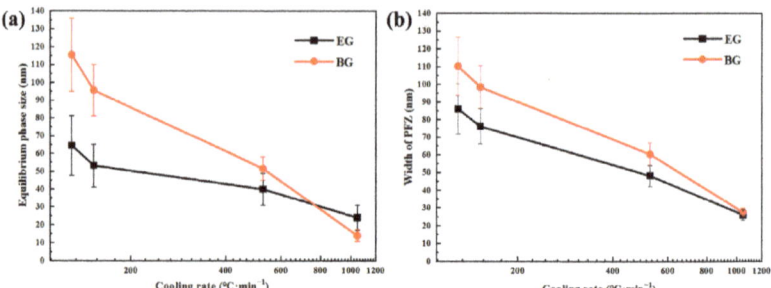

Figure 5. GBPs' size and PFZ width at different cooling rates: (**a**) GBPs size, (**b**) PFZ width.

The composition of GBPs at different cooling rates can be seen in Figure 6. The observation reveals a considerably higher content of Zn and Mg elements in the GBPs compared to Cu. As the cooling rate decreases, the Zn, Mg, and Cu content in the GBPs rise, with a much larger increase in Zn and Mg content compared to Cu content. At the same cooling rate, the GBPs in the BG sample have a slightly higher content of Zn and Mg compared to the EG sample, while the Cu content is similar. When the cooling rate descends from 1048 °C/min to 129 °C/min, the content of Zn, Mg, and Cu in the GBPs of the EG sample rises from 3.2%, 2.3%, and 0.9% to 12.3%, 11.2%, and 2.1%, respectively. Similarly, the content of Zn, Mg, and Cu in the GBPs of the BG sample increases from 3.4%, 2.5%, and 1.1% to 14.4%, 12.9%, and 2.3%, respectively. This indicates that there is not a significant difference in the composition of GBPs between the two grain structures.

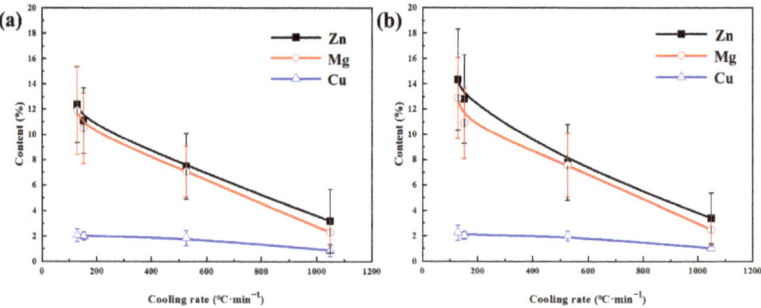

Figure 6. Compositions of GBPs at different cooling rates: (**a**) EG sample, (**b**) BG sample.

3.3. Exfoliation Corrosion Immersion Test Results

The SEM images of the sample with a cooling rate of 129 °C/min soaked in EXCO solution for 1 h are shown in Figure 7. The surface of the EG samples exhibits noticeable corrosion products, primarily distributed at GBs, as shown in Figure 7a. According to EDX analysis, the predominant corrosion product is Al_2O_3, as shown in Figure 7c. This is mainly due to the susceptibility of grain boundary quenching precipitates to corrosion. Since the grains are predominantly equiaxed, the corrosion follows the GBs, although the extent of exfoliation is not significant. The surface of the BG sample exhibits more severe corrosion, with abundant and larger-sized corrosion products. The presence of recrystallized grains with high aspect ratios and a large number of HAGBs in the BG sample leads to the generation of a substantial amount of corrosion products. This further exacerbates the occurrence of EXCO, indicating a higher sensitivity to EXCO, as shown in Figure 7b.

Figure 7. SEM image of the sample with a cooling rate of 129 °C/min soaked in EXCO solution for 1 h: (**a**) EG sample, (**b**) BG sample, (**c**) EDX.

The TEM images of the sample with a cooling rate of 129 °C/min soaked in EXCO solution for 1 h can be seen in Figure 8. It can be observed from the figure that the precipitates at the GBs of both the EG and BG samples have been completely corroded. Previous studies have shown that the potential of the grain boundary η phase is −1.05 V, the potential of the PFZ is −0.85 V, and the potential of the matrix inside the grain is −0.75 V [26]. Therefore, in the microgalvanic cells formed within the corrosive medium, GBPs and PFZs act as the anodes and are preferentially corroded, leading to GBs becoming the initiation points and propagation paths for corrosion.

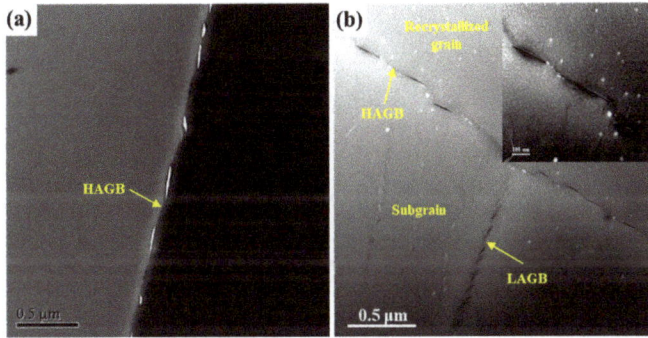

Figure 8. TEM image of the sample with a cooling rate of 129 °C/min soaked in EXCO solution for 1 h: (**a**) EG sample, (**b**) BG sample.

Figure 9 shows the corrosion morphology of the end-quenched samples at different soaking times in the EXCO solution. For the end-quenched samples of the EG alloy, uniform pitting corrosion is observed along the water spray direction. With increasing soaking time, slight "blistering" and "peeling" occur, but there is minimal corrosion product in the solution, and no significant evidence of EXCO is observed. According to the grading standard GB/T 22639-2008 for alloy EXCO, the corrosion rating at the low cooling rate after 48 h is classified as PC grade (Figure 10a). However, for the end-quenched samples of the BG alloy, as the cooling rate decreases, a larger number of bubbles are generated during the immersion process, indicating a more intense reaction with the corrosive solution. At the initial stage of immersion, a slight pitting corrosion appears on the alloy's surface, which intensifies over time. After 8 h, blistering becomes evident, and after 24 h, with further reduction in cooling rate, the alloy exhibits more pronounced blistering, peeling, and severe delamination on the surface, penetrating deep into the metal. The areas with smaller cooling rates exhibit more corrosion products and more severe EXCO (Figure 9h). After 48 h of immersion (Figure 9j), larger areas of corrosion are observed at locations with lower cooling rates. The surface blistering has completely ruptured and delaminated, penetrating deep into the interior of the metal. The solution contains a significant amount of detached corrosion products, and the corrosion rating has reached the EB grade (Figure 10b).

The corrosion ratings for the EG and BG samples are shown in Figure 10, indicating that corrosion becomes more severe with increasing immersion time, particularly at lower cooling rates (152–129 °C/min), but the difference in their ratings is not significant.

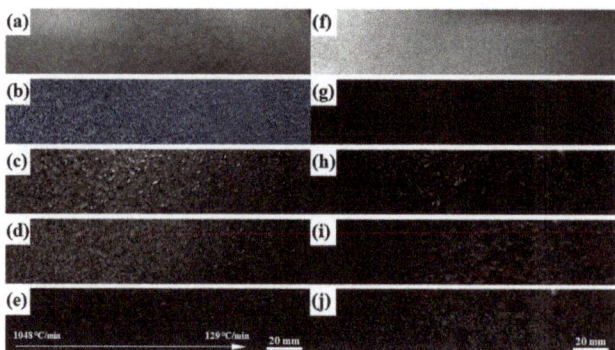

Figure 9. Corrosion morphology of the end-quenched samples at different soaking times in the EXCO solution, EG sample: (**a**) 2 h, (**b**) 12 h, (**c**) 24 h, (**d**) 36 h, (**e**) 48 h; BG sample: (**f**) 2 h, (**g**) 12 h, (**h**) 24 h, (**i**) 36 h, (**j**) 48 h.

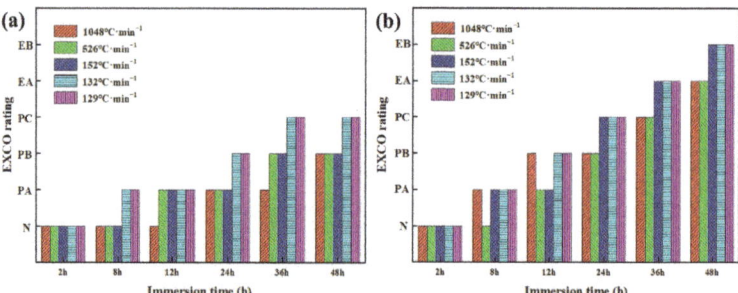

Figure 10. Corrosion ratings after soaking for different times: (**a**) EG sample, (**b**) BG sample.

The cross-sectional corrosion morphologies of EG and BG specimens at different cooling rates are shown in Figure 11. The extent of corrosion in the EG samples increases slightly as the cooling rate decreases. Corrosion occurs along the GBs, and in some areas, entire grains are corroded, resulting in the formation of corrosion pits. At a cooling rate of 1048 °C/min, corrosion is observed along the GBs, and significant corrosion is also present within the grains, as shown in Figure 11a. At 129 °C/min, almost the entire grain is corroded, and the maximum depth of corrosion is approximately the size of a single grain, as shown in Figure 11b. Similarly, the degree of EXCO in the BG samples increases significantly as the cooling rate decreases. At a cooling rate of 1048 °C/min, the EXCO is observed, and the maximum depth of corrosion is similar to that in the EG samples (Figure 11c). As shown in Figure 11e,f, at 129 °C/min, pronounced EXCO occurred. Simultaneously, significant corrosion cracks were observed and propagated along specific paths. Figure 11h is the inverse pole figure of corrosion cracks, where the black regions represent corroded areas, the red lines represent HAGBs, and the white lines represent LAGBs. It was observed that there were evident corrosion cracks along the HAGBs, and some of the SGBs were also corroded. Therefore, it can be inferred that the corrosion cracks primarily propagate along the HAGBs.

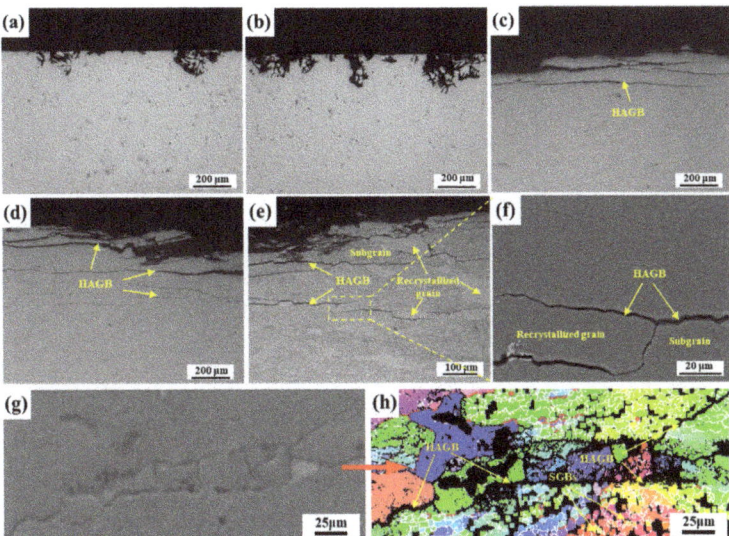

Figure 11. Cross-sectional corrosion morphologies of EG (**a**,**b**) and BG (**c**–**h**) samples at different cooling rates: (**a**,**c**) 1048 °C/min, (**b**,**d**,**e**–**h**) 129 °C/min.

Figure 12 illustrates the relationship between the corrosion depth and the cooling rate at different cooling rates. It can be observed that the corrosion depth of both samples increases with decreasing cooling rates. However, the variation in corrosion depth for the EG sample is relatively small. At a cooling rate of 1048 °C/min, the maximum and average corrosion depths of the BG sample are 30 μm and 28 μm greater than those of the EG sample, respectively. At a cooling rate of 526 °C/min, the maximum and average corrosion depths of the BG sample are 50 μm greater than those of the EG sample. At 129 °C/min, the maximum and average corrosion depths of the BG sample (315 μm and 267.5 μm, respectively) are 125 μm and 120 μm greater than those of the EG sample. This indicates that the peel corrosion of the BG sample becomes more severe at very low cooling rates. Additionally, the corrosion depth of the samples continuously increases with decreasing cooling rates. By performing linear regression analysis on the data, it is found that the average corrosion depth (H) has the following logarithmic relationship with the cooling rate (CR):

$$H_{EG} = 313.7 - 78.8 \lg(CR) \quad (1)$$

$$H_{BG} = 587.8 - 160.7 \lg(CR) \quad (2)$$

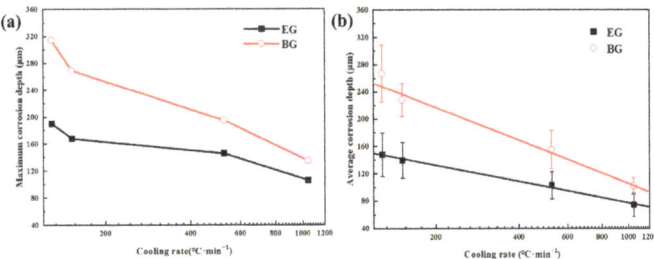

Figure 12. Relationship between the corrosion depth and the cooling rate at different cooling rates: (**a**) maximum corrosion depth, (**b**) average corrosion depth.

The linear correlation coefficients are 0.996 and 0.994, respectively, indicating a strong linear relationship between the logarithm of the average corrosion depth and the cooling rate within the studied range.

It is generally believed that the occurrence of EXCO in high-strength aluminum alloys is related to the characteristics of grain boundaries and the precipitation of phases at grain boundaries and PFZs [26,33]. On the one hand, the corrosion products formed on the aluminum alloy in the corrosive medium create an outward driving force. This outward driving force is related to the shape of the grains, and the more the grains are elongated, the greater the outward force that is generated. EXCO follows the stress corrosion cracking mechanism, where the corrosion product wedges generate tensile stress concentration at the crack tip, leading to the propagation of corrosion through the stress corrosion cracking (SCC) mechanism. As long as the tensile stress exists at the corrosion front, the EXCO will continue to propagate. After immersing different grain structure samples in an EXCO solution for 48 h, different types of localized corrosion occurred, as shown in Figure 9. In the corrosive solution, extensive corrosion products are generated after the sample undergoes corrosion, creating wedging forces that result in the delamination of the upper layer of the metal. Meanwhile, high aspect ratio grains lead to the gradual accumulation of surface strain, eventually resulting in the formation of larger blistering, delamination, and exfoliation morphologies [11]. Compared to EG samples, BG samples possess a significant amount of high-aspect ratio grain structure, which is more conducive to the propagation of corrosion cracks, thus exhibiting a higher sensitivity to EXCO.

On the other hand, in 7XXX-series aluminum alloys, EXCO typically initiates at GBs and propagates along them into the material's interior. This is attributed to the significant potential difference between GBPs and PFZs at GBs and the matrix. In the corrosive medium, GBPs and PFZs with lower potentials are preferentially dissolved as the anode, forming preferential pathways for anodic dissolution along GBs [34], resulting in corrosion cracks propagating along GBs, as illustrated in Figure 11. Based on the experimental results, the schematic diagrams of EXCO propagation for EG and BG samples under different cooling rates are shown in Figure 13. In EG samples, due to their smaller grain aspect ratio, corrosion propagates slower along HAGBs, resulting in lower EXCO sensitivity, as depicted in Figure 13a,b. In comparison to EG samples, BG samples exhibit reduced grain size after hot rolling, leading to a significant increase in the number of GBs, as shown in Figure 2. Additionally, hot rolling promotes the nucleation and growth of GBPs, forming wider PFZs. This provides more favorable conditions for the expansion of corrosion cracks, contributing to the higher EXCO sensitivity in BG samples.

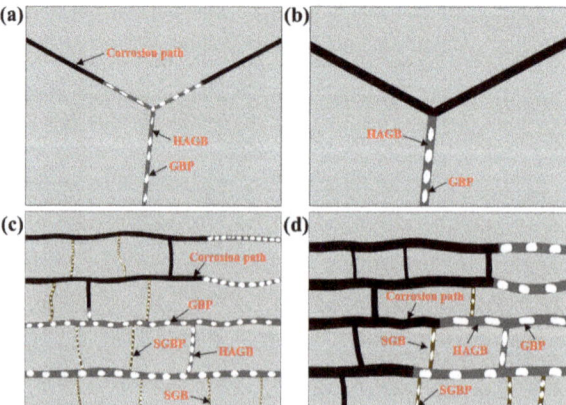

Figure 13. Schematic diagram of spalling corrosion expansion in EG (**a**,**b**) and BG (**c**,**d**) specimens at different cooling rates: (**a**,**c**) 1048 °C/min, (**b**,**d**) 129 °C/min.

Furthermore, the size, spacing, and chemical composition of GBPs and the width of PFZs significantly influence the alloy's sensitivity to EXCO [35]. At lower cooling rates, GBPs continue to absorb surrounding atoms and grow, leading to larger PFZ widths. Microstructural observations reveal that as cooling rates decrease, both EG and BG samples exhibit increased GBPs and PFZ widths. As shown in Figure 13b,d, this makes grain boundaries more susceptible to corrosion and accelerates the propagation of corrosion cracks. Simultaneously, the content of Zn and Mg elements in GBPs is notably higher than that of Cu, and with decreasing cooling rates, the increase in Zn and Mg content far exceeds that of Cu (Figure 5). This results in a greater potential difference between GBPs and the matrix, further accelerating the corrosion crack propagation rate along grain boundaries [36]. The results demonstrate that as cooling rates decrease, the EXCO sensitivity of both EG and BG samples gradually increases.

Existing research has indicated that for 7XXX-series aluminum alloys, EXCO primarily extends along HAGBs, while propagation along SGBs is challenging [11,31]. This is attributed to the higher interfacial energy of HAGBs, which serve as the primary diffusion pathways for atoms, facilitating the nucleation and growth of GBPs through continuous atom absorption [36,37]. Consequently, there are more and larger GBPs at HAGBs, accompanied by wider PFZs. In contrast, SGBs possess lower interfacial energy, hindering the nucleation and growth of GBPs and resulting in narrower PFZs that are less conducive to forming independent anodic regions, making it difficult to establish corrosion pathways. However, as cooling rates decrease, GBPs at SGBs also grow larger, and PFZs widen, providing favorable conditions for the expansion of corrosion cracks. As depicted in Figure 13d, when the cooling rate decreases to 129 °C/min, corrosion preferentially extends along both HAGBs and SGBs, resulting in increased corrosion depth and a wavy distribution of macroscopic corrosion morphology. This further elucidates that decreasing cooling rates reduce the resistance of the alloy to EXCO.

4. Conclusions

1. For the 7085 alloy, BG samples exhibit higher sensitivity to EXCO compared to EG samples, regardless of the cooling rate. This is because BG samples have larger aspect ratio grains, which lead to the accumulation of surface stress during the corrosion process, resulting in faster crack propagation along the GBs. Additionally, BG samples have a higher number of GBs, which results in a greater amount of GBPs and a higher sensitivity to EXCO.
2. With a decrease in cooling rate, both EG and BG samples show an increasing trend in EXCO sensitivity. This is attributed to the slower cooling rate, which leads to an increase in the size of GBPs and the width of the PFZ. Both the maximum corrosion depth and average corrosion depth significantly increase with decreasing cooling rates, with a higher rise observed in BG samples. In the meantime, a linear correlation can be established between the average depth of corrosion and the logarithm of the cooling rate.
3. GBPs and their PFZ have a lower potential compared to the matrix, causing them to act as anodes and preferentially dissolve during corrosion. Consequently, corrosion cracks propagate along the GBs. Corrosion cracks in both EG and BG samples primarily propagate along HAGBs. At lower cooling rates, crack propagation along SGBs is observed in BG samples.

Author Contributions: Conceptualization, C.L.; Data curation, C.Z.; Formal analysis, C.L.; Investigation, P.C.; Methodology, C.L.; Project administration, G.X.; Software, B.X.; Supervision, D.Z.; Visualization, P.C.; Methodology, G.X.; Writing—original draft, P.C.; Writing—review and editing, D.Z. All authors have read and agreed to the published version of the manuscript.

Funding: This research was funded by the National Natural Science Foundation of China (52205421); Guangxi Science and Technology Major Project (2022JBGS041); the Key laboratory open project of Guangdong Province (XF20230330-XT); the school-enterprise, industry–university–research cooperation project (2023XF-FW-32); and the science and technology innovation Program of Hunan Province, China (2021RC2087, and 2022JJ30570).

Institutional Review Board Statement: Not applicable.

Informed Consent Statement: Not applicable.

Data Availability Statement: Not applicable.

Acknowledgments: This study was supported by the National Natural Science Foundation of China (52205421); Guangxi Science and Technology Major Project (2022JBGS041); the Key laboratory open project of Guangdong Province (XF20230330-XT); the school-enterprise, industry–university–research cooperation project (2023XF-FW-32); and the science and technology innovation Program of Hunan Province, China (2021RC2087, and 2022JJ30570).

Conflicts of Interest: The authors declare no conflict of interest.

References

1. Wloka, J.; Virtanen, S. Influence of scandium on the pitting behaviour of Al-Zn-Mg-Cu alloys. *Acta Mater.* **2007**, *55*, 6666–6672. [CrossRef]
2. Sun, Y.W.; Pan, Q.L.; Lin, S.; Zhao, X.J.; Liu, Z.L.; Li, W.J.; Wang, G.Q. Effects of critical defects on stress corrosion cracking of Al-Zn-Mg-Cu-Zr alloy. *J. Mater. Res. Technol.* **2021**, *12*, 1303–1318. [CrossRef]
3. Ren, J.; Wang, R.C.; Wang, C.Q.; Feng, Y. Multistage aging treatment influenced precipitate characteristics improve mechanical and corrosion properties in powder hot-extruded 7055 Al alloy. *Mater. Charact.* **2020**, *170*, 110683. [CrossRef]
4. Rout, P.K.; Ghosh, M.M.; Ghosh, K.S. Effect of solution pH on electrochemical and stress corrosion cracking behaviour of a 7150 Al-Zn-Mg-Cu alloy. *Mater. Sci. Eng. A* **2014**, *604*, 156–165. [CrossRef]
5. Kittur, M.Y.; Kittur, M.I.; Reddy, A.R.; Baig, M.A.A.; Ridwan, S.A.K.; Faheem, M. Mechanical response of aluminum 7075 with heat treatment and exfoliation corrosion. *Mat. Today Proc.* **2021**, *47*, 6173–6179. [CrossRef]
6. Masoud, M.; Mahdieh, S.; Shigeru, K.; Tomohiko, H. Unraveling the effect of dislocations and deformation-induced boundaries on environmental hydrogen embrittlement behavior of a cold-rolled Al–Zn–Mg–Cu alloy. *Int. J. Hydrogen Energy* **2021**, *46*, 8285–8299.
7. Wloka, J.; Hack, T.; Virtanen, S. Influence of temper and surface condition on the exfoliation behavior of high strength Al-Zn-Mg-Cu alloys. *Corros. Sci.* **2007**, *49*, 1437–1449. [CrossRef]
8. Huang, T.S.; Frankel, G.S. Influence of grain structure on anisotropic localized corrosion kinetics of AA7xxx-T6 alloys. *Corros. Eng. Sci. Technol.* **2006**, *41*, 192–199. [CrossRef]
9. Marlaud, T.; Malki, B.; Deschamps, A.; Baroux, B. Electrochemical aspects of exfoliation corrosion of aluminum alloys: The effects of heat treatment. *Corros. Sci.* **2011**, *53*, 1394–1400. [CrossRef]
10. Lu, X.H.; Han, M.L.; Du, Z.H.; Wang, G.J.; Lu, L.Y.; Lei, J.Q.; Zhou, T.T. Effect of microstructure on exfoliation corrosion resistance in an Al-Zn-Mg alloy. *Mater. Charact.* **2018**, *135*, 167–174. [CrossRef]
11. Mcnaughtan, D.; Worsfold, M.; Robinson, M.J. Corrosion product force measurements in the study of exfoliation and stress corrosion cracking in high strength aluminum alloys. *Corros. Sci.* **2003**, *45*, 2377–2389. [CrossRef]
12. Keddam, M.; Kuntz, C.; Takenouti, H.; Schustert, D.; Zuili, D. Exfoliation corrosion of aluminum alloys examined by electrode impedance. *Electrochim. Acta* **1997**, *42*, 87–97. [CrossRef]
13. Chen, K.; Fang, H.; Zhang, Z.; Chen, X.; Liu, G. Effect of Yb, Cr and Zr additions on recrystallization and corrosion resistance of Al–Zn–Mg–Cu alloys. *Mater. Sci. Eng. A* **2008**, *497*, 426–431. [CrossRef]
14. Fan, X.; Jiang, D.; Li, Z.; Tao, W.; Ren, S. Influence of microstructure on the crack propagation and corrosion resistance of Al-Zn-Mg-Cu alloy 7150. *Mater. Characater.* **2007**, *58*, 24–28. [CrossRef]
15. Liu, S.; Zhong, Q.; Zhang, Y.; Liu, W.; Zhang, X.; Deng, Y. Investigation of quench sensitivity of high strength Al-Zn-Mg-Cu alloys by time- temperature-properties diagrams. *Mater. Des.* **2010**, *31*, 3116–3120. [CrossRef]
16. Dorward, R.C.; Beerntsen, D. Grain structure and quench-rate effects on strength and toughness of AA7050 Al-Zn-Mg-Cu-Zr alloy plate. *Metall. Trans. A* **1995**, *26*, 2481–2484. [CrossRef]
17. Tang, J.; Yang, Z.; Liu, S.; Wang, Q.; Chen, J.; Chai, W.; Ye, L. Quench sensitivity of AA 7136 alloy: Contribution of grain structure and dispersoids. *Metall. Mater. Trans. A* **2019**, *50*, 4900–4912. [CrossRef]
18. Sánchez-Amaya, J.M.; Bethencourt, M.; González-Rovira, L.; Botana, F.J. Noise resistance and shot noise parameters on the study of IGC of aluminium alloys with different heat treatments. *Electrochim. Acta* **2007**, *52*, 6569–6583. [CrossRef]
19. Marlaud, T.; Malki, B.; Henon, C.; Deschamps, A.; Baroux, B. Relationship between alloy composition, microstructure and exfoliation corrosion in Al-Zn-Mg-Cu alloys. *Corros. Sci.* **2011**, *53*, 3139–3149. [CrossRef]
20. Li, D.F.; Ming, Z.X.; Dan, L.S.; Wen, Y.B.; Yue, L. Effects of Quenching Rate on Exfoliation Corrosion of Al-5Zn-3Mg-1Cu Aluminum Alloy Thick Plate. *J. Hunan Univ.* **2015**, *42*, 47–52.

21. Liu, S.D.; Li, C.B.; Deng, Y.L.; Zhang, X.M. Influence of grain structure on quench sensitivity relative to localized corrosion of high strength aluminum alloy. *Mat. Chem. Phys.* **2015**, *167*, 320–329. [CrossRef]
22. Chen, M.Y.; Zheng, X.; He, K.H.; Liu, S.D.; Zhang, Y. Local corrosion mechanism of an Al-Zn-MgCu alloy in oxygenated chloride solution: Cathode activity of quenching induced η precipitates. *Corros. Sci.* **2021**, *191*, 109743. [CrossRef]
23. Ma, Z.M.; Liu, J.; Yang, Z.S.; Liu, S.D.; Zhang, Y. Effect of cooling rate and grain structure on the exfoliation corrosion susceptibility of AA7136 alloy. *Mater. Charact.* **2020**, *168*, 110533. [CrossRef]
24. Yuan, D.L.; Chen, K.H.; Chen, S.G.; Zhou, L.; Chang, J.Y.; Huang, L.P.; Yi, Y.P. Enhancing stress corrosion cracking resistance of low Cu-containing Al-Zn-Mg-Cu alloys by slow quench rate. *Mater. Design* **2019**, *164*, 107558. [CrossRef]
25. Song, F.X.; Zhang, X.M.; Liu, S.D.; Tan, Q.; Li, D.F. The effect of quench rate and over-ageing temper on the corrosion behaviour of AA7050. *Corros. Sci.* **2014**, *78*, 276–286. [CrossRef]
26. Li, C.B. Study on Quenching Sensitivity of 7085 Aluminum Alloy. Ph.D. Thesis, Central South University, Changsha, China, 2017.
27. GB/T 22639-2008; Test method of exfoliation corrosion for wrought aluminium and aluminium alloys. Standardization Administration of the People's Republic of China: Beijing, China, 2008.
28. Sun, Y.; Pan, Q.; Sun, Y.; Wang, W.; Huang, Z.; Wang, X.; Hu, Q. Localized corrosion behavior associated with Al_7Cu_2Fe intermetallic in Al-Zn-Mg-Cu-Zr alloy. *J. Alloy. Compd.* **2019**, *783*, 329–340. [CrossRef]
29. Li, C.; Deng, Y.; Tang, J. Effect of recrystallization fraction on quenching sensitivity of Al-Zn-Mg-Cu alloy. *J. Mater. Res.* **2018**, *32*, 881–888.
30. Liu, S.; Liu, W.; Yong, Z.; Zhang, X.; Deng, Y. Effect of microstructure on the quench sensitivity of Al-Zn-Mg-Cu alloys. *J. Alloy. Compd.* **2010**, *507*, 53–61. [CrossRef]
31. Liu, S.; Chen, B.; Li, C.; Dai, Y.; Deng, Y.; Zhang, X. Mechanism of low exfoliation corrosion resistance due to slow quenching in high strength aluminum alloy. *Corros. Sci.* **2015**, *91*, 203–212. [CrossRef]
32. Ogura, T.; Hirose, A.; Sato, T. Effect of PFZ and grain boundary precipitate on mechanical properties and fracture morphologies in Al-Zn-Mg (Ag) alloys. *Mater. Sci. Forum* **2010**, *638*, 297–302. [CrossRef]
33. Kelly, D.J.; Robinson, M.J. Influence of heat treatment and grain shape on exfoliation corrosion of Al-Li alloy 8090. *Corrosion* **1993**, *49*, 787–795. [CrossRef]
34. Yu, M.Y.; Zhang, Y.A.; Li, X.W.; Wen, K.B.; Xiong, Q.; Li, Z.H.; Yan, L.Z.; Yan, H.G.; Liu, H.W.; Li, Y.N. Effect of recrystallization on plasticity, fracture toughness and stress corrosion cracking of a high-alloying Al-Zn-Mg-Cu alloy. *Mater. Lett.* **2020**, *275*, 128074. [CrossRef]
35. Dan, L.S.; Chen, G.; Yin, Y.L.; Shen, Y.Z.; Lai, D.Y. Influence of Quenching Rate on Peel Corrosion Resistance of 7020 Aluminum Alloy Sheets. *J. Mater. Res.* **2018**, *32*, 423–431.
36. Fang, H.C.; Chen, K.H.; Chen, X.; Chao, H.; Peng, G.S. Effect of Cr, Yb and Zr additions on localized corrosion of Al-Zn-Mg-Cu alloy. *Corros. Sci.* **2009**, *51*, 2872–2877. [CrossRef]
37. Du, Y.; Chang, Y.A.; Huang, B.Y.; Gong, W.P.; Jin, Z.P.; Xu, H.H.; Yuan, Z.H.; Liu, Y.; He, Y.H.; Xie, F.Y. Diffusion coefficients of some solutes in fcc and liquid Al: Critical evaluation and correlation. *Mater. Sci. Eng. A* **2003**, *363*, 140–151. [CrossRef]

Disclaimer/Publisher's Note: The statements, opinions and data contained in all publications are solely those of the individual author(s) and contributor(s) and not of MDPI and/or the editor(s). MDPI and/or the editor(s) disclaim responsibility for any injury to people or property resulting from any ideas, methods, instructions or products referred to in the content.

Article

Microstructure Evolution of the Near-Surface Deformed Layer and Corrosion Behavior of Hot Rolled AA7050 Aluminum Alloy

Ergen Liu [1], Qinglin Pan [1,2], Bing Liu [1,*], Ji Ye [2] and Weiyi Wang [1]

[1] School of Materials Science and Engineering, Central South University, Changsha 410083, China; ergenliu@163.com (E.L.); weiyiwang@163.com (W.W.)
[2] Light Alloy Research Institute, Central South University, Changsha 410083, China
* Correspondence: bing.liu@csu.edu.cn

Abstract: The current study investigates the influence of hot rolling on the microstructure evolution of the near-surface region on AA7050 aluminum alloy and the corrosion performance of the alloy. It is revealed that hot rolling resulted in grain refinement in the near-surface region, caused by dynamic recrystallization, and equiaxed grains less than 500 nm can be clearly observed. Fibrous grains were evident in the hot rolled AA7050 aluminum alloy with relatively lower rolling temperature or larger rolling reduction, caused by the more severe elemental segregation at grain boundaries, which inhibited the progression of dynamic recrystallization. The density of the precipitates in the fibrous grain layer was higher, compared with those in the equiaxed grain layer, due to the increased dislocation density, combined with more severe elemental segregation, which significantly promoted the nucleation of precipitates. With the co-influence exerted by low density of precipitates and dislocations on the improvement of the corrosion performance of the alloy, the rolled AA7050 alloy with decreased density of precipitates and dislocations exhibited better corrosion resistance.

Keywords: AA7050 aluminum alloy; hot rolling; near-surface deformed layer; microstructure evlution; corrosion behavior

1. Introduction

Aluminum alloy is one of the most widely used nonferrous metal structural materials in industry. Due to its high power-to-weight ratio, low cost, high wear resistance and other good properties, aluminum alloy has been widely used in many structural parts of aviation, aerospace, automobile, machinery manufacturing, and shipbuilding [1–5]. Rolling is often employed during manufacture of such alloys. Thus, the microstructural evolution caused by rolling should be considered, as it will influence the properties and finally limit the better applications of such alloys. Guo et al., [6] found that the fraction of the low-angle grain boundaries (LAGBs) increased and S phase precipitates located at the grain boundaries were discontinuous when Al-Cu-Mg-Ag alloy were rolled at a lower temperature. Kumar et al., [7] proposed that dislocation cells, un-recrystallized grains and nano-sized precipitates were evident on cold rolled AA6082 aluminum alloy. Wang et al., [8] reported the presence of shear bands on cold rolled AA7050 aluminum alloy, and such shear bands became wide with the increase of the rolling reduction. Liu et al., [9] reported that the textures obviously changed in Al-Cu alloy due to the increase of the cold rolling reduction. Based on the previous study, extensive work has been done regarding the effect of rolling on the microstructure of aluminum alloys [6–12]. However, the near-surface region has rarely been discussed, which is directly in contact with the service environment and will determine the properties of rolled aluminum alloys to a large extent.

In recent years, it has been found that severe plastic deformation, including machining [13], orthogonal cutting [14], laser shock processing [15] and other processes significantly influence the near-surface microstructure of aluminum alloy, leading to the heavily deformed layers characterized by ultra-fine grains. Within such deformed layers, obvious residual stresses, high density of dislocations and element redistribution have also been observed which, in combination, affect the corrosion performance of aluminum alloys [16–21]. The near-surface deformed layer on the weld joint of Al-Cu-Li alloy is more active than the bulk alloy, and anodic dissolution usually occurs preferentially in such regions of the weld joint subjected to corrosive environment [22]. Liu et al., [23] found that element segregation in the near-surface deformed layer of AA7075-T6 aluminum alloy promoted the localized corrosion along the grain boundaries. Saklakoglu et al., [15] reported that high level of work-hardening and compressive residual stress were observed in the near-surface deformed layer on AA6061-T6 aluminum alloy treated by laser shock peening, which improved the resistance against pitting corrosion. Based on the previous study, the near-surface deformed layers have both advantages and disadvantages on the corrosion performance of aluminum alloys, depending on the microstructure formed during the severe deformation. Since hot rolling is normally employed during fabrication of aluminum alloys and corrosion which starts the surface of the alloy is always one of the main concerns of such alloys during service, it is necessary to investigate the corrosion behavior of hot rolled 7xxx series aluminum alloys, considering the impact of near-surface deformed layer.

The main purpose of the current work is to investigate the microstructure evolution of the near-surface deformed layer and the corrosion performance of the hot rolled AA7050 aluminum alloy, and to establish the correlation between them, which contributes to improving the performance and service life of hot rolled AA7050 aluminum alloy as a structural material.

2. Materials and Methods

2.1. Materials

AA7050 aluminum alloy ingot was cut into alloy samples with dimensions of 200 × 110 × 60 mm. The chemical composition of the alloy is shown in Table 1. After homogenization annealing at 465 °C for 24 h, the alloy ingots were hot rolled at the temperature of 380 °C (1#), 420 °C (3#), 450 °C (5#) with rolling reduction of 66.7%, and at the temperature of 420 °C with the rolling reductions of 83.3% (2#), 66.7% (3#), 50% (4#), respectively. Hot rolling of the alloy was carried out using the L6500 two rolling mill, of which the roll diameter and speed were 420 mm and 0.36 m/s, respectively. The reduction of each rolling pass was less than 5 mm. The rolling parameters of each sample are summarized in Table 2.

Table 1. Chemical composition of AA7050 aluminum alloy.

	Mg	Zn	Cu	Zr	Cr	Mn	Si	Ti	Fe	Al
Mass fraction/%	2.08	6.07	2.21	0.11	0.02	0.10	0.12	0.04	0.12	Bal.

Table 2. Hot rolling parameters.

Sample Number	Rolling Temperature/°C	Original Thickness/mm	Thickness After Rolling/mm	Rolling Reduction/%
1#	380	60	20	66.7
2#	420	60	10	83.3
3#	420	60	20	66.7
4#	420	60	30	50.0
5#	450	60	20	66.7

2.2. Mircrostructure Characterization

Transmission electron microscopy (TEM) samples were prepared by focused ion beam (FIB), using a FEI Helios NanoLab G3 UC dual-beam focused ion beam and a scanning electron microscope. The voltage and current during sample preparation were 5~30 kV and 41 pA~10 nA, respectively. A Tecnai G^2 20 transmission electron microscope and a Talos F200X field emission transmission electron microscope equipped with the EDS detector were employed to observe the microstructure of the near-surface deformed layers on the alloy samples subjected to hot rolling under different conditions. The morphology of the corroded surface was observed by a SIRION200 field emission scanning electron microscope equipped with the EDS detector.

2.3. Corrosion Test

Potentiodynamic polarization, open circuit potential (OCP) and electrochemical impedance spectroscopy (EIS) measurements were conducted in 3.5 wt.% NaCl solution on a Multi Autolab M204 electrochemical workstation. The alloy samples served as the working electrode, while the reference electrode (RE) and counter electrode (CE) were saturated calomel electrode (SCE) and platinum wire, respectively, in the three-electrode system. The testing area involved in the electrochemical measurements was 10×10 mm^2. The duration of the immersion tests were 6 h, 12 h, 18 h, and 24 h, respectively. The OCP test was carried out for 3600 s. The scan range selected for potentiodynamic polarization test was from -1.1 V (SCE) to -0.3 V (SCE), with a scan speed of 1 mV/s. The EIS measurement was carried out from 10^5 Hz to 0.01 Hz and the sinusoidal voltage amplitude was 10 mV.

3. Results

3.1. Microstructure of Near-Surface Deformed Layer on Hot Rolled AA7050 Aluminum Alloy

Figure 1 displays the High Angle Annular Dark Field (HAADF) micrographs of the near-surface region on AA7050 aluminum alloy subjected to hot rolling under different conditions. Figure 1a,b are the near-surface regions of 1# and 2# samples, respectively, indicating that the near-surface deformed layers on the two alloy samples were mainly composed of an equiaxed grain layer, which was near the rolled surface, and a fibrous grain layer beneath the equiaxed grain layer. The thickness of the equiaxed grain layer was approx. 3.5 µm and 2.5 µm, and the dimension of the equiaxed grains was approx. 400 nm and 500 nm, respectively, on 1# and 2# samples. The boundaries between the equiaxed grain layers and the fibrous grain layers are indicated by curved lines in Figure 1a,b. The density of precipitates in the fibrous grain layer was significantly higher than that in the equiaxed grain layer. Figure 1c–e are the near-surface deformed layers of 3#, 4# and 5# samples, respectively. It is revealed that the near-surface deformed layers were mainly characterized by equiaxed grains, with the dimensions of approx. 400 nm, 500 nm, and 500 nm, respectively. The thickness of the near-surface deformed layers was more than 5.5 µm, and the precipitates can be observed uniformly distributed in the near-surface deformed layers on such samples.

Figure 2 shows the HAADF micrographs, showing distribution of precipitates in the top regions of the near-surface deformed layers on the rolled AA7050 aluminum alloy under different conditions. It can be observed that such precipitates vary from 20–200 nm distributed at the equiaxed grain boundaries. It is also evident that the equiaxed grain boundaries were bright in the HAADF micrographs. Due to the high dependence of the brightness of the HAADF micrographs on atomic number, its brightness is the relative abundance of heavy alloy elements, suggesting the segregation of heavier elements such as Cu and Zn compared with aluminum.

Figure 3 exhibits the HAADF micrographs, showing distribution of precipitates in the bottom regions of the near-surface deformed layers on the rolled AA7050 aluminum alloy under different conditions. The precipitates can clearly be observed in the fibrous grain layers, indicated in Figure 3a,b. In addition, the bright bands also appeared at the fibrous grain boundaries, suggesting the presence of segregation of alloy elements at such locations. In the bottom regions of the near-surface deformed layers on the 3#, 4# and 5# samples (Figure 3c–e), the precipitates can be observed in the equiaxed grain boundaries. The brightest bands at the grain boundaries disappeared, indicating that the alloy elements segregation was not obvious at the grain boundaries on the rolled alloy samples under such conditions. From Figures 2 and 3, it can be observed that when the hot rolling parameters change, element segregation also changes, forming the near-surface deformed layer near the rolling surface to the entire near-surface deformed layer.

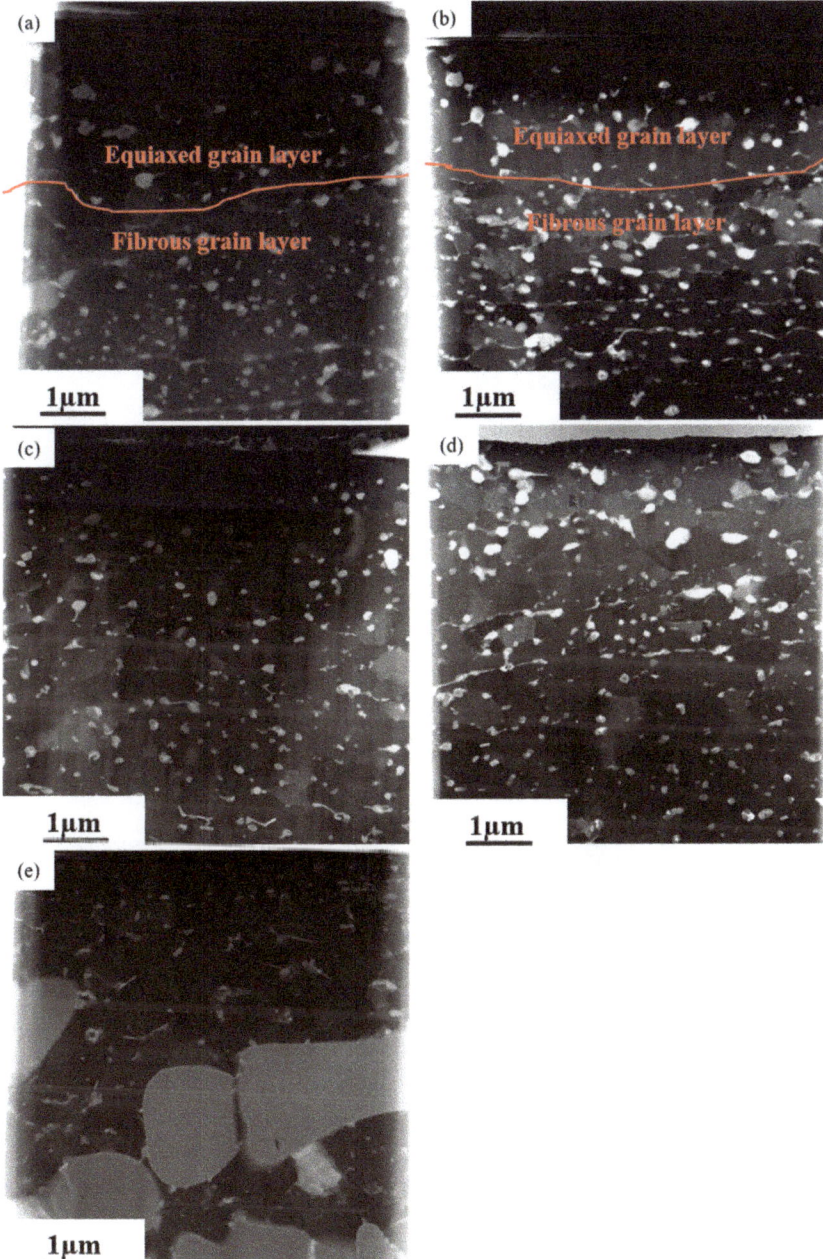

Figure 1. The HAADF micrographs of the near-surface deformed layers on AA7050 aluminum alloy subjected to hot rolling under different conditions: (**a**) 1# (380 °C, 66.7%); (**b**) 2# (420 °C, 83.3%); (**c**) 3 # (420 °C, 66.7%); (**d**) 4# (420 °C, 50%); (**e**) 5# (450 °C, 66.7%).

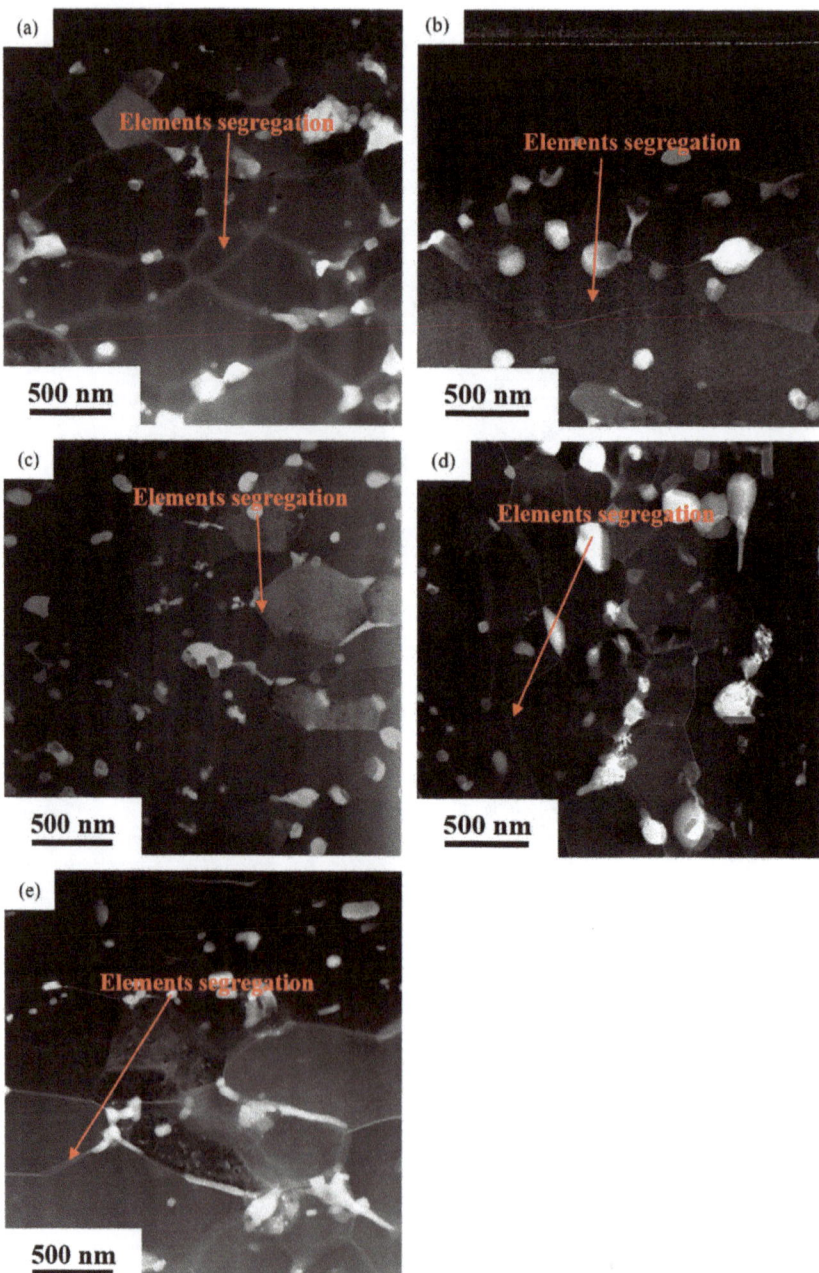

Figure 2. The HAADF micrographs showing the top regions of the near-surface deformed layers on hot rolled AA7050 aluminum alloy under different hot rolling conditions: (**a**) 1# (380 °C, 66.7%); (**b**) 2# (420 °C, 83.3%); (**c**) 3# (420 °C, 66.7%); (**d**) 4# (420 °C, 50%); (**e**) 5# (450 °C, 66.7%).

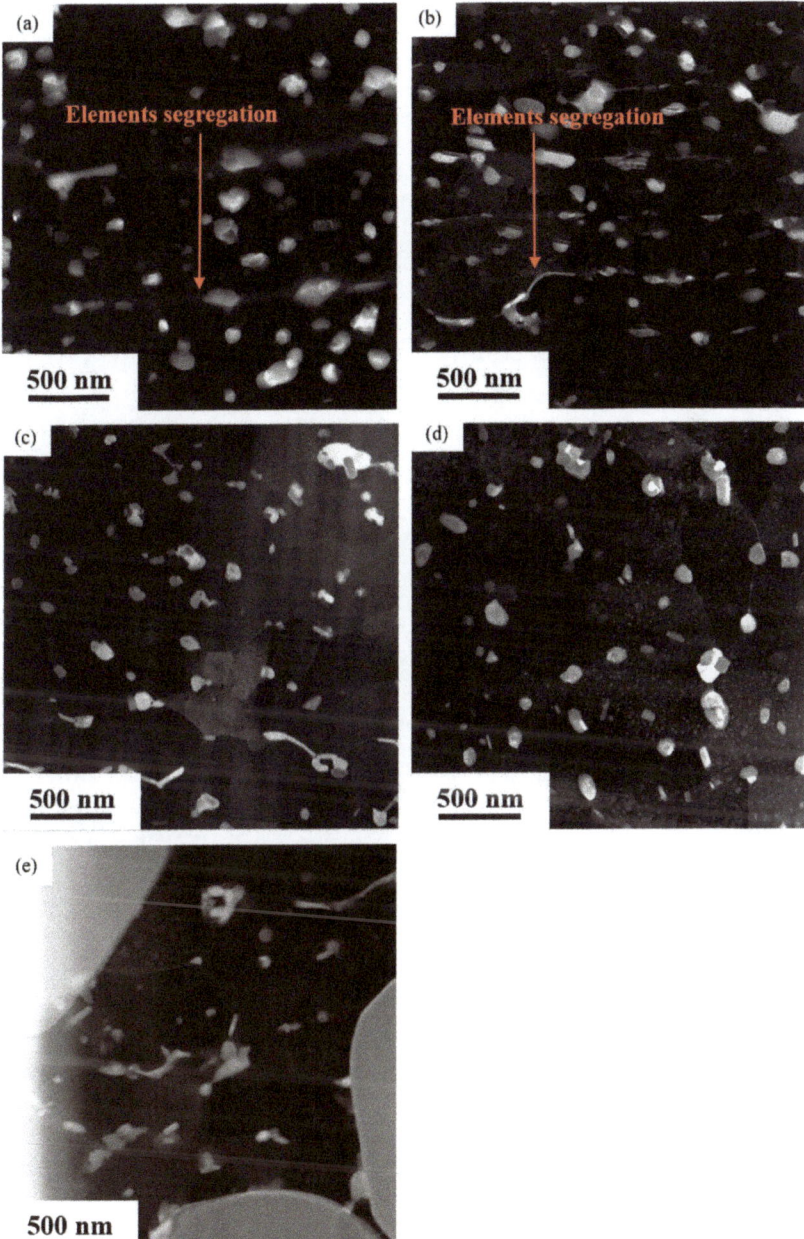

Figure 3. The STEM-HAADF micrographs showing the bottom regions of the near-surface deformed layers on hot rolled AA7050 aluminum alloy under different hot rolling conditions: (**a**) 1# (380 °C, 66.7%); (**b**) 2# (420 °C, 83.3%); (**c**) 3# (420 °C, 66.7%); (**d**) 4# (420 °C, 50%); (**e**) 5# (450 °C, 66.7%).

In order to further determine the composition of the elemental segregation, the 1# sample (380 °C, 66.7%) was selected for EDS mapping analysis, as shown in Figure 4. It is evident that the grain boundaries were rich in Cu, Zn and Mg in the near-surface deformed layer on the 1# sample, which also suggests that the bright bands were caused by the accumulation of Mg, Zn and Cu atoms. In addition, the precipitates observed in the near-surface deformed layer on 1# sample mainly contained Cu, Mg and Zn elements, as indicated in Figure 4, which might be the η (MgZn$_2$) phase enriched with Cu. The O element was mainly concentrated on the rolling surface, which was caused by the formation of oxides and promoted by the high temperature during hot rolling [24].

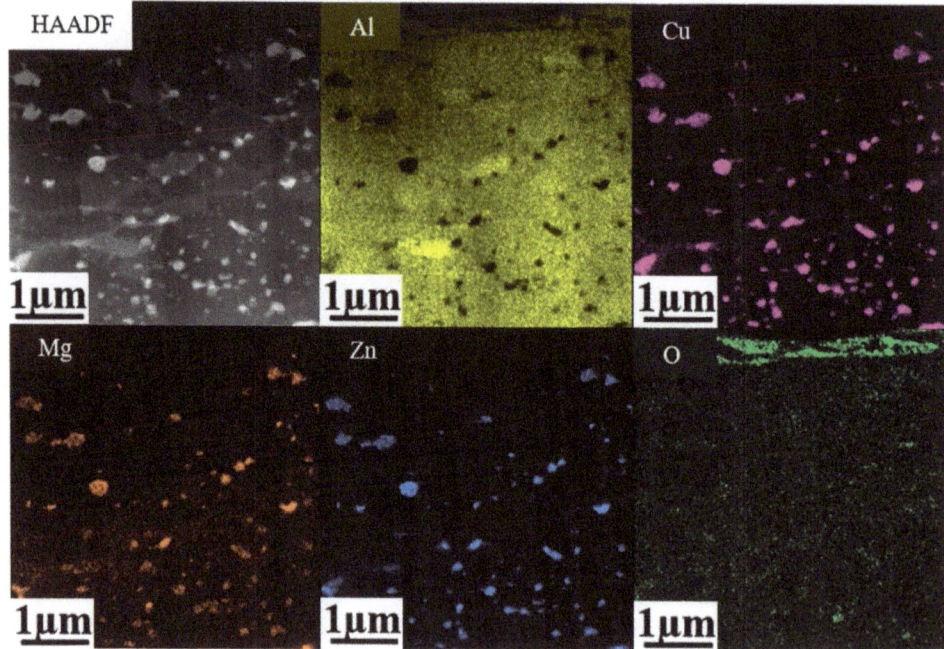

Figure 4. EDS mapping analysis of the near-surface deformed layer of 1# sample (380 °C, 66.7).

In order to further determine whether the precipitated phase was a η phase, we conducted a more in-depth study on it. Figure 5a displays the lattice image of the precipitates observed in the near-surface deformed layers along the $[\bar{1}12]_{Al}$ axis. The lattice parameters of the precipitate phase were measured as follows: a = 0.472 nm, c = 0.893 nm, respectively. Meanwhile, the lattice parameters of the η phase are as follows: a = 0.504 nm, c = 0.828 nm [25]. Figure 5b shows the enlarged image in Figure 5a. It can be found from the HAADF micrograph that the stacking structure was mainly R/R^{-1}, with occasional presence of an $R^{-1}R^{-1}$ stacking layer-structure. Previous studies pointed out that the R/R^{-1} stacking layer-structure often existed in η phase [26–29], and even an R/R^{-1} stacking layer-structure is presumed to exist in all types of the η phase [26]. Chung et al., [29] believed that the formation of the $R^{-1}R^{-1}$ stacking structure was typical in the η phase. Therefore, based on the experimental results, it can be determined that the precipitated phase is the η phase.

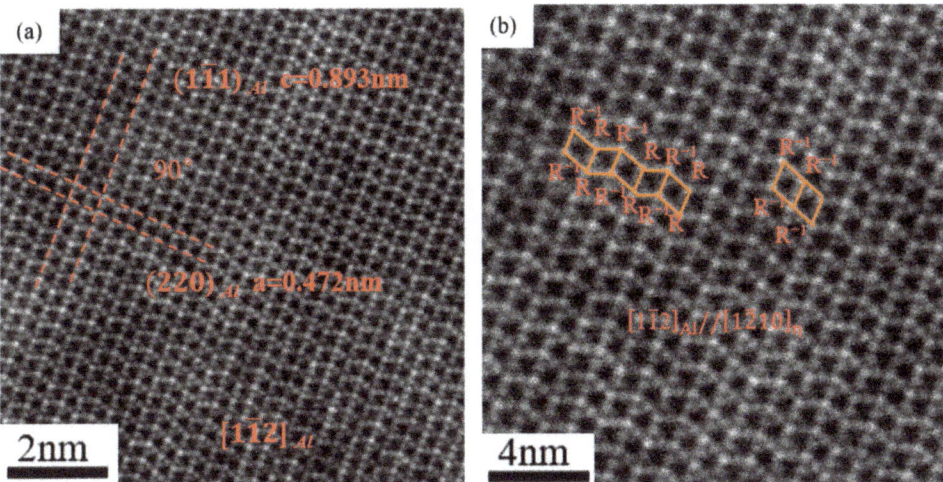

Figure 5. (a) The lattice image of the rounded plate precipitate along [1$\bar{1}$2]$_{Al}$ axis; (b) the enlarged image in (a).

In addition to precipitation of larger precipitates at grain boundaries (Figures 1–3), the fine precipitates are also precipitated inside the grains. Figure 6 displays the TEM micrographs showing the top regions of the near-surface layers, which are near the rolling surfaces on AA7050 aluminum alloy subjected to hot rolling under different conditions. It can be revealed that precipitate-free zone was not obviously adjacent to the grain boundaries in the near-surface deformed layers on the hot rolled AA7050 aluminum alloy under various conditions. At the same rolling temperature of 420 °C (2#, 3#, and 4# samples), the density of precipitates did not vary obviously with the increase of rolling reduction (Figure 6b–d). Meanwhile, the grain boundary precipitates of the three samples were discontinuous, and fine precipitates with dimensions of approx. 5–10 nm can be clearly observed. With the same rolling reduction of 66.7% (1#, 3#, and 5# samples), with the increase of rolling temperature, the density of such fine precipitates decreased in the order of 5# (450 °C) > 1# (380 °C) > 3# (420 °C), and the size of such fine precipitates also decreased, as indicated in Figure 6a,c,e. The grain boundary precipitates were distributed continuously at the lower rolling temperature (380 °C) and discontinuously at the higher rolling temperature (450 °C).

Figure 7 displays the TEM micrographs showing the top regions of the near-surface layers near the rolling surfaces on AA7050 aluminum alloy subjected to hot rolling under different conditions. It can be observed that dynamic recrystallization occurred, and the dislocations were annihilated and reordered by slipping and climbing in all samples, resulting in the disappearance of dislocations and the formation of dislocation cells in the structure. Compared with 3# sample (66.7%), the density of dislocations was reserved after dynamic recrystallization was higher in 2# (83.3%) and 4# samples (50%), as indicated by the arrows in Figure 7b,d. However, there was no significant difference in the dislocation density of sample 1#, 3# and 5# (Figure 7a,c,e).

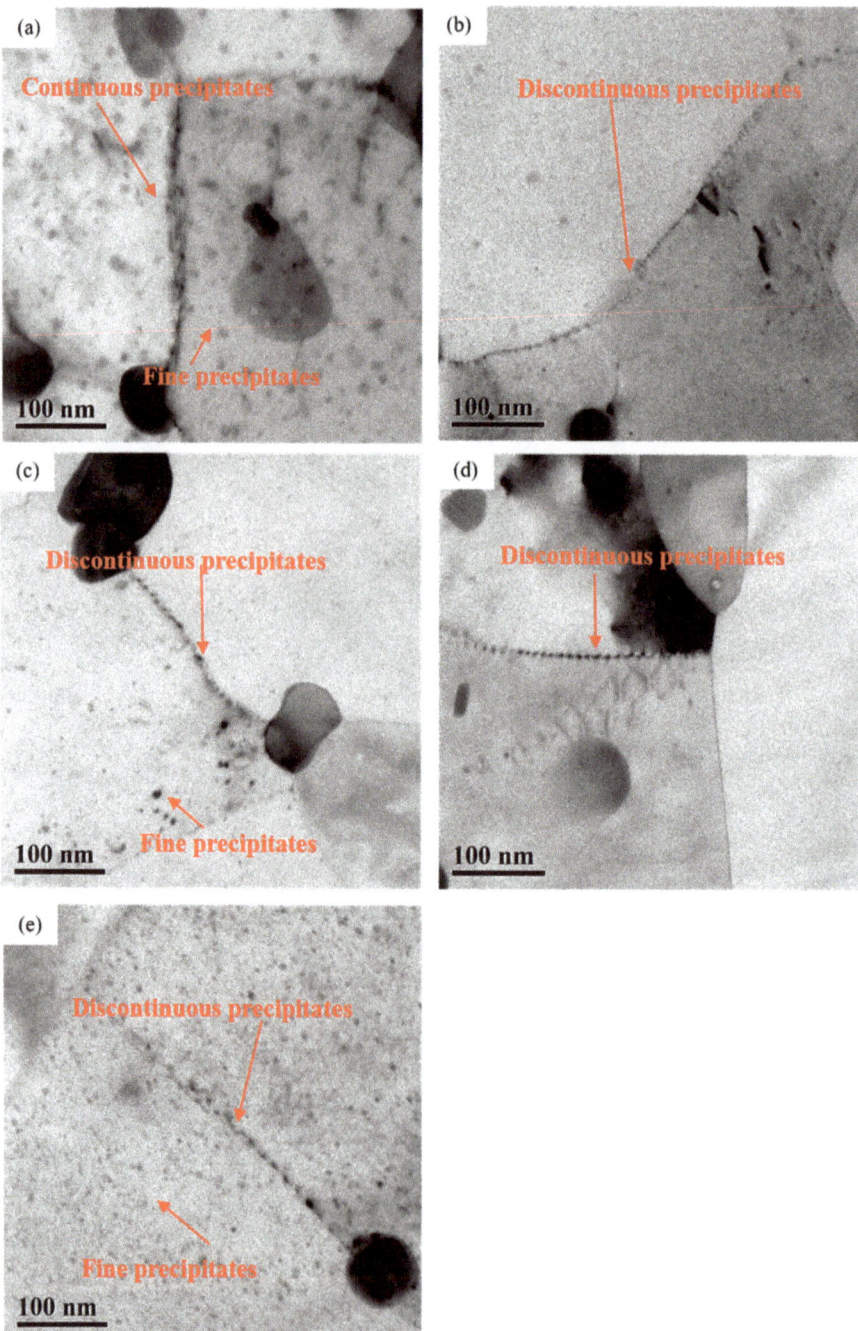

Figure 6. TEM micrographs showing the top regions of the near-surface deformed layers on hot rolled AA7050 aluminum alloy under different hot rolling conditions: (**a**) 1# (380 °C, 66.7%); (**b**) 2# (420 °C, 83.3%); (**c**) 3# (420 °C, 66.7%); (**d**) 4# (420 °C, 50%); (**e**) 5# (450 °C, 66.7%).

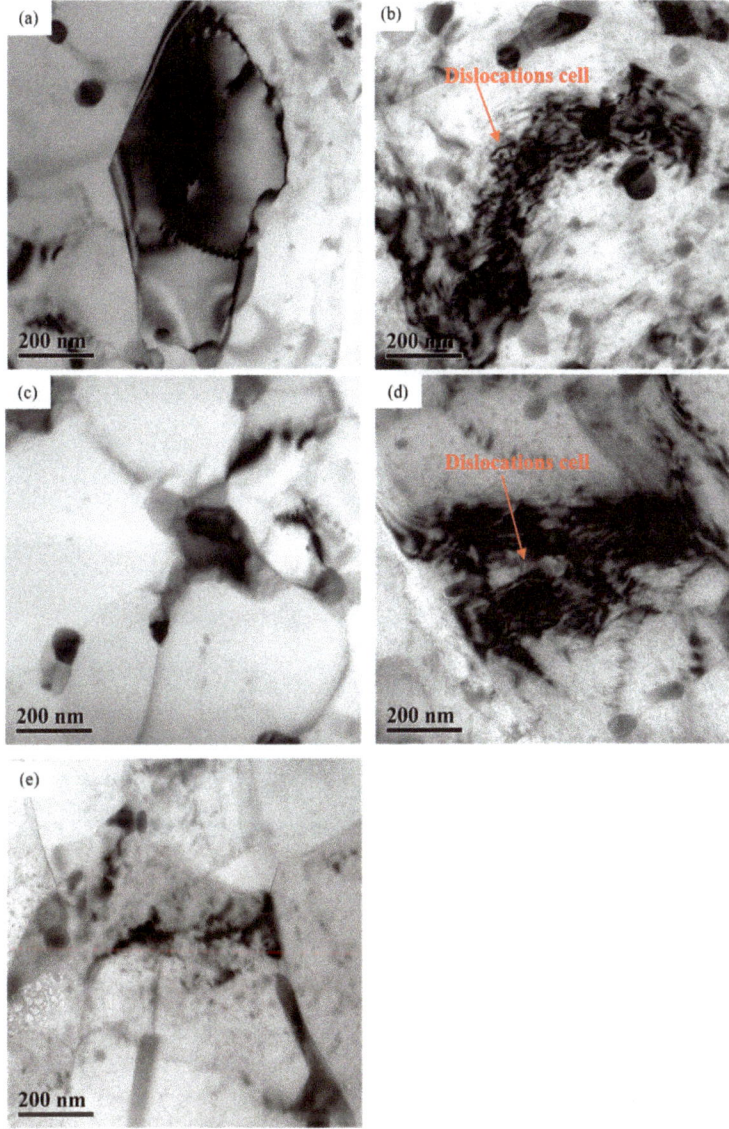

Figure 7. TEM micrographs showing the top regions of the near-surface deformed layers on hot rolled AA7050 aluminum alloy under different hot rolling conditions: (**a**) 1# (380 °C, 66.7%); (**b**) 2# (420 °C, 83.3%); (**c**) 3# (420 °C, 66.7%); (**d**) 4# (420 °C, 50%); (**e**) 5# (450 °C, 66.7%).

3.2. Electrochemical Corrosion Behavior of Hot Rolled AA7050 Aluminum Alloy

The OCP curve of hot rolled AA7050 aluminum alloy in 3.5 wt.% NaCl solution are displayed in Figure 8. According to Figure 8a, with the same rolling reduction of 66.7%, the OCP of 1# (380 °C), 3# (420 °C) and 5# (450 °C) samples varied obviously, with the values of −815.140 mV (SCE), −802.343 mV (SCE) and −844.650 mV (SCE), respectively, among which the OCP of 3# sample was the most positive, and the OCP of 5# sample was the most negative. The OCP values of such samples are listed in Table 3. It is clearly revealed that, with the same rolling reduction, the corrosion resistance of the aluminum alloy, rolled at different temperatures, was in the sequence: 3# > 1# > 5#. With the same hot rolling temperature of 420 °C, the OCP values of 2# (88.3%), 3# (66.7%) and 4# (50%) samples were −815.081 mV (SCE), −802.343 mV (SCE) and −804.718 mV (SCE), respectively, as evidenced in Figure 8b, indicating that the corrosion resistance of the hot rolled aluminum alloy was in the order of 3# > 4# > 2#.

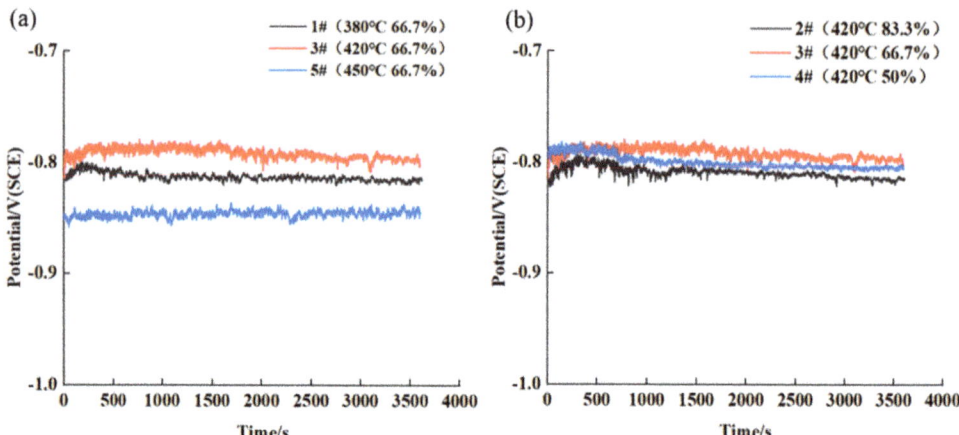

Figure 8. OCP of hot rolled AA7050 aluminum alloy in 3.5 wt.% NaCl solution: (**a**) alloy samples rolled with the same rolling reduction, at different rolling temperature; (**b**) alloy samples rolled at the same hot rolling temperature, with different rolling reduction.

Table 3. OCP, E_{corr} and I_{corr} of the rolled alloy samples.

Sample	1#	2#	3#	4#	5#
OCP/mV(SCE)	−815.140	−815.081	−802.343	−804.718	−844.650
E_{corr}/mV (SCE)	−803.528	−799.255	−778.503	−789.642	−826.569
I_{corr}/uA×cm^{-2} (SCE)	1.77	1.58	0.891	1.26	3.16

Figure 9 shows potentiodynamic polarization curves of hot rolled AA7050 aluminum alloy immersed in 3.5 wt.% NaCl solution. The corrosion potential and corrosion current density were calculated based on the results of the measurements, which are shown in Table 3. Figure 9a exhibits that with the same rolling reduction of 66.7%, the corrosion current density (I_{corr}) of the rolled alloy samples gradually decreased in the order of 5# (450 °C) > 1# (380 °C) > 3# (420 °C), and the corrosion potential gradually decreased in the order of 3# > 1# > 5#, suggesting that the corrosion resistance of the hot rolled aluminum alloy was in the order of 3# > 1# > 5#. Additionally, at the same hot rolling temperature of 420 °C, corrosion current density gradually decreased in the order of 2# (83.3%) > 4# (50%) > 3# (66.7%), and the corrosion potential gradually decreased in the order of 3# > 4# > 2#, as shown in Figure 9b, indicating that the corrosion resistance of the hot rolled aluminum alloy was in the order of 3# > 4# > 2#.

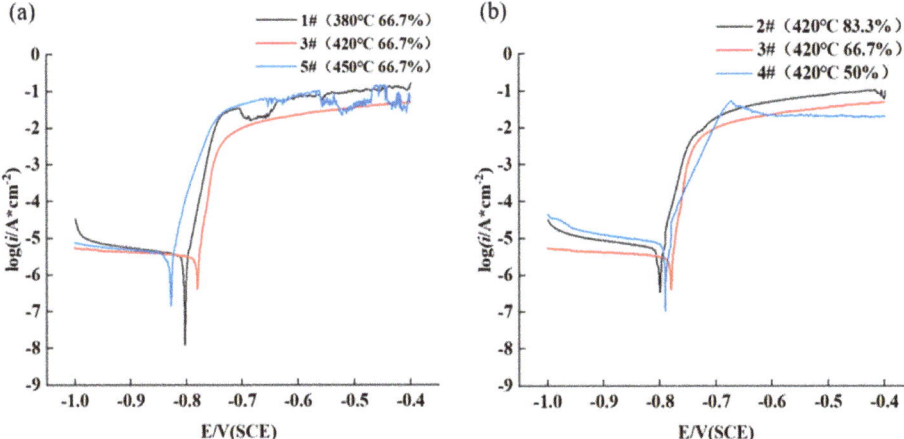

Figure 9. Potentiodynamic polarization curves of hot rolled AA7050 aluminum alloy in 3.5 wt.% NaCl solution: (**a**) the alloy samples rolled with the same rolling reduction, at different hot rolling temperature; (**b**) the alloy samples rolled at the same hot rolling temperature, with different rolling reduction.

Figure 10 shows the Nyquist diagrams of AA7050 aluminum alloy immersed in 3.5 wt.% NaCl solution. It is clearly revealed that the capacitive arc was evident at high frequency in the Nyquist diagram of the 1# sample, the inductive arc can be observed at the low frequency, and only the capacitive arc was obvious at high frequency in the Nyquist diagrams of the other four samples. The radius of the capacitive arc of the alloy samples rolled with the same reduction of 66.7% followed the order of 3# (420 °C) > 1# (380 °C) > 5# (450 °C), and that of the alloy samples rolled at the same temperature of 420 °C followed the order of 3# (66.7%) > 4# (50%) > 2# (83.3%), as evidenced in Figure 10a,b, respectively. Generally, the larger the radius of the capacitive arc is related to the better corrosion resistance of the alloy. Therefore, according to the experimental results, it is clearly evident that the 3# sample (420 °C, 66.7%) was the most corrosion resistant among the five samples.

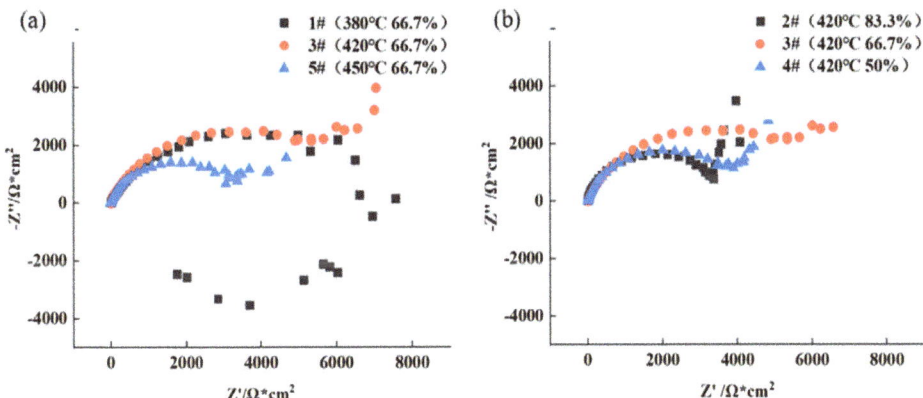

Figure 10. Nyquist diagrams of hot rolled AA7050 aluminum alloy in 3.5 wt.% NaCl solution: (**a**) the alloy samples rolled with the same rolling reduction, at different rolling temperature; (**b**) the alloy samples rolled at the same hot rolling temperature, with different rolling reduction.

Figures 11 and 12 show the corrosion morphology of rolled AA7050 aluminum alloy immersed in 3.5 wt.% NaCl solution for 6, 12, 18, and 24 h under different rolling conditions. It can be observed in Figures 11 and 12 that in the early stage of the immersion test (6 h), corrosion was not obvious on the rolled alloy surface of all samples. However, when the immersion time was extended to 12–24 h, compared with the 3# sample, the density of corrosion pits and corrosion products of the other samples significantly increased while corrosion was extended to increased areas on such samples. Based on the results, the 3# samples exhibited the best corrosion resistance, which was also consistent with the results of the potentiodynamic polarization and the EIS experiment.

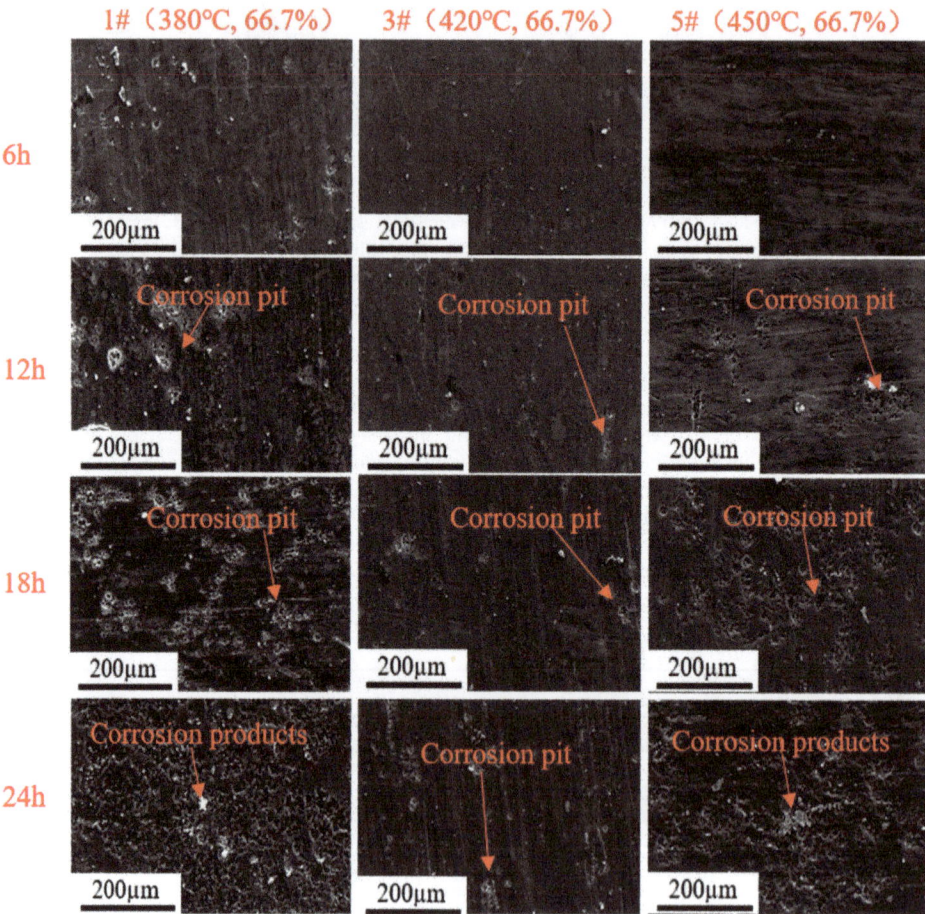

Figure 11. Corrosion morphology of the rolled AA7050 aluminum alloy immersed in 3.5 wt.% NaCl solution for 6, 12, 18, and 24 h (the samples were rolled with the same reduction at different temperatures).

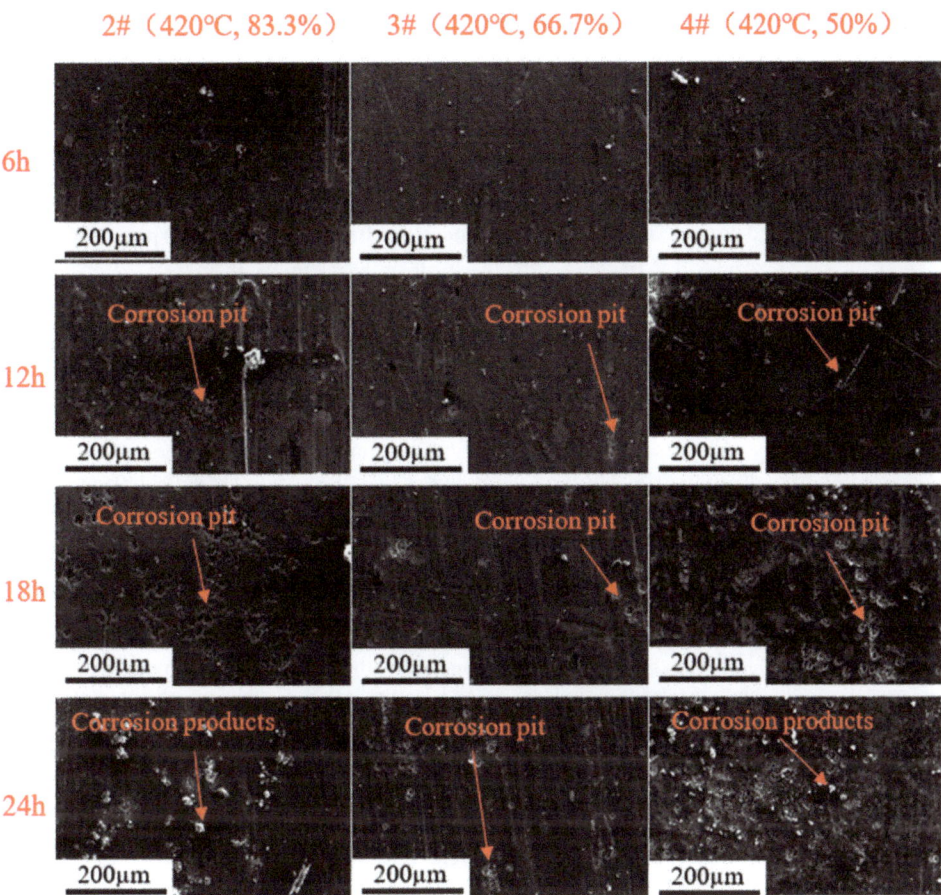

Figure 12. Corrosion morphology of the rolled AA7050 aluminum alloy immersed in 3.5 wt.% NaCl solution for 6, 12, 18, and 24 h (the samples were rolled at the same temperature with different reductions).

4. Discussion

4.1. Effect of Hot Rolling Parameters on the Near-Surface Deformed Layer

In summary, after hot rolling, the size of equiaxed grains in the near-surface regions on all the rolled samples did not exceed 500 nm, and the precipitates were uniformly distributed in the equiaxed grain layers. However, the density of precipitates in the fibrous grain layers was significantly higher than that in the equiaxed grain layers (1# and 2#). Additionally, variation in hot rolling conditions led to varying degrees of grain boundary segregation. The segregation in the near-surface regions on 3#, 4#, and 5# samples was mainly concentrated at the top of the near surface deformed layers, while in 1# and 2# samples, element segregation was distributed throughout the near-surface deformed layers.

The grain size of Al-Zn-Mg-Cu alloys treated by conventional thermo-mechanical processing is normally in the micron-scale [30]. Obviously, in the current study, under the co-influence of elevated temperature and high level of strain caused by severe deformation during hot rolling, dynamic recrystallization occurred in the alloy. Consequently, the deformation introduced dislocations with increased density which moved to form the sub-grain structures with relatively low-angle grain boundaries. Such sub-grains continuously rotated, and the large fibrous grains were eventually divided into fine equiaxed grains with dimensions less than 500 nm which were evident in the near-surface layers on the hot rolled AA7050 aluminum alloy under different conditions (Figures 1–3).

During the hot rolling process, grains in AA7050 aluminum alloy were elongated along the rolling direction, leading to the formation of fibrous grains. Such fibrous grains in the near-surface deformed layers were mostly subdivided into equiaxed grains caused by dynamic recrystallization during hot deformation, as evidenced in samples 3#, 4#, and 5#. However, fibrous grains were still evident at the bottom regions of the near-surface deformed layers in samples 1# and 2#. Such phenomena can be explained by the elemental segregation at grain boundaries in the near-surface deformed layers on samples 1# and 2# (Figures 2 and 3), and comparing that with the other samples. The solute atoms at grain boundaries significantly facilitated the occurrence of dynamic recovery and accelerated the annihilation and reordering of dislocations, thus reducing the deformation energy storage of aluminum alloys [30] which is the driving force of dynamic recrystallization. Consequently, dynamic recrystallization was not obvious in such regions, and fine equiaxed grains can hardly be observed.

Dynamic recrystallization introduced in equiaxed grain layers can also been observed adjacent to the fibrous grain layer in the 1# and 2# samples at the top of the near-surface regions near the rolling surfaces. Since the temperature and strain gradually decreased from the rolling surface to the bulk alloy, and the top region of the near-surface deformed layers experienced the highest level of temperatures and strains, the high density of dislocations led to the formation of sub-grains and finally the fibrous grains were divided into fine grains.

As evidenced in Figures 1–3, the density of the precipitates was higher in the fibrous grain layers in the near-surface regions with the presence of both the equiaxed grain layer and the fibrous grain layer (1# and 2# samples). Such phenomena can be explained by the higher density of defects, such as vacancies and dislocations, promoting the diffusion of the solute atoms, providing ideal locations for the nucleation of the precipitates and the segregation of elements at the grain boundaries in the fibrous grain layer. Such dislocations largely promoted the nucleation of precipitates, and the segregation of elements also contributed to the formation of the precipitates rich in alloy elements. While in the equiaxed grain layer, most dislocations were consumed by dynamic recrystallization, resulting in fewer locations for nucleation of precipitates. Additionally, during the hot rolling process, extensive heat was generated as a result of the elevation of the temperature near the rolling surface, leading to the further dissolution of the precipitates. Consequently, this was relatively far from the rolling surface compared with the equiaxed grain layer, as evidenced in 1# and 2# samples.

4.2. Effect of Microstructure of Near-Surface Deformed Layer on Corrosion

It is well known that the corrosion of aluminum alloys starts from the surface region and is closely related to the surface condition. Microstructure evolution of the near-surface region, including the feature of the grains, the density, dimension and distribution of precipitates, as well as the density and distribution of dislocations directly influence the corrosion behavior of the alloy. Generally, corrosion preferentially occurs in the locations with high energy [31–34], such as the locations with high density of dislocations.

At the same rolling temperature of 420 °C, with the increase of rolling reduction, the density, dimension and distribution of precipitates in the near-surface region of the rolled samples showed no obvious variation (Figure 6), but the presence of dislocations retained after dynamic recrystallization led to increased stored energy, as shown in Figure 7. This significantly promoted the corrosion susceptibility of the rolled alloy. Thus, the corrosion resistance of 2# (83.3%) and 4# (50%) samples with higher dislocation density decreased compared with 3# (66.7%).

With the same rolling reduction of 66.7%, the density of such fine precipitates decreased in the order of 5# (450 °C) > 1# (380 °C) > 3# (420 °C). The density of dislocations was similar among the samples, as evidenced in Figures 6 and 7. It is found that the corrosion potential of the Al matrix is -0.68 V (SCE), while that of the MgZn$_2$ phase is approx. $-1 \sim -1.07$ V (SCE) [30,35–38]. Consequently, the increase of the density of the precipitates results in the promotion of electrochemical inhomogeneity within the alloy, and the alloy was less corrosion resistant in the corrosion environment. At the same time, with the increase of the temperature, the distribution of grain boundary precipitates varied from continuous to discontinuous, indicating that the density of precipitates was an important factor affecting the corrosion behavior, compared with the distribution of precipitates.

5. Conclusions

In the current work, the hot rolling introduced near-surface deformed layer and its influence on corrosion behavior of the hot rolled AA7050 aluminum alloy was investigated. The main conclusions can be summarized as follows:

1. Under the action of elevated temperature and strain introduced during hot rolling, dynamic recrystallization occurred in the near-surface region, resulting in the generation of equiaxed grains and refinement of grains.
2. With the same rolling reduction of 66.7%, and with the reduction of rolling temperature or at the same rolling temperature of 420 °C, the increase of rolling reduction caused the segregation of Cu, Zn and Mg elements to become more serious. This occurred from the near-surface deformed layer near the rolling surface to the entire near-surface deformed layer.
3. The density of precipitates in the fibrous grain layer was much higher than that in the equiaxed grain layer due to the elemental segregation and higher density of dislocations in such layers, which provided increased locations as well as higher concentration of solute atoms for nucleation of precipitates.
4. The density of the precipitates and dislocations are the key factors affecting the corrosion properties of rolled alloys. With the same rolling reduction, the corrosion resistance mainly depends on the density of precipitates, since it significantly promotes the electrochemical inhomogeneity within the alloy. At the same rolling temperature, the corrosion resistance of the rolled AA7050 aluminum alloy is closely related to the density of dislocations, which also contributes to the initiation of corrosion.

Author Contributions: Methodology, Q.P. and W.W.; Formal analysis, E.L., J.Y. and W.W.; Investigation, E.L.; Writing—original draft, E.L.; Writing—review & editing, B.L.; Visualization, E.L.; Supervision, Q.P., B.L. and J.Y.; Project administration, Q.P.; Funding acquisition, B.L. All authors have read and agreed to the published version of the manuscript.

Funding: The current work was financially supported by the Natural Science Foundation of Hunan Province (Grant No. 2021JJ40748), and the authors also acknowledge the financial support from Hangzhou Win-Win Technology Co., Ltd. (Grant No. 2022430102001600) and Key-Area Research and Development Program of Foshan City. (Grant No. 2230032004640).

Institutional Review Board Statement: Not applicable.

Informed Consent Statement: Not applicable.

Data Availability Statement: Data will be made availability on request.

Conflicts of Interest: The authors declare that they have no known competing financial interest or personal relationship that could have appeared to influence the work reported in this paper.

References

1. Zou, Y.; Wu, X.; Tang, S.; Zhu, Q.; Song, H.; Guo, M.; Cao, L. Investigation on microstructure and mechanical properties of Al-Zn-Mg-Cu alloys with various Zn/Mg ratios. *J. Mater. Sci. Technol.* **2021**, *85*, 106–117. [CrossRef]
2. Heinz, A.; Haszler, A.; Keidel, C.; Moldenhauer, S.; Benedictus, R.; Miller, W. Recent development in aluminium alloys for aerospace applications. *Mater. Sci. Eng. A* **2000**, *280*, 102–107. [CrossRef]
3. Azarniya, A.; Taheri, A.; Taheri, K. Recent advances in ageing of 7xxx series aluminum alloys: A physical metallurgy perspective. *J. Alloys Compd.* **2019**, *781*, 945–983. [CrossRef]

4. Senkov, O.; Shagiev, M.; Senkova, S.; Miracle, D. Precipitation of Al3(Sc,Zr) particles in an Al–Zn–Mg–Cu–Sc–Zr alloy during conventional solution heat treatment and its effect on tensile properties. *Acta Mater.* **2008**, *56*, 3723–3738. [CrossRef]
5. Stojanovic, B.; Bukvic; Epler, I. Application of Aluminum and Aluminum Alloys in Engineering. *Appl. Eng. Lett. J. Eng. Appl. Sci.* **2018**, *3*, 52–62. [CrossRef]
6. Guo, S.; Liu, Z.; Bai, S.; Ou, L.; Zhang, J.; He, G.; Zhao, J. Effect of rolling temperature on mechanical properties and corrosion resistance of Al-Cu-Mg-Ag alloy. *J. Alloys Compd.* **2022**, *897*, 163168. [CrossRef]
7. Kumar, V.; Kumar, D. Investigation of tensile behaviour of cryorolled and room temperature rolled 6082 Al alloy. *Mater. Sci. Eng. A* **2017**, *691*, 211–217. [CrossRef]
8. Wang, D.; Ma, Z.; Gao, Z. Effects of severe cold rolling on tensile properties and stress corrosion cracking of 7050 aluminum alloy. *Mater. Chem. Phys.* **2009**, *117*, 228–233. [CrossRef]
9. Liu, F.; Liu, Z.; Jia, P.; Bai, S.; Yan, P.; Hu, Y. Dynamic dissolution and texture evolution of an Al–Cu–Mg–Ag alloy during hot rolling. *J. Alloys Compd.* **2020**, *827*, 154254. [CrossRef]
10. Liu, D.-Y.; Li, J.-F.; Liu, T.-L.; Ma, Y.-L.; Iwaoka, H.; Hirosawa, S.; Zhang, K.; Zhang, R.-f. Microstructure evolution and mechanical properties of Al-Cu-Li alloys with different rolling schedules and subsequent artificial ageing heat treatment. *Mater. Charact.* **2020**, *170*, 110676. [CrossRef]
11. Zuiko, I.; Mironov, S.; Kaibyshev, R. Microstructural evolution and strengthening mechanisms operating during cryogenic rolling of solutionized Al-Cu-Mg alloy. *Mater. Sci. Eng. A* **2019**, *745*, 82–89. [CrossRef]
12. Zhao, Q.; Liu, Z.; Hu, Y.; Li, S.; Bai, S. Evolution of Goss texture in an Al–Cu–Mg alloy during cold rolling. *Arch. Civ. Mech. Eng.* **2020**, *20*, 24. [CrossRef]
13. Liu, B.; Zhou, X.; Hashimoto, T.; Zhang, X.; Wang, J. Machining introduced microstructure modification in aluminium alloys. *J. Alloys Compd.* **2018**, *757*, 233–238. [CrossRef]
14. Liu, B.; Zhou, X.; Zhang, X. Orthogonal machining introduced microstructure modification in AA7150-T651 aluminium alloy. *Mater. Charact.* **2017**, *123*, 91–98. [CrossRef]
15. Saklakoglu, N.; Irizalp, S.G.; Akman, E.; Demir, A. Near surface modification of aluminum alloy induced by laser shock processing. *Opt. Laser Technol.* **2014**, *64*, 235–241. [CrossRef]
16. Liu, Y.; Hashimoto, T.; Zhou, X.; Thompson, G.; Scamans, G.; Rainforth, W.; Hunter, J. Influence of near-surface deformed layers on filiform corrosion of AA3104 aluminium alloy. *Surf. Interface Anal.* **2013**, *45*, 1553–1557. [CrossRef]
17. Scamans, G.; Frolish, M.; Rainforth, W.; Zhou, Z.; Liu, Y.; Zhou, X.; Thompson, G. The ubiquitous Beilby layer on aluminium surfaces. *Surf. Interface Anal.* **2010**, *42*, 175–179. [CrossRef]
18. Leth-Olsen, H.; Afseth, A.; Nisancioglu, K. Filiform corrosion of aluminium sheet. Ii. Electrochemical and corrosion behaviour of bare substrates. *Corros. Sci.* **1998**, *40*, 1195–1214. [CrossRef]
19. Leth-Olsen, H.; Nordlien, J.; Nisancioglu, K. Filiform corrosion of aluminium sheet. iii. Microstructure of reactive surfaces. *Corros. Sci.* **1998**, *40*, 2051–2063. [CrossRef]
20. Afseth, A.; Nordlien, J.; Scamans, G.; Nisancioglu, K. Effect of heat treatment on filiform corrosion of aluminium alloy AA3005. *Corros. Sci.* **2001**, *43*, 2093–2109. [CrossRef]
21. Afseth, A.; Nordlien, J.; Scamans, G.; Nisancioglu, K. Effect of thermo-mechanical processing on filiform corrosion of aluminium alloy AA3005. *Corros. Sci.* **2002**, *44*, 2491–2506. [CrossRef]
22. da Silva, R.; Izquierdo, J.; Milagre, M.; Araujo, J.; Antunes, R.; Souto, R.; Costa, I. Electrochemical characterization of alloy segregation in the near-surface deformed layer of welded zones of an Al−Cu−Li alloy using scanning electrochemical microscopy. *Electrochim. Acta* **2022**, *427*, 140873. [CrossRef]
23. Liu, Y.; Laurino, A.; Hashimoto, T.; Zhou, X.; Skeldon, P.; Thompson, G.; Scamans, G.; Blanc, C.; Rainforth, W.; Frolish, M. Corrosion behaviour of mechanically polished AA7075-T6 aluminium alloy. *Surf. Interface Anal.* **2010**, *42*, 185–188. [CrossRef]
24. Gali, O.; Shafiei, M.; Hunter, J.; Riahi, A. The influence of work roll roughness on the surface/near-surface microstructure evolution of hot rolled aluminum–magnesium alloys. *J. Mater. Process. Technol.* **2016**, *237*, 331–341. [CrossRef]
25. Chung, T.-F.; Yang, Y.-L.; Huang, B.-M.; Shi, Z.; Lin, J.; Ohmura, T.; Yang, J.-R. Transmission electron microscopy investigation of separated nucleation and in-situ nucleation in AA7050 aluminium alloy. *Acta Mater.* **2018**, *149*, 377–387. [CrossRef]
26. Bendo, A.; Matsuda, K.; Lee, S.; Nishimura, K.; Nunomura, N.; Toda, H.; Yamaguchi, M.; Tsuru, T.; Hirayama, K.; Shimizu, K.; et al. Atomic scale HAADF-STEM study of η′ and η1 phases in peak-aged Al–Zn–Mg alloys. *J. Mater. Sci.* **2018**, *53*, 4598–4611. [CrossRef]
27. Marioara, C.; Lefebvre, W.; Andersen, S.; Friis, J. Atomic structure of hardening precipitates in an Al–Mg–Zn–Cu alloy determined by HAADF-STEM and first-principles calculations: Relation to η-MgZn2. *J. Mater. Sci.* **2013**, *48*, 3638–3651. [CrossRef]
28. Xu, X.; Zheng, J.; Li, Z.; Luo, R.; Chen, B. Precipitation in an Al-Zn-Mg-Cu alloy during isothermal aging: Atomic-scale HAADF-STEM investigation. *Mater. Sci. Eng. A* **2017**, *691*, 60–70. [CrossRef]
29. Chung, T.-F.; Yang, Y.-L.; Shiojiri, M.; Hsiao, C.-N.; Li, W.-C.; Tsao, C.-S.; Shi, Z.; Lin, J.; Yang, J.-R. An atomic scale structural investigation of nanometre-sized η precipitates in the 7050 aluminium alloy. *Acta Mater.* **2019**, *174*, 351–368. [CrossRef]
30. Zuo, J.; Hou, L.; Shi, J.; Cui, H.; Zhuang, L.; Zhang, J. Enhanced plasticity and corrosion resistance of high strength Al-Zn-Mg-Cu alloy processed by an improved thermomechanical processing. *J. Alloys Compd.* **2017**, *716*, 220–230. [CrossRef]
31. Xiong, H.; Zhou, Y.; Yang, P.; Kong, C.; Yu, H. Effects of cryorolling, room temperature rolling and aging treatment on mechanical and corrosion properties of 7050 aluminum alloy. *Mater. Sci. Eng. A* **2022**, *853*, 143764. [CrossRef]

32. Kanta, P.L.M.; Srivastava, V.; Venkateswarlu, K.; Paswan, S.; Mahato, B.; Das, G.; Sivaprasad, K.; Krishna, K. Corrosion behavior of ultrafine-grained AA2024 aluminum alloy produced by cryorolling. *Int. J. Miner. Metall. Mater.* **2017**, *24*, 1293–1305. [CrossRef]
33. Zhang, X.; Zhou, X.; Nilsson, J.-O.; Dong, Z.; Cai, C. Corrosion behaviour of AA6082 Al-Mg-Si alloy extrusion: Recrystallized and non-recrystallized structures. *Corros. Sci.* **2018**, *144*, 163–171. [CrossRef]
34. Peng, Y.; Huang, B.; Zhong, Y.; Su, C.; Tao, Z.; Rong, X.; Li, Z.; Tang, H. Electrochemical corrosion behavior of 6061 Al alloy under high rotating speed submerged friction stir processing. *Corros. Sci.* **2023**, *215*, 111029. [CrossRef]
35. Sedriks, A.; Green, J.; Novak, D. Corrosion behavior of aluminum-boron composites in aqueous chloride solutions. *Metall. Trans.* **1971**, *2*, 871–875. [CrossRef]
36. Mattsson, E.; Gullman, L.-O.; Knutsson, L.; Sundberg, R.; Thundal, B. Mechanism of exfoliation (layer corrosion) of Al-5% Zn-1% Mg. *Br. Corros. J.* **1971**, *6*, 73–83. [CrossRef]
37. Ahmad, Z. *Recent Trends in Processing and Degradation of Aluminium Alloys*; BoD–Books on Demand: Norderstedt, Germany, 2011.
38. Chen, S.; Li, J.; Hu, G.-y.; Chen, K.; Huang, L. Effect of Zn/Mg ratios on SCC, electrochemical corrosion properties and microstructure of Al-Zn-Mg alloy. *J. Alloys Compd.* **2018**, *757*, 259–264. [CrossRef]

Disclaimer/Publisher's Note: The statements, opinions and data contained in all publications are solely those of the individual author(s) and contributor(s) and not of MDPI and/or the editor(s). MDPI and/or the editor(s) disclaim responsibility for any injury to people or property resulting from any ideas, methods, instructions or products referred to in the content.

Article

The Portevin–Le Chatelier Effect of Cu–2.0Be Alloy during Hot Compression

Daibo Zhu [1], Na Wu [1], Yang Liu [1,2,*], Xiaojin Liu [1], Chaohua Jiang [1], Yanbin Jiang [3], Hongyun Zhao [4], Shuhui Cui [4] and Guilan Xie [1,*]

[1] School of Mechanical Engineering and Mechanics, Xiangtan University, Xiangtan 411105, China; daibozhu@xtu.edu.cn (D.Z.); m17873820325@163.com (N.W.); liuxiaojin11225@126.com (X.L.); 202005501512@smail.xtu.edu.cn (C.J.)
[2] School of Materials Science and Engineering, Nanyang Technological University, Singapore 639798, Singapore
[3] School of Materials Science and Engineering, Central South University, Changsha 410083, China; jybin8113@163.com
[4] State Key Laboratory of Special Rare Metal Materials, Northwest Rare Metal Materials Research Institute Ningxia Co., Ltd., Shizuishan 753000, China; yunhongzhao2023@163.com (H.Z.); shuhuicui2023@163.com (S.C.)
* Correspondence: liuyang_225@126.com (Y.L.); xieguilan@xtu.edu.cn (G.X.); Tel.: +86-731-5289-8553 (Y.L.); +86-731-5829-2495 (G.X.)

Abstract: The Portevin–Le Chatelier effect of Cu–2.0Be alloy was investigated using hot isothermal compression at varying strain rates (0.01–10 s^{-1}) and temperature (903–1063 K). An Arrhenius-type constitutive equation was developed, and the average activation was determined. Both strain-rate-sensitive and temperature-sensitive serrations were identified. The stress–strain curve exhibited three types of serrations: type A at high strain rates, type B (mixed A + B) at medium strain rates, and type C at low strain rates. The serration mechanism is mainly affected by the interaction between the velocity of solute atom diffusion and movable dislocations. As the strain rate increases, the dislocations outpace the diffusion speed of the solute atoms, limiting their ability to effectively pin the dislocations, resulting in lower dislocation density and serration amplitude. Moreover, the dynamic phase transformation triggers the formation of nanoscale dispersive β phases, which impede dislocation and cause a rapid increase in the effective stress required for unpinning, leading to the formation of mixed A + B serrations at 1 s^{-1}.

Keywords: Cu–Be alloy; Portevin–Le Chaterlier effect; serrations; dynamic phase transformation

1. Introduction

The Portevin–Le Chatelier (PLC) effect, a well-known phenomenon first observed by Portevin and Le Chatelier [1,2], occurs in a large number of industrial alloys such as aluminum alloys, steel, and copper alloys [3–5]. During the plastic deformation process, the flow stress, work hardening, and ductility can be influenced by the PLC effect, which manifests as serrations on the stress–strain curve [5,6]. The extrinsic factors of the PLC effect are mainly the deformation temperature and strain rate, whereas the intrinsic causes include the size and concentration of atoms, solute diffusivity, precipitation characteristics, and dislocation patterns. Dynamic strain aging (DSA), which is defined as the interplay between movable dislocations and the diffusion of solute atoms, is the primary mechanism underlying the PLC effect [7–9]. DSA produces a variety of types of heterogeneous deformation, which are divided into three main stress–strain curve serration-types [6,10]:

- The type A serration is usually found at high strain rates and low deformation temperatures, exhibiting relatively small fluctuations in flow stress, and the fluctuations appear at a random frequency in the stress–strain curve. The main reason for this behavior is the locking of movable dislocations by the solute atoms.

- The type B serration, which usually occurs at medium to high strain rates, is characterized by rapid continuous oscillations above and below the average stress value, obtaining higher amplitudes than those of type A serrations. Under certain conditions, the type B serration could exhibit a minor drop in stress within a certain interval.
- The type C serration is commonly found at relatively low strain rates and exhibits high amplitude and frequency stress oscillations. Type C serration oscillations are above and below the mean value, with larger amplitudes than those of type B serrations.

The PLC effect in copper alloys has been studied in Cu–Al [11], Cu–Cr–Nb [12], Cu–Ga [13], Cu–Sn [14], Cu–Si [15], Cu–Ti [15], Cu–P [15], Cu–Zn [16], and Cu–Ni–Zn [17] copper alloy systems. For Cu–Be alloys [18,19], researchers have focused on the serrations under various aging temperatures (593 K—733 K) and strain rates, due to the precipitation process, which is well-known as G.P. zones—γ'' ordered phase—γ' phase—equilibrium γ phase [20–22]. Previous work on the PLC effects is beneficial for the shape forming of Cu–Be sheets. However, the wide application of Cu–Be alloys in industrial applications is not limited to manufacturing various geometric and near-net shapes, It, also includes the production of sheet/foil/bar/wire materials [23,24], the processing temperature of which is higher than 893 K. At the deformation temperatures above 893 K, the β phases are precipitated through dynamic phase transformation, the diffusion rate of which is higher than that of the traditional aging precipitations, and the effect of β phases on serrations of the Cu–Be alloy is not clear at present. Hence, comprehensively understanding the serrated flow behavior and serration mechanism of Cu–Be alloys at high deformation temperatures is of great significance. To investigate the PLC effect and serration types of Cu–2.0Be alloy at various deformation conditions, hot isothermal compression deformation was employed in this work. The Arrhenius model was chosen to calculate the thermal activation energy and decouple the constitutive relationship under various deformation conditions. In addition, the serration mechanism under various deformation conditions was illustrated.

2. Experimental Process

Table 1 lists the component (in wt.%) of the Cu–Be alloy used in this study. A homogenization heat treatment at 1073 K for 24 h was used to prepare all the samples, and then cooled with the furnace for 36 h. For hot compression testing, cylindrical samples with a diameter of 8 mm and height of 12 mm were selected using a Gleeble-3500 thermomechanical simulator with a heating rate of 5 K/s. To achieve consistent heat deformation, all samples were maintained at the necessary temperature for 5 min. In order to reduce the effect of friction between the sample and the processing platform, graphite lubricant with a thickness of approximately 1 mm was applied to coat on the flat end of the sample. The experimental parameters were as follows: for a true strain of 0.7, the deformation temperatures ranged from 903 to 1063 K with a 40 K interval, whereas the strain rates were $0.01\ s^{-1}$, $0.1\ s^{-1}$, $1\ s^{-1}$, and $10\ s^{-1}$. All compression tests were carried out in a vacuum, and all samples were promptly water-quenched to maintain the deformed microstructure following the hot compression tests. The samples for microstructural analysis were cut along the longitudinal plane following the hot compression tests. Figure 1 depicts the hot compression experimentation process.

Table 1. Components of Cu–2.0Be alloy.

Component	Be	Fe	Al	Si	Pb	Others	Cu
Content (wt.%)	2.02	0.045	0.009	0.02	0.0018	0.23	Bal.

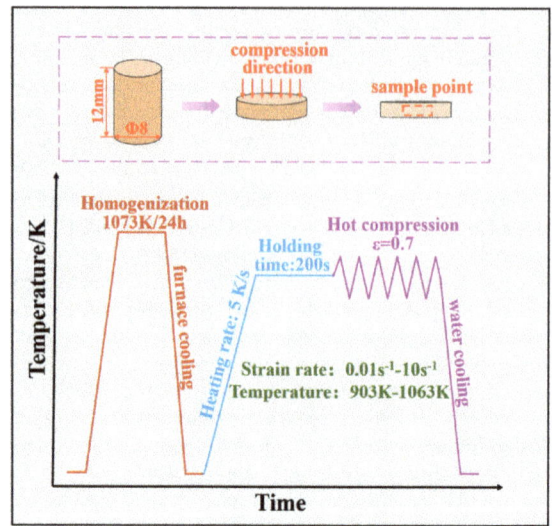

Figure 1. Schematic diagram of hot compression experiment process.

Samples for electron backscatter diffraction pattern (EBSD) analysis were mechanically and vibratory polished for 0.5 h by a Buehler VibroMet2 Vibratory Polisher. A field emission gun-equipped FEI scanning electron microscope was used to characterize the grain structure. The HKL Channel 5 software (VMware Workstation 15.5.5 Pro) was chosen to analyze the EBSD data using a step size of 0.6 um. Low-angle grain boundaries (LAGBs) with orientations between 2° and 10° in the EBSD maps are depicted by red lines, whereas the high-angle grain boundaries (HAGBs) with orientations >10° are depicted by blue lines. Transmission electron microscopy (TEM) images were collected using a JEM-2100F transmission electron microscope at 200 kV. The TEM sample was formed into a disc with a 50 μm thickness and 3 mm diameter, and twin-jet thinning was applied in a mixed solution of $HNO_3:CH_3OH = 1:3$. The experimental voltage and temperature were 30 V and −30 °C respectively.

3. Result and Discussion

3.1. The Influence of the Temperature and Strain Rate on the Serrated Flow

The true flow stress curves of hot isothermal deformation are shown in Figure 2 for strain rates of $0.01~s^{-1}$ (Figure 2a), $0.1~s^{-1}$ (Figure 2c), $1~s^{-1}$ (Figure 2e), and $10~s^{-1}$ (Figure 2g), with magnified views of the red dashed box given in Figure 2b,d,f,h, respectively. According to the results, the compression temperature and strain rate were closely related to the flow stress change. At a certain strain rate, the flow stress decreased with rising temperature. Additionally, the flow stress rose with an increasing strain rate at a certain deformation temperature. The development of the flow stress curves can be summarized as occurring in three stages. In the first stage, the flow stress increased practically linearly with increasing strain (the occurrence of work hardening), which was caused by the development and propagation of dislocations. In the second stage, the increasing rate of flow stress slowed down and reached its peak value, indicating the emergence of dynamic restoration (flow softening) processes, such as dynamic recovery (DRV) and dynamic recrystallization (DRX). In the third stage, DRX appeared, and the flow stress decreased and eventually began to level off.

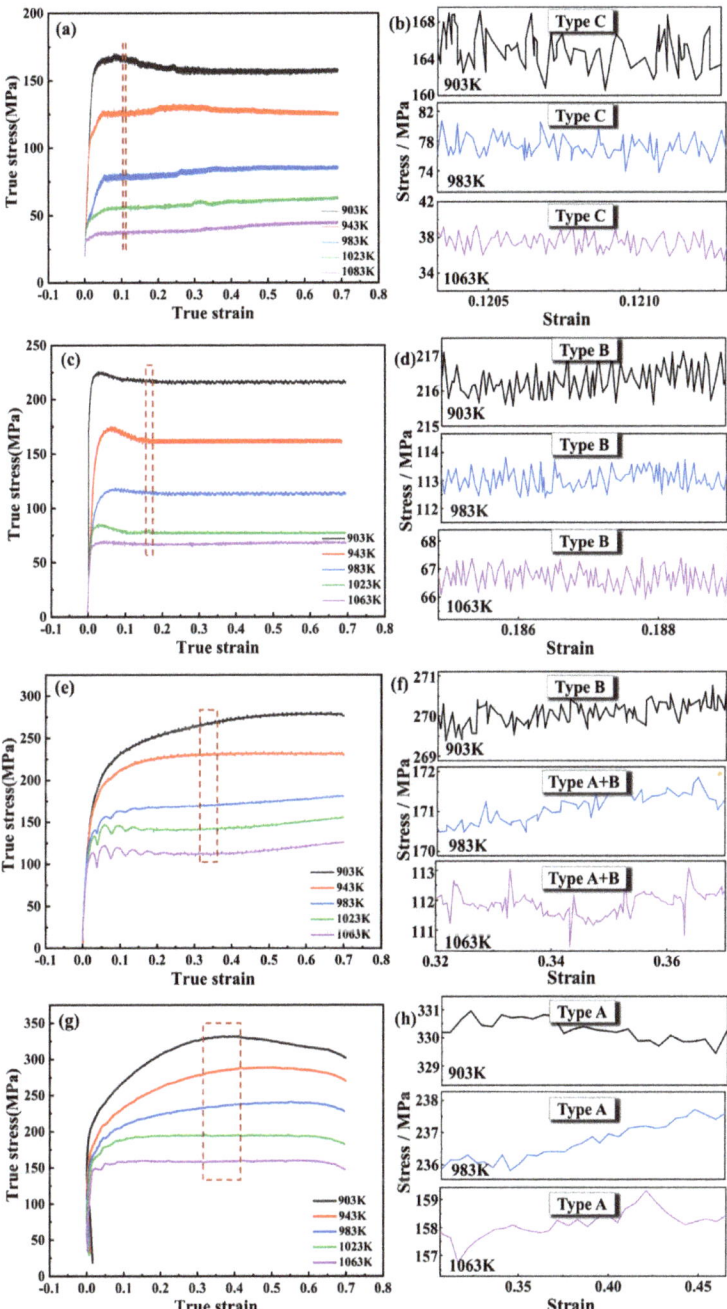

Figure 2. Flow stress curves of the Cu–2.0Be alloy at various strain rates and temperatures: (**a**) 0.01 s^{-1}, (**c**) 0.1 s^{-1}, (**e**) 1 s^{-1}, and (**g**) 10 s^{-1}. (**a,c,e,g**) are magnified views of the red dashed boxes, respectively, in (**b,d,f,h**).

Based on the magnified views of flow stress in Figure 2, with increasing strain rate, the serration changed from type C (Figure 2a) to type B (Figure 2c,e) and eventually to type A (Figure 2g). Table 2 displays the distribution of serration types in accordance with the temperature and strain rate. The decisive factor affecting the serration was obviously the strain rate. Furthermore, the temperature also affected the serration type under a certain strain rate (1 s^{-1} in Figure 2e). When the compression temperature was above than 943 K, the type of serration changed from type B to type A + B. The serration transformation rules of the Cu–2.0Be alloy are not affected by the deformation temperature in the same manner as other copper alloy works with regard to hot compression [4,25–29].

Table 2. Summary of serrations types observed on stress–strain data.

$\dot{\varepsilon}$ (s^{-1})	Temperature (K)				
	903	943	983	1023	1063
0.01	C	C	C	C	C
0.1	B	B	B	B	B
1	B	B	A + B	A + B	A + B
10	A	A	A	A	A

The amplitudes of the serrations represent the degree to which the stress deviates away from the average level. The average serration amplitudes of the various serration types at the corresponding temperatures are shown in Figure 3. The type C serration had the greatest amplitude (7.75 MPa), and the ordinary serration amplitudes of type A and type B (type A + B) serrations were approximately 2.38 MPa and 1.52 MPa, respectively. Low strain rate causes high concentration of clustered solutes at dislocations during the effective waiting time, resulting in type C serrations. Therefore, a high average strength will be needed for unpinning, and larger amplitudes could be obtained compared with those of type A or B serrations [30].

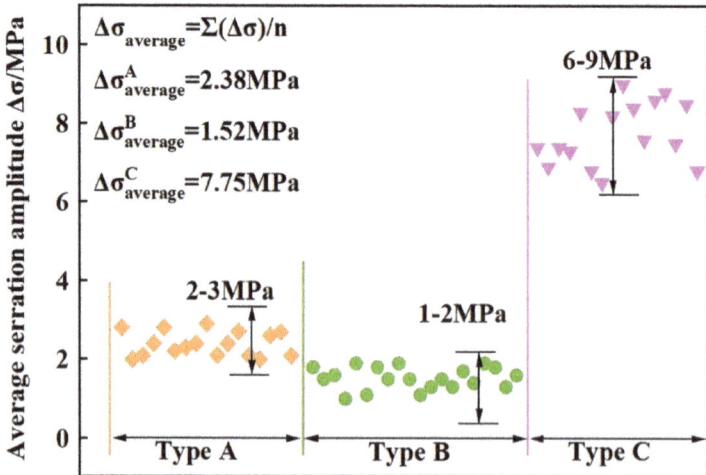

Figure 3. Statistics of variation in the average serration amplitude of various serration types. Serrations of types A, B, and C have average amplitudes of 2~3 MPa, 1~2 MPa, and 6~9 MPa, respectively.

3.2. Constitutive Equation and Activation Energy

In order to obtain the activation energy under different strain rates, the Arrhenius equation was employed. The plastic flow behavior during the deformation processes was described using constitutive equations connected to different deformation situations. Under both low- and high-stress circumstances, the hyperbolic sine function of the Arrhenius equation may typically be reduced to an exponential and power function. Equations (1) and (2) illustrate how the deformation circumstances affect the thermal working behavior of the Cu–2.0Be alloy using the Arrhenius-type constitutive model [31].

$$Z = A[\sinh(a\sigma)]^S = \dot{\varepsilon}\exp(Q/RT) \quad (1)$$

$$\dot{\varepsilon} = \begin{cases} A[\sinh(a\sigma)]^n \exp\left(-\frac{Q}{RT}\right), & \text{for all values of } a\sigma \\ A_1\sigma^{n_1} \exp\left(-\frac{Q}{RT}\right), & \text{if } a\sigma < 0.8 \\ A_2 \exp(\beta\sigma)\exp\left(-\frac{Q}{RT}\right), & \text{if } a\sigma > 1.2 \end{cases} \quad (2)$$

where A, A_1, A_2, n, n_1, a and β are material constants, $\dot{\varepsilon}$ is the strain rate (s^{-1}), R is the gas constant, σ is the flow stress (MPa), T is the absolute temperature (K), and Q is the thermal activation energy needed during plastic deformation. The α can be defined as $a = \beta/n_1$.

The logarithm of the above equation results in [32,33]:

$$\ln\dot{\varepsilon} = \begin{cases} \ln A - \frac{Q}{RT} + n\ln[\sinh(\sigma)] \\ \ln A_1 - \frac{Q}{RT} + n_1 \ln(\sigma) \\ \ln A_2 - \frac{Q}{RT} + \beta\sigma \end{cases} \quad (3)$$

Thus, through analyzing the slope of the curve in the Figure 4a–c, the value of $1/n = 0.19455$, $1/n_1 = 0.148258$, and $1/\beta = 15.00435$ can be obtained, respectively. The value of $\ln A = 34.76345$ can be calculated by taking the logarithm of Equation (1) and fitting it to the intercept of the line in Figure 4d.

$$\ln Z = \ln A + s\ln[\sinh(a\sigma)] \quad (4)$$

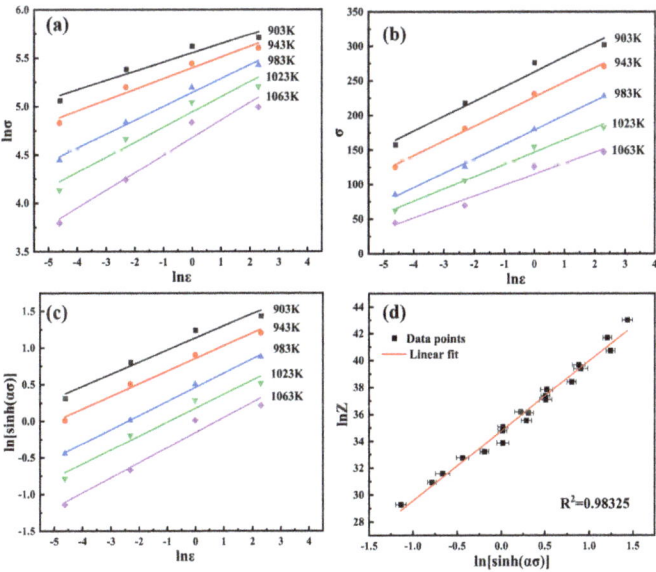

Figure 4. The flow stress and strain rate relationship: (a) $\ln\sigma\sim\ln\dot{\varepsilon}$, (b) $\sigma\sim\ln\dot{\varepsilon}$, (c) $\ln[\sinh(a\sigma)]\sim\ln\dot{\varepsilon}$, (d) $\ln Z\sim\ln[\sinh(a\sigma)]$.

The value of Q can be obtained by linear fitting in Figure 5, which can be expressed by deducing Equation (3):

$$Q = Rs \frac{d\{\ln[\sinh(\alpha\sigma)]\}}{d(1/T)} \tag{5}$$

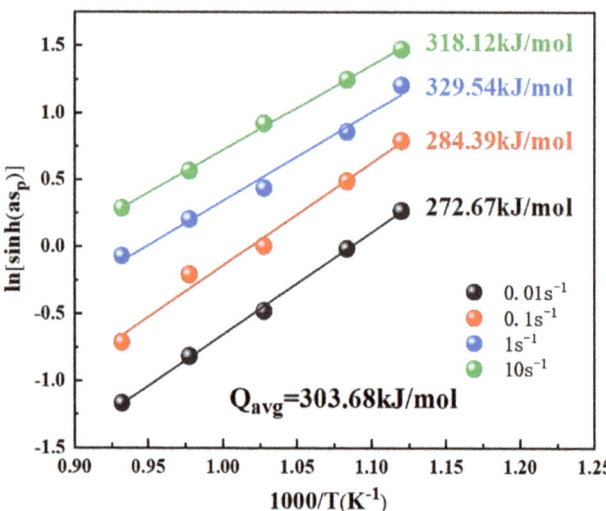

Figure 5. Calculation of the activation energy for serrated flow Plot of $\ln[\sinh(a\sigma)]$ versus $1/T$.

The activation energy decreases with strain rate enhancement, and the activation energy (Q) is 303.68 kJ/mol on average. According to the above values, the constitutive equation of hot compression can be expressed as follows:

$$\dot{\varepsilon} = e^{34.76}[\sin h(0.0071\sigma)]^{5.14}\exp\left(-\frac{303.68}{RT}\right) \tag{6}$$

3.3. Microstructure Evolution

The EBSD diagrams of the deformed Cu–2.0Be alloy samples are shown in Figure 6. When the deformation conditions were 903 K/1 s^{-1}, large grains dominated, as shown in Figure 6a. A small amount of recrystallization occurred at the grain boundaries of the large grains, and the original grains of the alloy elongated. When the temperature was increased from 903 to 1063 K, dynamic recrystallization occurred, the region became larger, and the grain growth direction changed from an "item chain" distribution to a spherical distribution, as shown in Figure 6c.

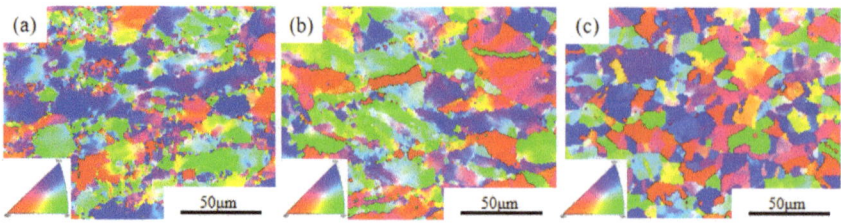

Figure 6. EBSD micrographs of the Cu–2.0Be alloy deformed at: (a) 903 K, 1 s^{-1}; (b) 983 K, 1 s^{-1}; (c) 1063 K, 1 s^{-1}.

The LAGBs and HAGBs distribution maps and misorientation angle distributions of the Cu–2.0Be alloy samples are shown in Figure 7. Due to the accumulation of dislocations in the deformed grains, many LAGBs formed at 903 K, as seen in Figure 7a. The development of recrystallized grains caused LAGBs to become HAGBs as the deformation temperature rose. Additionally, in Figure 7d,f, the average misorientation angle value rose from 12.02 to 17.35. Based on the statistical data in Figure 8a, the volume percentage of HAGBs increased from 16.4 to 35.4% with increasing temperature, indicating that the degree of recrystallization was further enhanced with increasing HAGBs [34]. Particularly, the fraction of HAGBs close to 60° has increased can be seen in Figure 7d–f, which indicates the presence of annealing twin boundaries (TBs). Further, the distribution of the percentage of TBs is shown in Figure 8b. It can be found that the percentage of TBs increased from 1.5% to 11% with the temperature increase. The recrystallized grain distribution of the hot compressed samples is shown in Figure 9. At 903 K (Figure 9a), the DRX is suppressed due to the low deformation temperature, and there are more red inhomogeneous deformed tissues. With the temperature enhancement, the deformed grains gradually transform to the recrystallized grains through the nucleation and growth of new grains, which results in the increase the proportion of the blue DRXed grains and yellow Sub-structured grains. The kernel average misorientation (KAM) maps in Figure 10 can be used to determine the dislocation density variations and distributions. The average dislocation density value fell as the deformation temperature rose, indicating the presence and coarsening of DRX sacrifice dislocations [35]. Furthermore, the region of high dislocation density is primarily localized close to the grain boundaries shown in Figure 10. The dislocation density was substantially higher in the grain boundary area than within the grains, indicating the preferential occurrence of DRX in the grain boundary area.

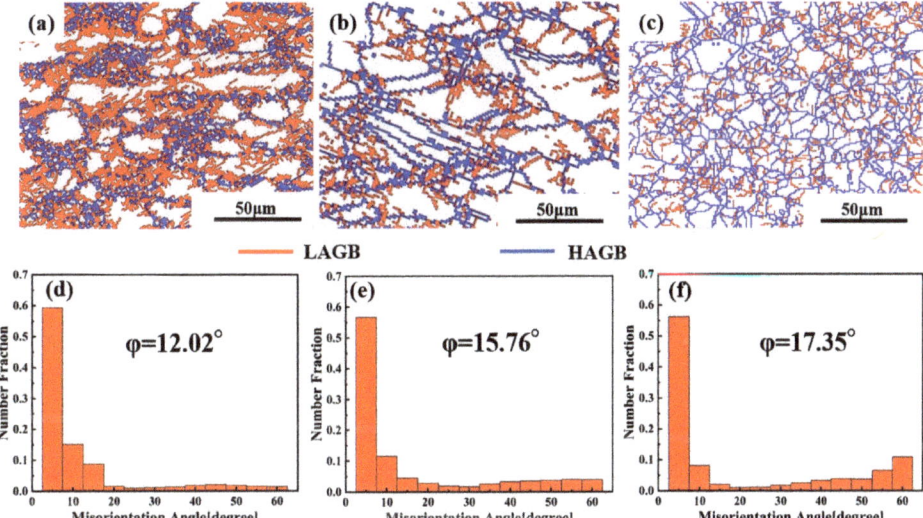

Figure 7. HAGBs and LAGBs distribution maps and misorientation angle distributions of the Cu–2.0Be alloy deformed at: (**a**,**d**) 903 K, 1 s^{-1}; (**b**,**e**) 983 K, 1 s^{-1}; (**c**,**f**) 1063 K, 1 s^{-1}. LAGBs are depicted as red lines, while HAGBs are displayed as blue lines.

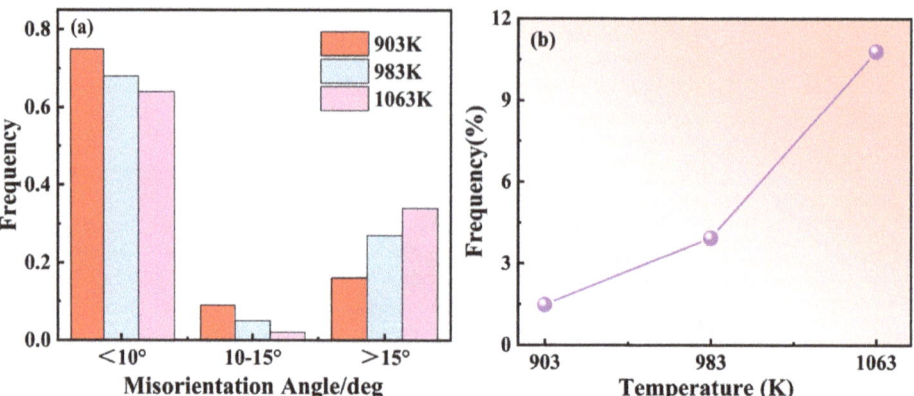

Figure 8. Diagram of grain boundary characteristics: (**a**) misorientation angle scope variations; (**b**) the distribution of TBs.

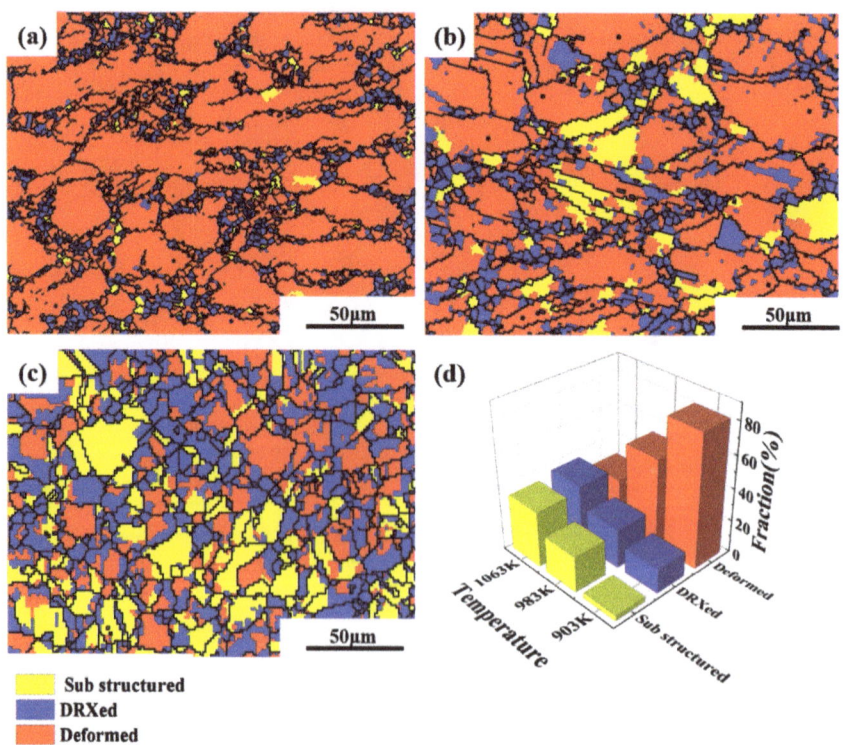

Figure 9. Recrystallization grain distribution of the Cu–2.0Be alloy deformed at: (**a**) 903 K, 1 s^{-1}; (**b**) 983 K, 1 s^{-1}; (**c**) 1063 K, 1 s^{-1}; (**d**) distribution statistics chart.

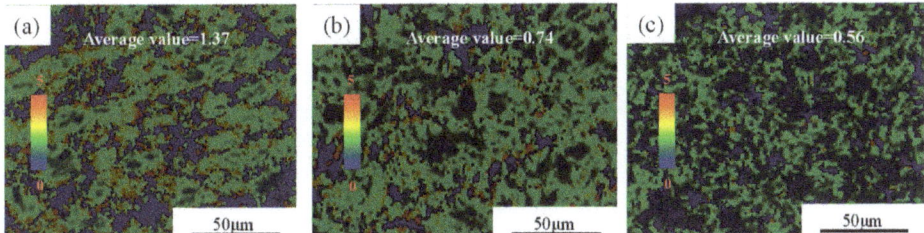

Figure 10. The kernel average misorientation (KAM) of the Cu–2.0Be alloy deformed at: (**a**) 903 K, 1 s^{-1}; (**b**) 983 K, 1 s^{-1}; (**c**) 1063 K, 1 s^{-1}.

To clarify the effect of the compression rate and temperature on the serration types, compressed samples under different conditions were analyzed by TEM/HRTEM. The results are shown in Figures 11 and 12.

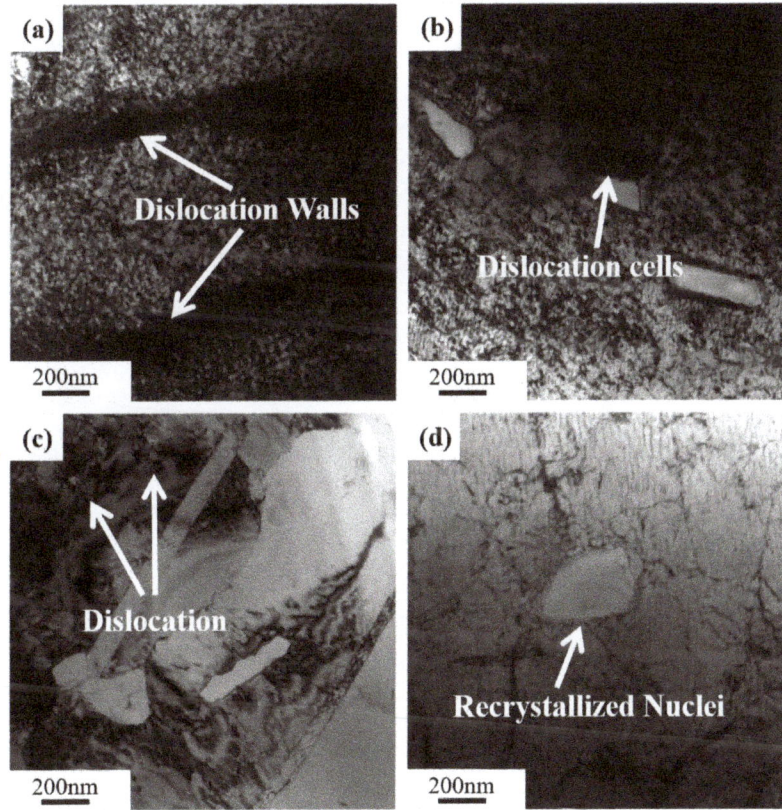

Figure 11. TEM images along with the compressive temperature of 903 K at various strain rate: (**a**) 0.01 s^{-1}; (**b**) 0.1 s^{-1}; (**c**) 1 s^{-1}; (**d**) 10 s^{-1}.

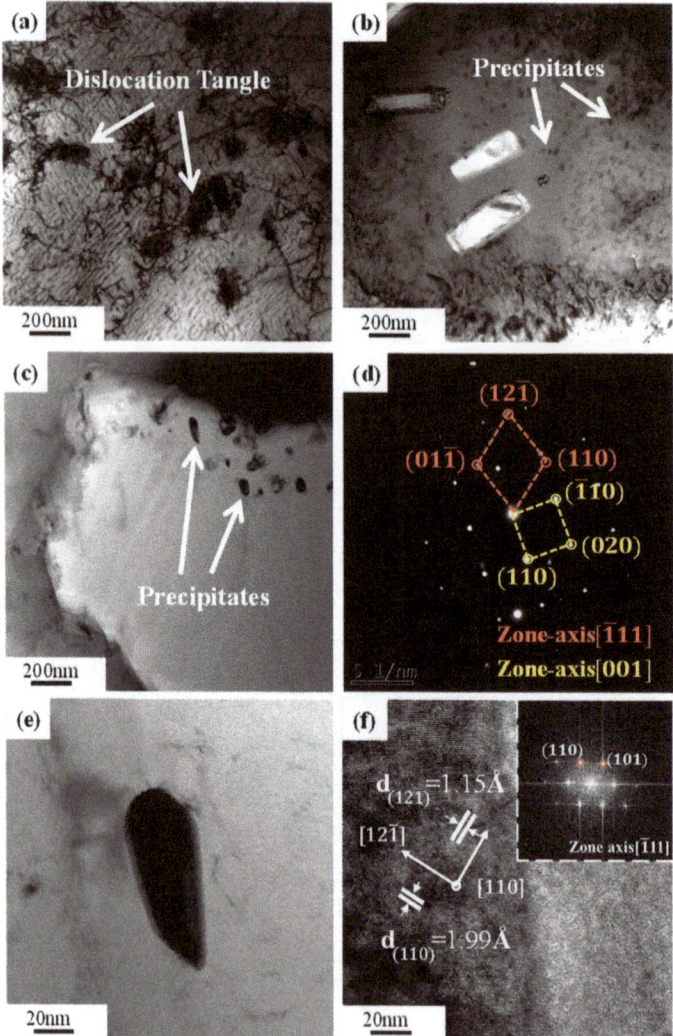

Figure 12. TEM images along with the compressive strain rate of 1 s^{-1} at different temperature: (**a**) 943 K; (**b**) 983 K; (**c**) 1063 K, (**d**) represents the selected area electron diffraction (SAED) patterns of (**c**), (**e**) is a magnified view of (**c**), and (**f**) is the HRTEM of (**e**).

Figure 11 is images of samples with different strain rates at 903 K. The dislocations were gradually unlocked as the strain rate increased, leading to the emergence of the different serration mechanisms shown in Figure 2. When the strain rate was relatively low in Figure 11a, the dislocation movement velocity was lower than the diffusion velocity of solute atoms. Hence, the dislocations were effectively pinned, resulting in a higher amplitude and frequency of serrations in the flow stress curves (type C serrations) [36]. As strain rates rose, the dislocation movement velocity gradually outpaced the rate at which solute atoms diffused, leading to the unlocking of dislocations. At medium strain rates, type B serrated flow was achieved [37]. When the strain rate was 10 s^{-1}, the dislocation movement speed was faster than the rate of solute atom diffusion, and a lower dislocation density was obtained Figure 11d, generating the minor, random undulation of type A serrations in the stress–strain curves [30]. Moreover, with the gradual unlocking of dislocations, the typical dislocation-induced recrystallization can be seen in Figure 11b–d.

TEM images of hot compressed samples at a strain rate of 1 s^{-1} and temperatures of 943 K, 983 K, and 1063 K are shown in Figure 12. At 943 K, the precipitated phases had a small size and a large number of dislocation tangles (Figure 12a). The size of the precipitated phases steadily grew from 5–15 nm (Figure 12b) to 80–160 nm (Figure 12c) as the temperature rose, while the density of dislocations gradually decreased. Furthermore, Figure 12d shows the SAED of the precipitate in Figure 12c, demonstrating a body center cubic (BCC) structure. Figure 12e displays the block-shaped precipitate's high-resolution TEM (HRETEM) micrograph and accompanying Fast Fourier Transform (FFT) pattern in Figure 12f. The distance between the layered precipitates, called d-spacing, is around 1.99 Å, which is close to that of the (hkl) = (110) plane of the BCC structure found in the diffraction pattern database of the $β$ phase [38–40].

The mechanism of serration (Figure 13) is essentially the interaction between the dislocation motion (grain boundary motion) and the diffusion of solute atoms [41]. As illustrated in Figure 13a, when the strain rate is large enough, such as 10 s^{-1}, the amounts of solute atoms caught by excessive dislocation movement speed to form entanglement is greatly reduced under the influence of external force. As a result, the dislocation density is relatively low in Figure 11d, only a small amount of serration occurs randomly, and the serration frequency is greatly reduced in Figure 2g,h. Furthermore, insufficient time passes to generate the $β$ phase, which results in an extremely small serration amplitude (type A). The same phenomenon is also well verified in the articles on copper alloys [42], nickel-based superalloys [36], magnesium alloys [30] and steels [43]. However, as illustrated in Figure 13b–d, when the strain rate is lower than 10 s^{-1}, the reduced strain rate allows enough time for the mobile dislocation to be effectively pinned and generate the $β$ phase, which can pin the mobile dislocation ulteriorly. The effective stress needed for unpinning is enhanced, increasing the amplitude of the serration and dislocation density. The lower the strain rate, the higher the required effective stress and the larger the serration amplitude. As a result, the frequency and amplitude of type C serrations are the largest among all the stress–strain curves in Figure 2a.

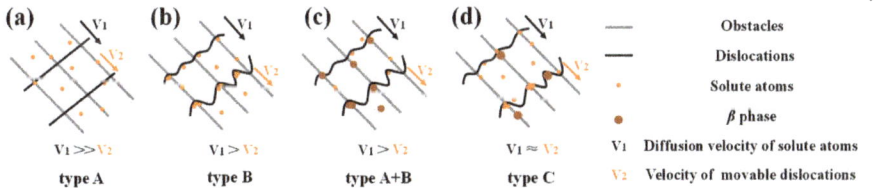

Figure 13. The sketch map of the serration mechanism process in type (**a**) A, (**b**) B, (**c**) A + B and (**d**) C.

Normally, despite the fact that the speed of solute atom diffusion increases as the temperature rises, the development of grains by DRX significantly reduces the number of dislocations in the grain border and matrix (Figures 6–10), which results in a serration amplitude that is not sensitive to temperature at certain strain rate. Meanwhile, at a particular strain rate, such as 1 s^{-1}, when the temperature is higher than 903 K, a mixed type of serration between types B and A emerges, which is characterized by an abruptly large serration in a small serration. This mixed A + B type of serration combines the characteristics of type A, the serration amplitude of which is large, unstable, and random, and type B, the serration amplitude of which moves up and down steadily with a specific number continuously. As shown in Figure 12, nanoscale dispersive $β$ phases will be readily formed at high temperatures (983 K and 1063 K) at 1 s^{-1}, which can hinder the movement of the dislocations and lead to the rapid increase in the effective stress required for unpinning. As a result, mixed A + B type serrations are generated, as illustrated in Figure 13c.

4. Conclusions

To investigate the PLC effect of Cu–2.0Be alloy, the hot compression method was employed at temperatures ranging from 903 to 1063 K with strain rates of 0.01 to 10 s^{-1}, and the characteristics of serrated flow were studied. In addition, EBSD and TEM microscopic analyses were performed to explore the mechanisms of different serration types. The main conclusions are as follows:

(1) The PLC effect of Cu–2.0Be alloy is sensitive to the strain rate. Types A, B and C serrations occur at high, medium and low strain rates, respectively. In particular, at a strain rate of 1 s^{-1}, temperature-sensitive serrations were discovered, and mixed type A + B serrations were produced when the temperature exceeded 983 K.

(2) The Arrhenius-type constitutive equation was created using the stress–strain data in the following manner: $\dot{\varepsilon} = e^{34.76}[\sin h(0.0071\sigma)]^{5.14} \exp\left(-\frac{303.68}{RT}\right)$. The activation energy increased with increasing strain rate and the ordinary activation energy Q = 303.68 kJ/mol.

(3) The serrations type is determined by the interaction between the dislocation motion (grain boundary motion) and the diffusion of solute atoms. The flow stress serration is very sensitive to the pinning and unpinning of the mobile dislocations, which can be mainly affected by the solute atom diffusion velocity.

(4) Under appropriate deformation conditions, the formation of fine β phases can hinder dislocation movement, resulting in temperature-sensitive serrations and the formation of the mixed type A + B serrations.

Author Contributions: Conceptualization, D.Z.; Methodology, Y.L. and G.X.; Software, N.W. and C.J.; Formal analysis, H.Z. and S.C.; Investigation, D.Z. and G.X.; Data curation, X.L.; Writing—original draft, N.W.; Writing—review & editing, D.Z.; Supervision, Y.J., Y.L. and G.X.; Project administration, Y.L. and Y.J.; Funding acquisition, D.Z. All authors have read and agreed to the published version of the manuscript.

Funding: This work was finally supported by the science and technology innovation Program of Hunan Province, China (Grant No. 2021RC2087, and 2021JJ30673), and the Project of the Education Department of Hunan Province, China (Grant No. 20B569).

Institutional Review Board Statement: Not applicable.

Informed Consent Statement: Not applicable.

Data Availability Statement: The data are available from the corresponding author upon reasonable request.

Conflicts of Interest: The authors declare no conflict of interest.

References

1. Zhang, P.; Liu, G.; Sun, J. A critical review on the Portevin-Le Chatelier effect in aluminum alloys. *J. Cent. South Univ.* **2022**, *29*, 744–766. [CrossRef]
2. Yilmaz, A. The Portevin-Le Chatelier effect: A review of experimental findings. *Sci. Technol. Adv. Mater.* **2011**, *12*, 063001. [CrossRef] [PubMed]
3. Chandravathi, K.S.; Laha, K.; Parameswaran, P.; Mathew, M.D. Effect of microstructure on the critical strain to onset of serrated flow in modified 9Cr–1Mo steel. *Int. J. Press. Vessel. Pip.* **2012**, *89*, 162–169. [CrossRef]
4. Shukla, A.K.; Murty, S.N.; Sharma, S.C.; Mondal, K. The serrated flow and recrystallization in dispersion hardened Cu–Cr–Nb alloy during hot deformation. *Mater. Sci. Eng. A* **2016**, *673*, 135–140. [CrossRef]
5. Halim, H.; Wilkinson, D.S.; Niewczas, M. The Portevin–Le Chatelier (PLC) effect and shear band formation in an AA5754 alloy. *Acta Mater.* **2007**, *55*, 4151–4160. [CrossRef]
6. Estrin, Y.; Kubin, L.P. Plastic instabilities: Phenomenology and theory. *Mater. Sci. Eng. A* **1991**, *137*, 125–134. [CrossRef]
7. Mulford, R.A.; Kocks, U.F. New observations on the mechanisms of dynamic strain aging and of jerky flow. *Acta Metall.* **1979**, *27*, 1125–1134. [CrossRef]
8. Beukel, A.V.D. On the mechanism of serrated yielding and dynamic strain ageing. *Acta Metall.* **1980**, *28*, 965–969. [CrossRef]
9. Kubin, L.P.; Chihab, K.; Estrin, Y. The rate dependence of the portevin-Le chatelier effect. *Acta Metall.* **1988**, *36*, 2707–2718. [CrossRef]

10. Jiang, H.F.; Zhang, Q.C.; Chen, X.D.; Chen, Z.Z.; Jiang, Z.Y.; Wu, X.P.; Fan, J.H. Three types of Portevin–Le Chatelier effects: Experiment and modelling. *Acta Mater.* **2007**, *55*, 2219–2228. [CrossRef]
11. Onodera, R.; Ishibashi, T.; Koga, M.; Shimizu, M. The relation between the Portevin-Le chatelier effect and the solid solubility in some binary alloys. *Acta Metall.* **1983**, *31*, 535–540. [CrossRef]
12. Jovanovic, M.; Djuric, B.; Drobnjak, D. Serrated yielding in commercial Cu-Be-Co alloy. *Scr. Metall.* **1981**, *15*, 469–473. [CrossRef]
13. Era, H.; Ohura, N.; Onodera, R.; Shimizu, M. Precipitation and the Portevin-Le Chatelier effect in Cu-5.5, 11.6 and 14.2 at.% Ga alloy. *Scr. Metall.* **1984**, *18*, 1041–1044. [CrossRef]
14. Qian, K.W.; Reed-Hill, R.E. A model for the flow stress and strain rate sensitivity of a substitutional alloy–Cu-3.1 at.% Sn. *Acta Metall.* **1983**, *31*, 87–94. [CrossRef]
15. Onodera, R.; Ishibashi, T.; Era, H.; Shimizu, M. The portevin-le chatelier effects in Cu-Ti, Cu-P and Cu-Si alloys. *Acta Metall.* **1984**, *32*, 817–822. [CrossRef]
16. Adams, S.M. Serrated flow in alpha-beta brass. *Scr. Metall.* **1973**, *7*, 173–177. [CrossRef]
17. Mayer, M.; Vohringer, O.; Macherauch, E. Portevin-Le chatelier effect during tensile deformation in Cu-Zn and Cu-Ni-Zn alloys. *Scr. Metall.* **1975**, *9*, 1333–1339. [CrossRef]
18. Jovanović, M.; Drobnjak, D. Onset of serrated flow in Cu-Be alloy. *Scr. Metall.* **1973**, *7*, 997–1002. [CrossRef]
19. Hayes, R.W. The disappearance of serrated flow in two copper alloys. *Mater. Sci. Eng. A* **1986**, *82*, 85–92. [CrossRef]
20. Zhou, Y.J.; Song, K.X.; Xing, J.D.; Zhang, Y.M. Precipitation behavior and properties of aged Cu-0.23Be-0.84Co alloy. *J. Alloy. Compd.* **2016**, *658*, 920–930. [CrossRef]
21. Yagmur, L.; Duygulu, O.; Aydemir, B. Investigation of metastable γ' precipitate using HRTEM in aged Cu–Be alloy. *Mater. Sci. Eng. A* **2011**, *528*, 4147–4151. [CrossRef]
22. Xie, G.L.; Wang, Q.S.; Mi, X.J.; Xiong, B.Q.; Peng, L.J. The precipitation behavior and strengthening of a Cu–2.0wt% Be alloy. *Mater. Sci. Eng. A* **2012**, *558*, 326–330. [CrossRef]
23. Zhu, D.B.; Liu, C.M.; Yu, H.J.; Han, T. Effects of Changing Hot Rolling Direction on Microstructure, Texture and Mechanical Properties of Cu-2.7Be Sheets. *J. Mater. Eng. Perform.* **2018**, *27*, 3532–3543. [CrossRef]
24. Zhu, D.B.; Liu, C.M.; Liu, Y.D.; Han, T.; Gao, Y.H.; Jiang, S.N. Evolution of the texture, mechanical properties, and microstructure of Cu-2.7Be alloys during hot cross-rolling. *Appl. Phys. A* **2015**, *120*, 1605–1613. [CrossRef]
25. Hui, J.; Feng, Z.X.; Xue, C.; Gao, W.F. Effects of hot compression on distribution of $\Sigma3^n$ boundary and mechanical properties in Cu-Sn alloy. *J. Mater. Sci.* **2018**, *53*, 15308–15318. [CrossRef]
26. Zhang, Y.; Liu, P.; Tian, B.H.; Liu, Y.; Li, R.Q.; Xu, Q.Q. Hot deformation behavior and processing map of Cu–Ni–Si–P alloy. *Trans. Nonferrous Met. Soc. China* **2013**, *23*, 2341–2347. [CrossRef]
27. Zhang, H.; Zhang, H,g.; Peng, D.S. Hot deformation behavior of KFC copper alloy during compression at elevated temperatures. *Trans. Nonferrous Met. Soc. China* **2006**, *16*, 562–566. [CrossRef]
28. Yang, J.Y.; Kim, W.J. The effect of addition of Sn to copper on hot compressive deformation mechanisms, microstructural evolution and processing maps. *J. Mater. Res. Technol.* **2020**, *9*, 749–761. [CrossRef]
29. Zhang, H.; Zhang, H.G.; Li, L.X. Hot deformation behavior of Cu–Fe–P alloys during compression at elevated temperatures. *J. Mater. Process. Technol.* **2009**, *209*, 2892–2896. [CrossRef]
30. Wang, W.H.; Wu, D.; Shah, S.S.A.; Chen, R.S.; Lou, C.S. The mechanism of critical strain and serration type of the serrated flow in Mg–Nd–Zn alloy. *Mater. Sci. Eng. A* **2016**, *619*, 214–221. [CrossRef]
31. Sellars, C.M.; McTegart, W.J. On the mechanism of hot deformation. *Acta Metall.* **1966**, *14*, 1136–1138. [CrossRef]
32. Huang, K.; Logé, R.E. A review of dynamic recrystallization phenomena in metallic materials. *Mater. Des.* **2016**, *111*, 548–574. [CrossRef]
33. Doherty, R.D.; Szpunar, J.A. Kinetics of sub-grain coalescence-A reconsideration of the theory. *Acta Metall.* **1984**, *32*, 1789–1798. [CrossRef]
34. Kabir, A.S.H.; Sanjari, M.; Su, J.; Jung, I.H.; Yue, S. Effect of strain-induced precipitation on dynamic recrystallization in Mg–Al–Sn alloys. *Mater. Sci. Eng. A* **2014**, *616*, 252–259. [CrossRef]
35. Momeni, A.; Dehghani, K. Characterization of hot deformation behavior of 410 martensitic stainless steel using constitutive equations and processing maps. *Mater. Sci. Eng. A* **2010**, *527*, 5467–5473. [CrossRef]
36. Cai, Y.L.; Tian, C.G.; Fu, S.H.; Han, G.M.; Cui, C.Y.; Zhang, Q.C. Influence of γ' precipitates on Portevin–Le Chatelier effect of NI-based superalloys. *Mater. Sci. Eng. A* **2015**, *638*, 314–321. [CrossRef]
37. Jiang, Z.W.; Zhu, L.L.; Yu, L.X.; Sun, B.A.; Cao, Y.; Zhao, Y.H.; Zhang, Y. The mechanism for the serrated flow induced by Suzuki segregation in a Ni alloy. *Mater. Sci. Eng. A* **2021**, *820*, 141575. [CrossRef]
38. Zhu, D.B.; Liu, C.M.; Han, T.; Liu, Y.D.; Xie, H.P. Effects of secondary β and γ phases on the work function properties of Cu–Be alloys. *Appl. Phys. A* **2015**, *120*, 1023–1026. [CrossRef]
39. Montecinos, S.; Cuniberti, A.; Sepúlveda, A. Grain size and pseudoelastic behaviour of a Cu–Al–Be alloy. *Mater. Charact.* **2008**, *59*, 117–123. [CrossRef]
40. Castro, M.L.; Fornaro, O. Formation of dendritic precipitates in the beta phase of Cu-based alloys. *J. Mater. Sci.* **2009**, *44*, 5829–5835. [CrossRef]
41. Li, Z.G.; Zhang, L.T.; Sun, N.R.; Fu, L.M.; Shan, A.D. Grain size dependence of the serrated flow in a nickel based alloy. *Mater. Lett.* **2015**, *150*, 108–110. [CrossRef]

42. Peng, G.W.; Gan, X.P.; Jiang, Y.X.; Li, Z.; Zhou, K.C. Effect of dynamic strain aging on the deformation behavior and microstructure of Cu-15Ni-8Sn alloy. *J. Alloys Compd.* **2017**, *718*, 182–187. [CrossRef]
43. Choudhary, B.K. Influence of strain rate and temperature on serrated flow in 9Cr–1Mo ferritic steel. *Mater. Sci. Eng. A* **2013**, *564*, 303–309. [CrossRef]

Disclaimer/Publisher's Note: The statements, opinions and data contained in all publications are solely those of the individual author(s) and contributor(s) and not of MDPI and/or the editor(s). MDPI and/or the editor(s) disclaim responsibility for any injury to people or property resulting from any ideas, methods, instructions or products referred to in the content.

Article

Thermal Deformation Behavior and Dynamic Softening Mechanisms of Zn-2.0Cu-0.15Ti Alloy: An Investigation of Hot Processing Conditions and Flow Stress Behavior

Guilan Xie [1], Zhihao Kuang [1], Jingxin Li [1], Yating Zhang [1], Shilei Han [1], Chengbo Li [1], Daibo Zhu [1,*] and Yang Liu [1,2,3,*]

[1] School of Mechanical Engineering and Mechanics, Xiangtan University, Xiangtan 411105, China; xieguilan@xtu.edu.cn (G.X.); 17373241475@163.com (Z.K.); 202005501510@smail.xtu.edu.cn (J.L.); 202005501530@smail.xtu.edu.cn (Y.Z.); 16673234565@163.com (S.H.); csulicb@163.com (C.L.)
[2] School of Materials Science and Engineering, Nanyang Technological University, Singapore 639798, Singapore
[3] Zhuzhou Smelter Group Co., Ltd., Zhuzhou 412005, China
* Correspondence: daibozhu@xtu.edu.cn (D.Z.); liuyang_225@126.com (Y.L.); Tel.: +86-731-5289-8553 (Y.L.)

Abstract: Through isothermal hot compression experiments at various strain rates and temperatures, the thermal deformation behavior of Zn-2.0Cu-0.15Ti alloy is investigated. The Arrhenius-type model is utilized to forecast flow stress behavior. Results show that the Arrhenius-type model accurately reflects the flow behavior in the entire processing region. The dynamic material model (DMM) reveals that the optimal processing region for the hot processing of Zn-2.0Cu-0.15Ti alloy has a maximum efficiency of about 35%, in the temperatures range (493–543 K) and a strain rate range (0.01–0.1 s^{-1}). Microstructure analysis demonstrates that the primary dynamic softening mechanism of Zn-2.0Cu-0.15Ti alloy after hot compression is significantly influenced by temperature and strain rate. At low temperature (423 K) and low strain rate (0.1 s^{-1}), the interaction of dislocations is the primary mechanism for the softening Zn-2.0Cu-0.15Ti alloys. At a strain rate of 1 s^{-1}, the primary mechanism changes to continuous dynamic recrystallization (CDRX). Discontinuous dynamic recrystallization (DDRX) occurs when Zn-2.0Cu-0.15Ti alloy is deformed under the conditions of 523 K/0.1 s^{-1}, while twinning dynamic recrystallization (TDRX) and CDRX are observed when the strain rate is 10 s^{-1}.

Keywords: Zn-Cu-Ti alloy; hot compression; dynamic material model (DMM); flow stress behavior; softening mechanism

Citation: Xie, G.; Kuang, Z.; Li, J.; Zhang, Y.; Han, S.; Li, C.; Zhu, D.; Liu, Y. Thermal Deformation Behavior and Dynamic Softening Mechanisms of Zn-2.0Cu-0.15Ti Alloy: An Investigation of Hot Processing Conditions and Flow Stress Behavior. *Materials* **2023**, *16*, 4431. https://doi.org/10.3390/ma16124431

Academic Editor: Frank Czerwinski

Received: 21 May 2023
Revised: 10 June 2023
Accepted: 11 June 2023
Published: 16 June 2023

Copyright: © 2023 by the authors. Licensee MDPI, Basel, Switzerland. This article is an open access article distributed under the terms and conditions of the Creative Commons Attribution (CC BY) license (https:// creativecommons.org/licenses/by/ 4.0/).

1. Introduction

Zinc-based alloys are highly sought after due to their exceptional mix of qualities, such as excellent ductility, weldability, outstanding corrosion resistance [1], and surface aspect, making them suitable for various applications in the medical implantation, building industry, and anti-corrosive fields [2–4]. Moreover, these alloys possess strong creep resistance [5], which further increases their demand. The favorable hot workability of these high-strength alloys also contributes to their popularity. Hot extrusion/rolling is the main processing technique used for zinc-based alloys. However, the thermal deformation behavior of the alloys can be affected by several factors [6,7], including temperature, strain rates, and strain [8]. Thus, it is difficult to achieve the parameter specification of zinc-based alloys [9–11]. Up to now, the thermal deformation behavior of Zn-Cu-Ti has not been studied in depth. Little consideration has been given to the flow stress evolution and constitutive relationship of the Zn-Cu-Ti alloy during hot deformation.

The constitutive equations and the processing maps are important methods to study the hot workability characteristics and the deformation mechanisms of zinc-based alloys [12,13]. Unfortunately, the traditional Arrhenius equations and 2D processing maps cannot reflect the influence of strain, which has an obvious impact on flow stress [14]. In order to take the strain effect into account, modified Arrhenius equations are used to improve the accuracy of the numerical simulation, overcoming the disadvantages of being time-consuming and labor-intensive [15]. Meanwhile, it can also provide theoretical guidance for the optimization and the predictions of numerical simulations. The 3D processing map established by the dynamic material model (DMM) can reveal the reasonable processing region to optimize the thermal working process [16,17] from which stable and unstable regions can be found. Stable regions are accompanied by dynamic recrystallization [18] (DRX) and dynamic recovery (DRV), which can uniform microstructures and improve processability [19]. However, few systematic studies have revealed the evolution of the microstructure of Zn-Cu-Ti alloys and optimized its forming parameters by 3D processing maps. Therefore, it is of great significance to study flow stress behavior, constitutive equation, processing maps (3D), and dynamic recrystallization (DRX) behavior of Zn-Cu-Ti alloys during hot compression deformation.

The aim of this work is to investigate the thermal deformation behavior of Zn-2.0Cu-0.15Ti alloy under various strains, strain rates, and deformation temperatures. The 3D processing maps of Zn-2.0Cu-0.15Ti alloy are created to optimize the hot deformation parameters and find out the unstable deformation factors. The effects of thermal deformation parameters on the DRX mechanism are analyzed by the electron backscatter diffraction pattern (EBSD) and transmission electron microscopy (TEM), which provides scientific guidance for the thermal deformation behavior and dynamic softening mechanism of Zn-Cu-Ti alloy.

2. Experimental Procedure

Table 1 lists the component (in wt.%) of Zn-Cu-Ti alloys. In our experiments, using pure Zn (purity 99.995%), high-purity copper foil (99.99%), and high-purity titanium tablet as starting materials for casting samples with a composition of Zn-2.0Cu-0.15Ti (wt.% hereafter). The raw ingredients were melted at 973 K with the protection of argon (Ar) gas in a graphite crucible. The alloy was poured (pouring temperature: 823 K) into the steel mold preheated to 493 K. After natural cooling and solidification, an ingot was obtained. The ingot underwent homogenization at 653 K for 10 h and was used to prepare all the samples. The shape of each specimen for the compression testing is a cylinder (diameter: 8 mm, height: 12 mm). The samples were heated to the specified temperature (423, 473, 523, and 573 K) by using a Gleeble-3500 thermomechanical simulator (Dynamic Systems Inc. America) with a heating rate of 10 K/s. The other experimental parameters were as follows: the strain rates were $0.01~s^{-1}$, $0.1~s^{-1}$, $1~s^{-1}$, and $10~s^{-1}$, and the height of each sample was compressed by 60%. All compression tests were carried out in a vacuum, and in order to achieve a homogeneous temperature ahead of deformation, all samples were maintained at the necessary temperature for 200 s. Additionally, after the compression test, each sample was promptly water-quenched so that the microstructures could be preserved. Figure 1 depicts the hot compression experimentation process.

Table 1. Chemical composition of Zn-2.0Cu-0.15Ti alloy.

Component	Cu	Ti	Impurity	Zn
Content (wt.%)	2.03	0.152	0.045	Bal.

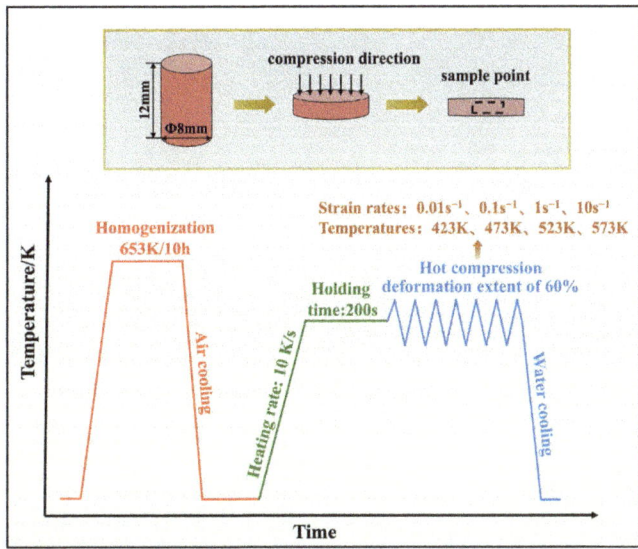

Figure 1. Schematic diagram of hot compression experiment process.

Samples for electron backscatter diffraction pattern (EBSD) analysis were prepared into round pieces with a thickness of 100 μm and a diameter of 3 mm. The round pieces were mechanically and vibrationally polished for 0.5 h using a Buehler VibroMet2 Vibratory Polisher (ITW, America), and then the Angle was reduced using a plasma thinning instrument. SEM with a 200 Sirion field emission gun was used for the EBSD studies. OIM7.3 software was used to analyze the EBSD data. Transmission electron microscope (TEM) images were collected using a Tecnai G220 transmission (FEI, America) electron microscope at 200 kV. The shape of TEM samples was similar to EBSD samples, and twin-jet thinning was applied in a mixed solution of HNO3:CH3OH = 1:3. The experimental parameters were 30 V and −243 K, respectively.

3. Results and Discussion
3.1. Analysis of Flow Stress Curves

Figure 2 displays the true stress–strain curves of the Zn-2.0Cu-0.15Ti alloy obtained from the compression test conducted at strain rates 0.01, 0.1, 1, and $10\ \text{s}^{-1}$ and deformation temperatures 423, 473, 523, and 573 K. The evolution of flow stress curves is roughly divided into three stages. In the first stage, flow stress increases with increasing strain resulting from an increase in dislocation density (work hardening). In the second stage, the increasing rate of flow stress slows down and reaches its peak value, indicating the occurrence of dynamic recovery (DRV) and DRX. In the third stage, flow stress gradually drops until reaching the final steady value. This phenomenon in hot working is usually caused by dynamic recrystallization (DRX) [20,21].

As shown in Figure 2, on the one hand, it can be seen that flow stress decreases with the increase in deformation temperature at a given strain rate. This is because higher temperatures can promote the occurrence of DRX, which causes a decrease in dislocation density. On the other hand, flow stress increases with the increase in strain rates at a given temperature. This is due to the fact that with the limited time for dislocation annihilation to manifest, the dislocation density gradually rises.

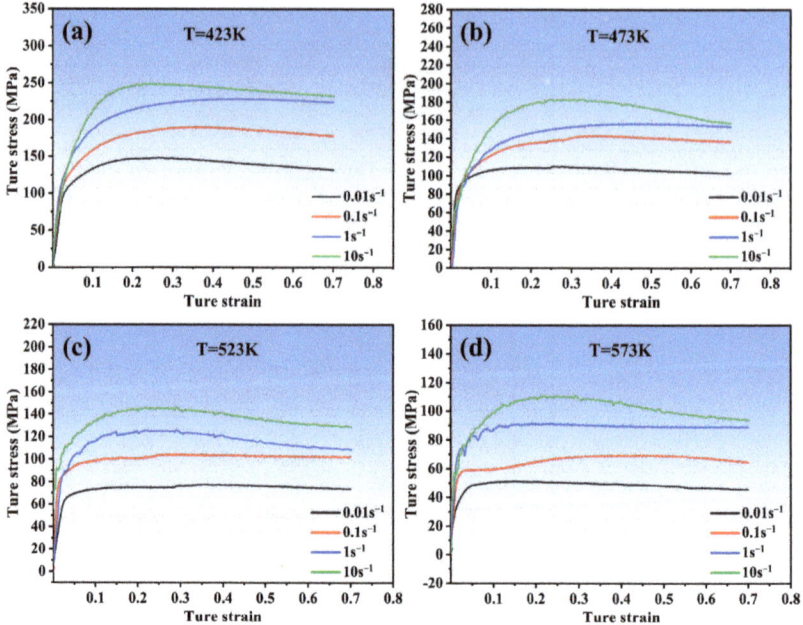

Figure 2. The true stress–strain curves in the hot compression tests of Zn-2.0Cu-0.15Ti alloy at 0.01–10 s^{-1} with a deformation temperature of (**a**) 423 K, (**b**) 473 K, (**c**) 523 K, (**d**) 573 K.

3.2. Development of Constitutive Equation

The flow stress caused by deformation conditions can be evaluated using the Arrhenius equations, which are presented in Equations (1) and (2) and are commonly utilized for describing thermal deformation behavior [22,23].

$$\dot{\varepsilon} = AF(\sigma)\exp\left(\frac{-Q}{RT}\right) \quad (1)$$

$$F(\sigma) = \begin{cases} \sigma^{n_1}, & \alpha\sigma < 0.8 \\ \exp(\beta\sigma), & \alpha\sigma > 1.2 \\ \sinh(\alpha\sigma)^n & \text{for all } \sigma \end{cases} \quad (2)$$

For a certain strain, the true stress (MPa) represented by σ, where the strain rate (s^{-1}) is represented by $\dot{\varepsilon}$, is calculated by some material constants, including A, n_1, n, α, and β, where $\alpha = \beta/n_1$, as well as gas constant R (8.314 J mol^{-1}K^{-1}), the activation energy Q (kJ mol^{-1}), and absolute temperature T (K).

Meanwhile, combining Equations (1) and (2) leads to the Zener–Hollomon (Z) parameter [17,24]:

$$\dot{\varepsilon} = A[\sinh(\alpha\sigma)]^n \exp\left(-\frac{Q}{RT}\right) \quad (3)$$

$$Z = \dot{\varepsilon}\exp\left(\frac{Q}{RT}\right) = A[\sinh(\alpha\sigma)]^n \quad (4)$$

By substituting Equation (2) into Equation (1) and taking the natural logarithm of both sides, the following equation was obtained:

$$\ln\sigma = \frac{1}{n_1}\ln\dot{\varepsilon} - \frac{1}{n_1}\ln A_1 + \frac{Q}{n_1 RT} \quad (5)$$

$$\sigma = \frac{1}{\beta}\ln\dot{\varepsilon} - \frac{1}{\beta}\ln A_2 + \frac{Q}{\beta RT} \tag{6}$$

Assuming the material's activation energy for the deformation is a fixed value unaffected by temperature, according to the true stress–strain curve, the peak stress corresponding to the alloy under various deformation circumstances is determined, and then the $\ln\sigma$-$\ln\dot{\varepsilon}$ and σ-$\ln\dot{\varepsilon}$ curves (Figure 3) can be plotted by linear regression processing according to Equations (5) and (6). Here, the average slopes of the $\ln\sigma$-$\ln\dot{\varepsilon}$ and σ-$\ln\dot{\varepsilon}$ curves are shown by n_1 and β, respectively, $n_1 = 10.601$ and $\beta = 0.086$ MPa^{-1} can be obtained, so that $\alpha = \beta/n_1 = 0.007$ MPa^{-1}.

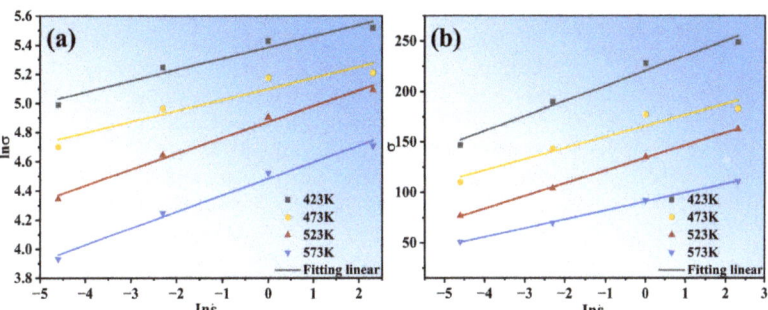

Figure 3. Linear relationship of (**a**) $\ln\sigma$-$\ln\dot{\varepsilon}$ and (**b**) σ-$\ln\dot{\varepsilon}$ at different temperatures.

Equations (7) and (8) are presented using the partial derivative method applied to the logarithm of Equation (3) provided:

$$n = \frac{\partial \ln\dot{\varepsilon}}{\partial[\sinh(\alpha\sigma)]} \tag{7}$$

$$\frac{Q}{Rn} = \frac{\partial \ln[\sinh(\alpha\sigma)]}{\partial(1/T)} \tag{8}$$

The value of both parameters ($\sigma, \dot{\varepsilon}$) is used for the plots of $\ln[\sinh(\alpha\sigma)]$-$\ln\dot{\varepsilon}$ and $\ln[\sinh(\alpha\sigma)]$-$1000/T$ (Figure 4) by linear regression processing according to Equations (7) and (8). By using the linear fitting method, the average values of n and b are calculated to be 7.70001 and 2.534, respectively. Q is calculated to be 130.69 kJ mol^{-1}. A is calculated to be 3.33×10^{12} s^{-1} by substituting the corresponding data into Equation (5).

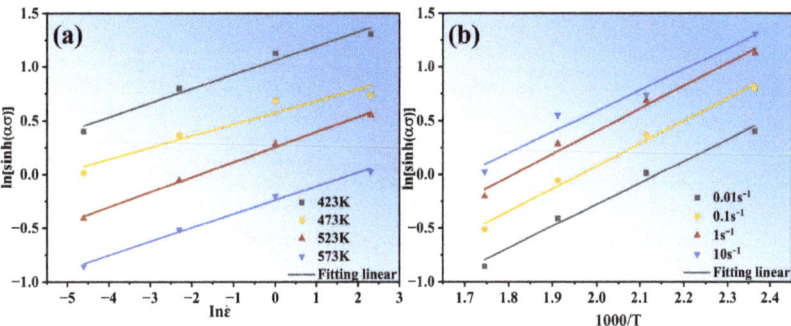

Figure 4. (**a**) Relationships between $\ln[\sinh(\alpha\sigma)]$ and $\ln\dot{\varepsilon}$ at different temperatures; (**b**) Relationships between $\ln[\sinh(\alpha\sigma)]$ and $1000/T$ at different strain rates.

Then, the relationship of σ-Z can be defined by Equation (9), and the formula can be written as:

$$\sigma = \frac{1}{\alpha} \ln \left\{ \left(\frac{Z}{A}\right)^{\frac{1}{n}} + \left[\left(\frac{Z}{A}\right)^{\frac{2}{n}} + 1\right]^{\frac{1}{2}} \right\} \tag{9}$$

The effect of strain on thermal deformation behavior is disregarded based on the aforementioned constitutive equations. Nevertheless, it can be found from Figure 2 that the strain played a considerable role in the true stress. This means that taking the impact of strain into account on the constitutive equations is essential.

The material constants (A, n, Q, and a) are known to be functions of the strain and can be described using polynomial functions [25,26]. To obtain the values of these constants, the strain is varied from 0.05 to 0.7 in increments of 0.05, and the material constants are computed at each strain. The relationship between the material constants and strain is then established through polynomial fitting techniques, as shown in Figure 5. It was found that a six-order polynomial function can accurately depict the impact of strain on material constants, as shown in Equation (10). Figure 5 summarizes the relationship between the material constants and strain. The validity of the seventh-order polynomial model is confirmed by the results of polynomial fitting, which are presented in Table 2.

$$\begin{aligned} a &= h_0 + h_1\varepsilon + h_2\varepsilon^2 + h_3\varepsilon^3 + h_4\varepsilon^4 + h_5\varepsilon^5 + h_6\varepsilon^6 \\ n &= i_0 + i_1\varepsilon + i_2\varepsilon^2 + i_3\varepsilon^3 + i_4\varepsilon^4 + i_5\varepsilon^5 + i_6\varepsilon^6 \\ Q &= j_0 + j_1\varepsilon + j_2\varepsilon^2 + j_3\varepsilon^3 + j_4\varepsilon^4 + j_5\varepsilon^5 + j_6\varepsilon^6 \\ \ln A &= k_0 + k_1\varepsilon + k_2\varepsilon^2 + k_3\varepsilon^3 + k_4\varepsilon^4 + k_5\varepsilon^5 + k_6\varepsilon^6 \end{aligned} \tag{10}$$

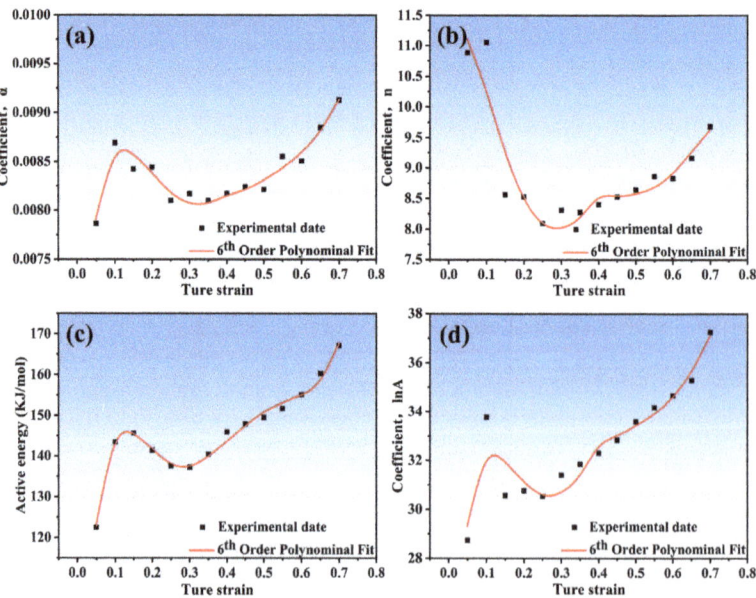

Figure 5. Relationship between the fitted parameters and the strain: (a) a; (b) n; (c) Q; (d) $\ln A$.

Table 2. Coefficients of the polynomial equations.

a	n	Q	lnA
$h_0 = 0.0025$	$i_0 = 9.0261$	$j_0 = 25.071$	$k_0 = 14.882$
$h_1 = 0.1749$	$i_1 = 85.08$	$j_1 = 3023$	$k_1 = 454.91$
$h_2 = -1.6323$	$i_2 = -1147.2$	$j_2 = -27,214$	$k_2 = -4230.1$
$h_3 = 6.8839$	$i_3 = 5542.6$	$j_3 = 113,308$	$k_3 = 17,877$
$h_4 = -14.666$	$i_4 = -12,569$	$j_4 = -238,415$	$k_4 = -37,778$
$h_5 = 15.409$	$i_5 = 13,604$	$j_5 = 246,903$	$k_5 = 39,081$
$h_6 = -6.3395$	$i_6 = -5665$	$j_6 = -99,939$	$k_6 = -15,752$

3.3. Performance Evaluations

Figure 6 displays that the Arrhenius-type model has good predictability of the flow behavior of Zn-2.0Cu-0.15Ti alloy under various deformation conditions. However, to ensure the reliability of the predictions of the Arrhenius-type model in our work, the mean absolute percentage error (MAPE), correlation coefficient (R), and relative error parameters are used to evaluate the reliability. The formula can be written as:

$$R = \frac{\sum_{i=1}^{n}(A_i - \overline{A})(B_i - \overline{B})}{\sqrt{\sum_{i=1}^{n}(A_i - \overline{A})^2 \sum_{i=1}^{n}(B_i - \overline{B})^2}} \quad (11)$$

$$\text{MAPE}(\%) = \frac{1}{n}\sum_{i=1}^{n}\left|\frac{A_i - B_i}{A_i}\right| \times 100 \quad (12)$$

where n is all the data; A_i and B_i indicate the measured data and the predicted data, respectively. $\overline{A}$ and $\overline{B}$ represent the average values of A_i and B_i, respectively.

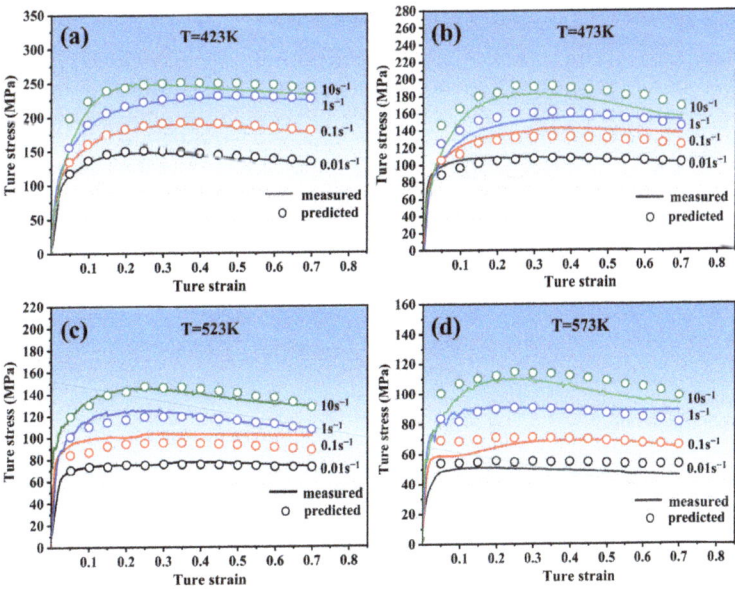

Figure 6. Comparisons between the measured and the predicted true stress at different temperatures: (a) 423 K; (b) 473 K; (c) 523 K; and (d) 573 K.

Generally, R is used to reflect the closeness of the correlation between variables [27]. In our study, the closer the R is to one, the more accurate the prediction of the model. Furthermore, the MAPE is typically used to reflect the practical prediction errors. The smaller the value of MAPE, the better performance for the predicted model. It can be observed from Figure 7a that R was calculated at 0.986. Moreover, the MAPE was 6.48%, and the relative error percentage was used to investigate the predictability performance of the model. One can therefore obtain the following:

$$\text{Relative error} = \left(\frac{A_i - B_i}{A_i}\right) \times 100\% \tag{13}$$

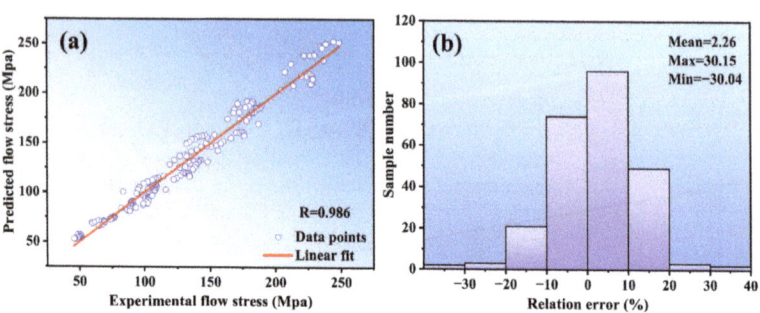

Figure 7. (a) Correlation between the experimental and predicted flow stress values obtained from Arrhenius-Type constitutive model; (b) Statistical analysis of the relative error by Arrhenius-Type constitutive model.

From Figure 7b, the value of the relative error parameter ranges from 30.04 to 30.15%. Combining the above calculation parameters, the result shows that predicted and measured values of the Arrhenius-type model have good agreement in our study.

3.4. Processing Maps

At present, Prased proposed hot processing maps by studying the theory of DMM [28,29] and continuum mechanics. In general, the hot processing maps can be obtained by combining the power dissipation map and the instability map. Additionally, in accordance with the theory of DMM, the total power (P) of the input system is made up of two components: dissipated co-content J and dissipated content G [20], which are produced by, respectively, structural change and plastic deformation. The equation appears to be the following:

$$P = \sigma\dot{\varepsilon} = G + J = \int_0^{\dot{\varepsilon}} \sigma d\dot{\varepsilon} + \int_0^{\dot{\varepsilon}} \dot{\varepsilon} d\sigma \tag{14}$$

where σ and $\dot{\varepsilon}$ represent the flow stress and the strain rate, the sensitivity of strain rate (m) can be expressed using the following equation:

$$m = \frac{dJ}{dG} = \frac{\partial \ln \sigma}{\partial \ln \dot{\varepsilon}} \tag{15}$$

Typically, the power dissipation efficiency (η) is used to assess an alloy's capacity for power dissipation, which is as follows [30]:

$$\eta = \frac{J}{J_{max}} = \frac{2m}{m+1} \tag{16}$$

When $m \leq 0$, there is no power dissipation in the entire system. When $0 < m < 1$, it is considered to be in a steady-state flow regime. When m = 1, the value of J is equal to its peak value (J_{max}). A 3D contour map of power dissipation can express the changes in η under various deformation temperatures, strain, and strain rates. Generally, the higher the power consumption efficiency is, the better the performance of alloy processing can be [31], which demonstrates that DRX may occur. However, processes such as adiabatic shear banding and crack growth usually result in structural instability. According to the extreme principle of irreversible thermodynamics proposed by Ziegler [32], the following is an expression for the unstable criterion:

$$\xi(\dot{\varepsilon}) = \frac{\partial \ln(\frac{m}{m+1})}{\partial \ln \dot{\varepsilon}} + m < 0 \tag{17}$$

The flow behavior of the material can become unstable when the values of the parameters (ξ) mentioned above fall below zero. Figures 8 and 9 illustrate, respectively, the 3D maps of power dissipation and flow instability for the strain range (0.1–0.7), temperature range (423–573 K), and strain rate range (0.01–10 s^{-1}). The colored grids of the power dissipation map represent η, and the shaded regions of the flow instability map can be used to identify the unstable area. The results suggest that the power dissipation efficiency (η) and the flow instability zone change significantly under different deformation conditions. With the increase in strain (from 0.1 to 0.7), the value of η declines (Figure 8). Moreover, it can be found that the peak values of the efficiency region are the strain rate range (0.01–0.1 s^{-1}) and the temperature range (493–543 K) (Figure 8b,c).

From Figure 9a, the shaded regions increase rapidly at lower strain levels (0.1–0.3). Meanwhile, it can be seen that the shaded regions are mostly presented at low temperatures (423–473 K) and strain rates (0.1–10 s^{-1}) (Figure 9b,c). To ensure excellent processability, the shaded regions should be avoided. According to the analysis based on the power dissipation map and the unstable map, the optimal processing region for the hot processing of Zn-2.0Cu-0.15Ti alloy is the temperature range (493–543 K) and the strain rate range (0.01–0.1 s^{-1}).

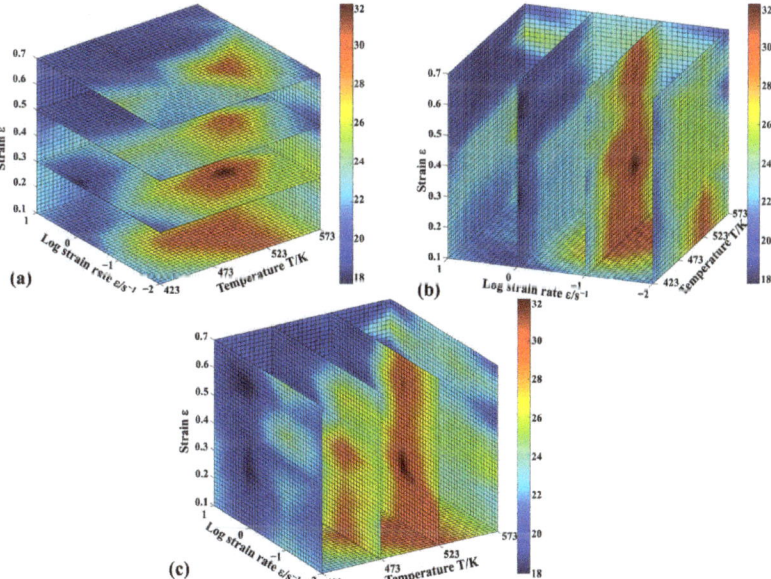

Figure 8. The 3D power dissipation maps of Zn-2.0Cu-0.15Ti alloy: (**a**) strain section, (**b**) strain rate section, and (**c**) temperature section.

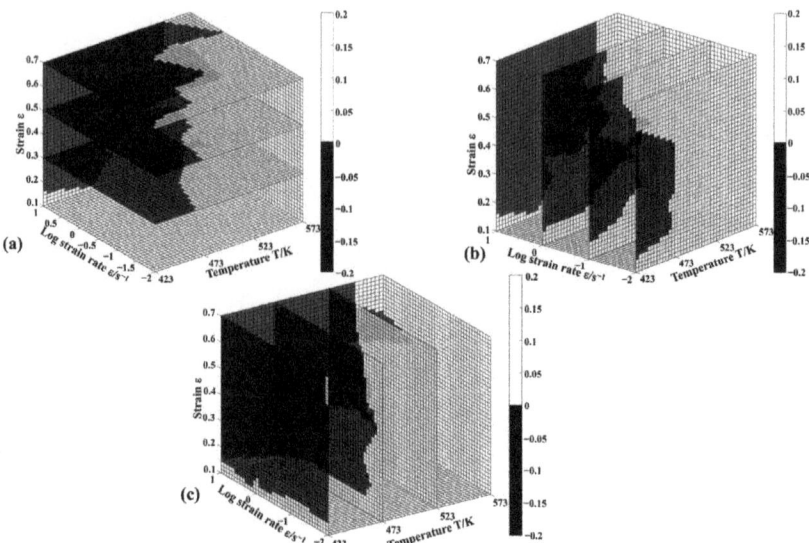

Figure 9. The 3D flow-instability maps of Zn-2.0Cu-0.15Ti alloy: (**a**) strain section, (**b**) strain rate section, and (**c**) temperature section.

3.5. Microstructure Evolution

Figure 10(a1,a4) displays the EBSD maps of the specimens deformed at 1 s^{-1} and the deformation temperature of 423–573 K. In Figure 10(a1,a4), the low-angle grain boundaries (LAGBs, 2–15°) are described as fine black lines, while the high-angle grain boundaries (HAGBs, ≥15°) are described as thick black lines. It is dominated by elongated grains when the specimen deformed at 423 k (Figure 10(a1)). With increasing deformation temperature (from 423 to 473 K), elongated grains gradually disappear and are replaced by smaller rounded grains or equiaxed grains (Figure 10(a2)). As the deformation temperature rises, it shows that the DRX begins to occur. When T = 523 K, an increase in the amount of recrystallization fraction was observed, and some sub-grain boundaries from the original grains can be observed (Figure 10(a3)). Meanwhile, some fine grains develop along the original grain boundaries. The phenomenon is in line with the typical feature of discontinuous-type DRX (DDRX) induced by strain-induced boundary migration [33]. During deformation, the increase in HAGBs will hinder the continuity of dislocation slip [34] and cause stress concentration. In order to reduce stress concentration, the grain boundary bulges and migrates locally to form a distortion-free recrystallized nucleation. When T = 573 K, the recrystallized grains further grow, tending to equiaxed grain formation. The complete processing of the DRX is depicted in Figure 10(a4).

In Figure 10(b1,b4), blue denotes the fully recrystallized grains, yellow symbolizes the subgrains, and red represents the deformed microstructures. Figure 10(b1) shows that the microstructure underwent deformation with only a few subgrains and recrystallized grains. Low deformation temperatures cause inadequate recrystallization time. When T = 523 K (Figure 10(b3)), the deformed structure is replaced by the sub-grains and disappears. When the temperature exceeds 523 K (Figure 10(b4)), full recrystallized grains formed along the original grain boundaries can be detected. Therefore, the DRX volume percent is enhanced with increasing deformation temperature [34,35].

From Figure 10(c1,c4), it can be found that the proportion of LAGBs declines, and the proportion of HAGBs increases gradually as the deformation temperature increases [36,37]. The proportion of HAGBs increases from 48.5% to 68.1% when the deformation temperature rises from 423 to 573 K. This is because higher deformation temperature favors DRX, which causes subgrain boundaries to absorb dislocations and merge into large subgrains [38]. In addition, when T = 423 K, the maximum pole intensity of the basal texture in Figure 10(c1) is 27.068. However, when T = 573 K, the maximum pole intensity suddenly decreases to 5.947. The result indicates that the deformation temperature plays a significant role in the dynamic softening of the alloy [39].

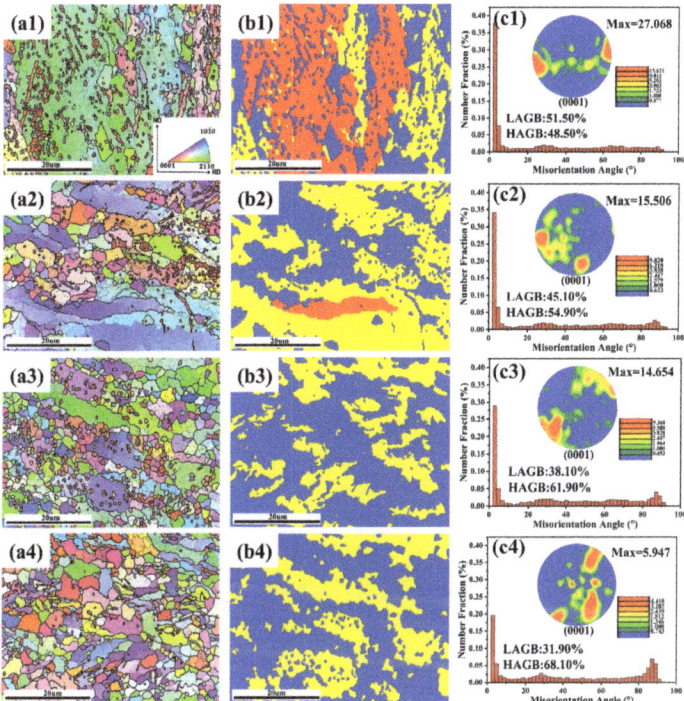

Figure 10. EBSD maps (**a1–a4**), recrystallization grain distribution maps (**b1–b4**) and misorientation angle (**c1–c4**) of the specimens deformed with strain rates of 1 s^{-1} at (**a1–c1**) 423 K, (**a2–c2**) 473 K, (**a3–c3**) 523 K, and (**a4–c4**) 573 K.

Figure 11 displays the TEM micrographs of Zn-2.0Cu-0.15Ti alloy under various deformation conditions. According to the results, the deformation temperature and strain rate are closely related to DRX [40,41]. Dislocation tangling and walls can be observed in the sample, along with original subgrains forming inside without obvious dynamic recrystallization grains in Figure 11a in the case of 423 K and 0.1 s^{-1}. The softening of the alloy was mainly achieved through dislocation slip, cross-slip, climb, and recombination inside the grain [42]. When T = 423 K and $\dot{\varepsilon}$ = 1 s^{-1}, a polygonal structure appeared clearly inside the grain due to the segmentation effect of dislocation grids or walls (Figure 11b), indicating the occurrence of continuous dynamic recrystallization (CDRX) inside the alloy. Within the strain rate range of 0.01–10 s^{-1}, the elevated strain rate can enhance the effect of CDRX, due to the decline in the CRSSnon-basal/CRSSbasal ratio [12,43], which leads to CDRX occurring easily, weakening DDRX. At 523 K/0.1 s^{-1} (Figure 11c), an obvious curved grain boundary was observed on the original grain boundary. These bow-outs were the nucleation sites of discontinuous dynamic recrystallization (DDRX) [34,36]. Additionally, the elevated temperature increases dislocation mobility, which makes more dislocations tend to accumulate near grain boundaries (GBs). The sliding and moving rates of the GB are correspondingly increased due to higher driving forces at high temperatures. As a result, flow stress decreases (Figure 2c). However, as the strain rate increases to 10 s^{-1}, twinning dynamic recrystallization (TDRX) and CDRX are observed (Figure 11d). This indicates that twinning is generated during the thermal deformation process, and twins can form much easier than slip at a high strain rate (10 s^{-1}), due to the highly effective interface velocity [13] of the twin. A twin boundary (TB) will hinder the dislocation motion, thus increasing the strain energy around the twinning and providing the driving force for the nucleation of TDRX on TB. Meanwhile, the high strain rate increases deformation inhomogeneity, making dislocation slip more difficult, which leads to a significant increase in flow stress (Figure 2c) and the weakening of DDRX [13,44].

Figure 11. TEM micrographs of Zn-2.0Cu-0.15Ti alloy deformed at (**a**) 423 K/0.1 s^{-1}; (**b**) 423 K/1 s^{-1}; (**c**) 523 K/0.1s^{-1}; (**d**) 523 K/10 s^{-1}.

4. Conclusions

The hot deformation behavior of Zn-2.0Cu-0.15Ti alloy was investigated at strain rates 0.01, 0.1, 1, and 10 s^{-1} and deformation temperatures 423, 473, 523, and 573 K, using the Arrhenius model and 3D processing maps. The following are the primary conclusions:

1. The Arrhenius-type model is utilized to forecast flow stress behavior. The results show the Arrhenius model can accurately predict the flow stress behavior of Zn-2.0Cu-0.15Ti alloy.
2. Three-dimensional processing maps are generated at different strains based on DDM theory. The ideal processing domain for Zn-2.0Cu-0.15Ti alloy is the temperature range from 493 to 543 K and strain rate range from 0.01 to 0.1 s^{-1}.
3. The softening mechanism of Zn-2.0Cu-0.15Ti alloy has diversification, including CDRX, DDRX, and TDRX, and is activated at T = 423–573 K and $\dot{\varepsilon}$ = 0.01–10 s^{-1}. CDRX is activated at low deformation temperature (423 K) and high strain rate (1 s^{-1}) and is inhibited with increasing deformation temperature and decreasing strain rate. However, when the deformation temperature increases (523 K) and the strain rate declines (0.1 s^{-1}), DDRX becomes the primary softening mechanism, and is weakened with increasing strain rate. CDRX and TDRX become the main softening mechanisms when the strain rate is 10 s^{-1}.

Author Contributions: Conceptualization, G.X.; Methodology, G.X.; Software, J.L. and Y.Z.; Formal analysis, C.L.; Investigation, C.L.; Data curation, S.H.; Writing—original draft, Z.K.; Project administration, D.Z. and Y.L.; Funding acquisition, D.Z. and Y.L. All authors have read and agreed to the published version of the manuscript.

Funding: This research was funded by Natural Science Foundation of Hunan Province, China (Grant Nos. 2021JJ30673 and 2022JJ30570), the Research Foundation of Zhuzhou Smelter Group Co. (grant number ZYKJ202309), and the Project of the Education Department of Hunan Province (Grant No. 20B569).

Institutional Review Board Statement: Not applicable.

Informed Consent Statement: Not applicable.

Data Availability Statement: The data are available from the corresponding author upon reasonable request.

Acknowledgments: This work was supported by the Natural Science Foundation of Hunan Province, China (Grant Nos. 2021JJ30673 and 2022JJ30570), the Research Foundation of Zhuzhou Smelter Group Co (grant number ZYKJ202309), and the Project of the Education Department of Hunan Province (Grant No. 20B569).

Conflicts of Interest: The authors declare no conflict of interest.

References

1. Gao, Z.; Zhang, X.; Huang, H.; Chen, C.; Jiang, J.; Niu, J.; Dargusch, M.; Yuan, G. Microstructure evolution, mechanical properties and corrosion behavior of biodegradable Zn-2Cu-0.8Li alloy during room temperature drawing. *Mater. Charact.* **2022**, *185*, 111722. [CrossRef]
2. Sun, J.; Zhang, X.; Shi, Z.-Z.; Gao, X.-X.; Li, H.-Y.; Zhao, F.-Y.; Wang, J.-Q.; Wang, L.-N. Development of a high-strength Zn-Mn-Mg alloy for ligament reconstruction fixation. *Acta Biomater.* **2021**, *119*, 485–498. [CrossRef] [PubMed]
3. Zhou, H.; Hou, R.; Yang, J.; Sheng, Y.; Li, Z.; Chen, L.; Li, W.; Wang, X. Influence of Zirconium (Zr) on the microstructure, me-chanical properties and corrosion behavior of biodegradable zinc-magnesium alloys. *J. Alloys Compd.* **2020**, *840*, 155792. [CrossRef]
4. Lu, X.Q.; Wang, S.R.; Xiong, T.Y.; Wen, D.S.; Wang, G.Q.; Du, H. Study on Corrosion Resistance and Wear Resistance of Zn-Al-Mg/ZnO Composite Coating Prepared by Cold Spraying. *Coatings* **2019**, *9*, 505. [CrossRef]
5. Chuvil'Deev, V.; Nokhrin, A.; Kopylov, V.; Gryaznov, M.; Shotin, S.; Likhnitskii, C.; Kozlova, N.; Shadrina, Y.; Berendeev, N.; Melekhin, N.; et al. Investigation of mechanical properties and corrosion resistance of fine-grained aluminum alloys Al-Zn with reduced zinc content. *J. Alloys Compd.* **2022**, *891*, 162110. [CrossRef]
6. Liu, Y.; Geng, C.; Zhu, Y.K.; Peng, J.F.; Xu, J.R. Hot Deformation Behavior and Intrinsic Workability of Carbon Nano-tube-Aluminum Reinforced ZA27 Composites. *J. Mater. Eng. Perform.* **2017**, *26*, 1967–1977. [CrossRef]
7. Zhu, D.; Chen, D.; Liu, Y.; Du, B.; Li, S.; Tan, H.; Zhang, P. Hot Deformation Behavior of Cu-2.7Be Alloy during Isothermal Compression. *J. Mater. Eng. Perform.* **2021**, *30*, 3054–3067. [CrossRef]
8. Liu, Y.; Wang, X.; Zhu, D.; Chen, D.; Wang, Q.; Li, X.; Sun, T. Hot Workability Characteristics and Optimization of Processing Parameters of 7475 Aluminum Alloy Using 3D Processing Map. *J. Mater. Eng. Perform.* **2020**, *29*, 787–799. [CrossRef]
9. Li, C.; Huang, T.; Liu, Z. Effects of thermomechanical processing on microstructures, mechanical properties, and biodegradation behaviour of dilute Zn–Mg alloys. *J. Mater. Res. Technol.* **2023**, *23*, 2940–2955. [CrossRef]
10. Wang, X.; Meng, B.; Han, J.; Wan, M. Effect of grain size on superplastic deformation behavior of Zn-0.033 Mg alloy. *Mater. Sci. Eng. A-Struct.* **2023**, *870*, 144877. [CrossRef]
11. Zhao, R.; Ma, Q.; Zhang, L.; Zhang, J.; Xu, C.; Wu, Y.; Zhang, J. Revealing the influence of Zr micro-alloying and hot extrusion on a novel high ductility Zn–1Mg alloy. *Mater. Sci. Eng. A-Struct.* **2021**, *801*, 140395. [CrossRef]
12. Bednarczyk, W.; Wątroba, M.; Kawałko, J.; Bała, P. Can zinc alloys be strengthened by grain refinement? A critical evaluation of the processing of low-alloyed binary zinc alloys using ECAP. *Mater. Sci. Eng. A-Struct.* **2019**, *748*, 357–366. [CrossRef]
13. Li, M.; Shi, Z.-Z.; Wang, Q.; Cheng, Y.; Wang, L.-N. Zn–0.8Mn alloy for degradable structural applications: Hot compression behaviors, four dynamic recrystallization mechanisms, and better elevated-temperature strength. *J. Mater. Sci. Technol.* **2023**, *137*, 159–175. [CrossRef]
14. Guo, S.; Shen, Y.; Guo, J.; Wu, S.; Du, Z.; Li, D. An investigation on the hot workability and microstructural evolution of a novel dual-phase Mg–Li alloy by using 3D processing maps. *J. Mater. Sci. Technol.* **2023**, *23*, 5486–5501. [CrossRef]
15. Ahmedabadi, P.M.; Kain, V. Constitutive models for flow stress based on composite variables analogous to Zener–Holloman parameter. *Mater. Today Commun.* **2022**, *33*, 104820. [CrossRef]
16. Ding, Z.; Jia, S.; Zhao, P.; Deng, M.; Song, K. Hot deformation behavior of Cu–0.6Cr–0.03Zr alloy during compression at elevated temperatures. *Mater. Sci. Eng. A-Struct.* **2013**, *570*, 87–91. [CrossRef]
17. Prasad, Y.; Rao, K. Processing maps for hot deformation of rolled AZ31 magnesium alloy plate: Anisotropy of hot workability. *Mater. Sci. Eng. A-Struct.* **2008**, *487*, 316–327. [CrossRef]
18. Lin, H.-B. Dynamic recrystallization behavior of 6082 aluminum alloy during hot deformation. *Adv. Mech. Eng.* **2021**, *13*, 16878140211046107. [CrossRef]

19. Galiyev, A.; Kaibyshev, R.; Gottstein, G. Correlation of plastic deformation and dynamic recrystallization in magnesium alloy ZK60. *Acta Mater.* **2001**, *49*, 1199–1207. [CrossRef]
20. Wang, Z.; Liu, X.; Xie, J. Constitutive relationship of hot deformation of AZ91 magnesium alloy. *Acta Metall. Sin.* **2008**, *44*, 1378–1383.
21. Kong, F.; Yang, Y.; Chen, H.; Liu, H.; Fan, C.; Xie, W.; Wei, G. Dynamic recrystallization and deformation constitutive analysis of Mg–Zn-Nd-Zr alloys during hot rolling. *Heliyon* **2022**, *8*, e09995. [CrossRef] [PubMed]
22. Wan, M.P.; Zhao, Y.Q.; Zeng, W.D.; Gang, C. Ambient Temperature Deformation Behavior of Ti-1300 Alloy. *Rare Met. Mat. Eng.* **2015**, *44*, 2519–2522.
23. Wang, B.; Yi, D.; Fang, X.; Liu, H.; Wu, C. Thermal Simulation on Hot Deformation Behavior of ZK60 and ZK60 (0.9Y) Magne-sium Alloys. *Rare Met. Mat. Eng.* **2010**, *39*, 106–111.
24. Zener, C.; Hollomon, J.H. Effect of Strain Rate Upon Plastic Flow of Steel. *J. Appl. Phys.* **1944**, *15*, 22–32. [CrossRef]
25. Lin, Y.; Chen, M.-S.; Zhong, J. Constitutive modeling for elevated temperature flow behavior of 42CrMo steel. *Comput. Mater. Sci.* **2008**, *42*, 470–477. [CrossRef]
26. Lin, Y.; Chen, X.-M. A critical review of experimental results and constitutive descriptions for metals and alloys in hot working. *Mater. Des.* **2011**, *32*, 1733–1759. [CrossRef]
27. Williams, S. Coefficient, Pearson's correlation coefficient. *N. Z. Med. J.* **1996**, *109*, 38.
28. Deng, Y.; Yin, Z.; Huang, J. Hot deformation behavior and microstructural evolution of homogenized 7050 aluminum alloy during compression at elevated temperature. *Mater. Sci. Eng. A-Struct.* **2011**, *528*, 1780–1786. [CrossRef]
29. Prasad, Y.V.R.K.; Gegel, H.L.; Doraivelu, S.M.; Malas, J.C.; Morgan, J.T.; Lark, K.A.; Barker, D.R. Modeling of dynamic material behavior in hot deformation: Forging of Ti-6242. *Metall. Trans. A* **1984**, *15*, 1883–1892. [CrossRef]
30. Murty, S.V.S.N.; Sarma, M.S.; Rao, B.N. On the evaluation of efficiency parameters in processing maps. *Met. Mater. Trans. A* **1997**, *28*, 1581–1582. [CrossRef]
31. Seshacharyulu, T.; Medeiros, S.C.; Frazier, W.G.; Prasad, Y.V. Hot working of commercial Ti-6Al-4V with an equiaxed a-b microstructure: Materials modeling considerations. *Mater. Sci. Eng. A-Struct.* **2000**, *284*, 184–194. [CrossRef]
32. Ziegler, H. An Introduction to Thermomechanics. *J. Appl. Mech.-Trans. ASME* **1978**, *45*, 996. [CrossRef]
33. Huang, W.; Yang, X.; Yang, Y.; Mukai, T.; Sakai, T. Effect of yttrium addition on the hot deformation behaviors and micro-structure development of magnesium alloy. *J. Alloys Compd.* **2019**, *786*, 118–125. [CrossRef]
34. Fan, D.-G.; Deng, K.-K.; Wang, C.-J.; Nie, K.-B.; Shi, Q.-X.; Liang, W. Hot deformation behavior and dynamic recrystallization mechanism of an Mg-5wt.%Zn alloy with trace SiCp addition. *J. Mater. Res. Technol.* **2021**, *10*, 422–437. [CrossRef]
35. Mollaei, N.; Fatemi, S.; Aboutalebi, M.; Razavi, S.; Bednarczyk, W. Dynamic recrystallization and deformation behavior of an extruded Zn-0.2 Mg biodegradable alloy. *J. Mater. Res. Technol.* **2022**, *19*, 4969–4985. [CrossRef]
36. Liu, C.; Barella, S.; Peng, Y.; Guo, S.; Liang, S.; Sun, J.; Gruttadauria, A. Modeling and characterization of dynamic recrystalli-zation under variable deformation states. *Int. J. Mech. Sci.* **2023**, *238*, 107838. [CrossRef]
37. Xu, C.; Huang, J.; Jiang, F.; Jiang, Y. Dynamic recrystallization and precipitation behavior of a novel Sc, Zr alloyed Al-Zn-Mg-Cu alloy during hot deformation. *Mater. Charact.* **2022**, *183*, 111629. [CrossRef]
38. Huang, T.; Liu, Z.; Wu, D.; Yu, H. Microstructure, mechanical properties, and biodegradation response of the grain-refined Zn alloys for potential medical materials. *J. Mater. Res. Technol.* **2023**, *15*, 226–240. [CrossRef]
39. Li, H.; Huang, Y.; Liu, Y. Dynamic recrystallization mechanisms of as-forged Al–Zn–Mg-(Cu) aluminum alloy during hot comp.ression deformation. *Mater. Sci. Eng. A-Struct.* **2023**, *878*, 145236. [CrossRef]
40. Shi, G.; Zhang, Y.; Li, X.; Li, Z.; Yan, L.; Yan, H.; Liu, H.; Xiong, B. Dynamic Recrystallization Behavior of 7056 Aluminum Alloys during Hot Deformation. *J. Wuhan Univ. Technol.* **2022**, *37*, 90–95. [CrossRef]
41. Liao, Q.; Jiang, Y.; Le, Q.; Chen, X.; Cheng, C.; Hu, K.; Li, D. Hot deformation behavior and processing map development of AZ110 alloy with and without addition of La-rich Mish Metal. *J. Mater. Sci. Technol.* **2020**, *61*, 1–15. [CrossRef]
42. Xu, D.; Zhou, M.; Zhang, Y.; Tang, S.; Zhang, Z.; Liu, Y.; Tian, B.; Li, X.; Jia, Y.; Volinsky, A.A.; et al. Microstructure and hot deformation behavior of the Cu-Sn-Ni-Zn-Ti(-Y) alloy. *Mater. Charact.* **2023**, *196*, 112559. [CrossRef]
43. Zhao, J.; Deng, Y.; Tang, J.; Zhang, J. Influence of strain rate on hot deformation behavior and recrystallization behavior under isothermal compression of Al-Zn-Mg-Cu alloy. *J. Alloys Compd.* **2019**, *809*, 151788. [CrossRef]
44. Ji, S.; Liang, S.; Song, K.; Li, H.; Li, Z. Evolution of microstructure and mechanical properties in Zn–Cu–Ti alloy during severe hot rolling at 300 °C. *J. Mater. Res.* **2017**, *32*, 3146–3155. [CrossRef]

Disclaimer/Publisher's Note: The statements, opinions and data contained in all publications are solely those of the individual author(s) and contributor(s) and not of MDPI and/or the editor(s). MDPI and/or the editor(s) disclaim responsibility for any injury to people or property resulting from any ideas, methods, instructions or products referred to in the content.

Article

Effect of Boronizing on the Microstructure and Mechanical Properties of CoCrFeNiMn High-Entropy Alloy

Mingyu Hu [1,2], Xuemei Ouyang [1,2,*], Fucheng Yin [1,2], Xu Zhao [1,2], Zuchuan Zhang [1,2] and Xinming Wang [1,2]

1. School of Materials Science and Engineering, Xiangtan University, Xiangtan 411105, China; 202021001605@smail.xtu.edu.cn (M.H.); fuchengyin@xtu.edu.cn (F.Y.); 202121551594@smail.xtu.edu.cn (X.Z.); 202005720811@smail.xtu.edu.cn (Z.Z.); wangxm@xtu.edu.cn (X.W.)
2. Key Laboratory of Materials Design and Preparation Technology of Hunan Province, Xiangtan University, Xiangtan 411105, China
* Correspondence: ouyangxuemei@xtu.edu.cn

Abstract: The CoCrFeNiMn high-entropy alloys were treated by powder-pack boriding to improve their surface hardness and wear resistance. The variation of boriding layer thickness with time and temperature was studied. Then, the frequency factor D_0 and diffusion activation energy Q of element B in HEA are calculated to be 9.15×10^{-5} m^2/s and 206.93 kJ/mol, respectively. The diffusion behavior of elements in the boronizing process was investigated and shows that the boride layer forms with the metal atoms diffusing outward and the diffusion layer forms with the B atoms diffusing inward by the Pt-labeling method. In addition, the surface microhardness of CoCrFeNiMn HEA was significantly improved to 23.8 ± 1.4 Gpa, and the friction coefficient was reduced from 0.86 to 0.48~0.61.

Keywords: high entropy alloy; boronizing; growth kinetics; boronizing mechanism; wear resistance

Citation: Hu, M.; Ouyang, X.; Yin, F.; Zhao, X.; Zhang, Z.; Wang, X. Effect of Boronizing on the Microstructure and Mechanical Properties of CoCrFeNiMn High-Entropy Alloy. Materials 2023, 16, 3754. https://doi.org/10.3390/ma16103754

Academic Editor: Frank Czerwinski

Received: 20 April 2023
Revised: 13 May 2023
Accepted: 14 May 2023
Published: 16 May 2023

Copyright: © 2023 by the authors. Licensee MDPI, Basel, Switzerland. This article is an open access article distributed under the terms and conditions of the Creative Commons Attribution (CC BY) license (https://creativecommons.org/licenses/by/4.0/).

1. Introduction

The equiatomic CoCrFeNiMn alloy with an fcc structure shows high ductility and high fracture toughness even at significantly lower temperatures [1–3]. However, more slip systems in fcc solid solution [4–6] caused lower hardness of CoCrFeNiMn [7,8] and worse wear resistance [9]. Alloying is a crucial method to improve strength and wear resistance. Some expensive and refractory alloying elements such as W, Ta, Nb, and Mo are added to CoCrFeNiMn HEA. This method, however, reduces its plasticity [10–13]. Thus, preparing a reinforced layer on the surface of CoCrFeNiMn HEA is a vital method to solve this problem. In many applications, the service life of materials depends highly on their surface properties, which can be extended by improving the wear resistance of the surface. Thus, preparing a reinforced layer on the surface of CoCrFeNiMn HEA is a vital method to solve this problem.

Boronizing is a thermo-chemical surface treatment method used to enhance the surface hardness and wear resistance of ferrous materials [14–17]. This method is used to enhance the surface properties of ferrous materials, while the boriding behavior of various steels has been extensively studied. The boride layer (BL) formed on the steel surface by boriding treatment can significantly improve its surface properties [18–20]. The principal elements in CoCrFeNiMn HEA, such as Fe, Co, and Cr, have a strong tendency to form stable borides [18,21]. It is reported that the wear resistance of boronized samples exceeds two times the wear resistance value of carburized, carbo-nitrided, or chromium-plated samples [22,23]. Hou et al. [24] obtained the boronized layer on the Al$_{0.25}$CoCrFeNi HEA by the solid-boronizing method. A boronized layer with the phases (Ni, Co, Fe)$_2$B and CrB has been prepared for holding for 9 h at 900 °C. The microhardness of the boronized sample surface approximately approaches 1136 HV$_5$, which is six times that of the untreated HEA. Hiroaki Nakajo et al. [25] obtained ceramic layers on the surface of CoCrFeMnNi HEA using the SPS method, and a powdered mixture of B_4C and KBF_4 was used as the boron source. M_2B, MB, and Mn_3B_4 type borides were formed on the surface of HEA, and the surface hardness reached 2000~2500 $Hv_{0.1}$. Most studies are focused on the microstructure and mechanical properties of borided bulk HEAs after boriding, but there are few reports on the boronizing mechanism of HEA [26,27].

In this study, a reinforced layer was prepared on the surface of HEA by powder-pack boriding and focused on the effect of temperature and duration on the boronizing of CoCrFeNiMn HEA. The boronized samples were characterized using various methods. Furthermore, the properties of boronized CoCrFeNiMn alloys and diffusion behavior were investigated. This study could provide a reference for high hardness and high wear resistance HEA that could be used in a variety of industrial applications, such as the mold and crucible used in the cast aluminum industry [28,29], because the borides formed on the surface have high hardness, excellent corrosion resistance, and wear resistance [30,31], which allow HEA to be applied in special environments to increase service life.

2. Materials and Methods

2.1. Alloys Preparation

Alloy ingots with a nominal composition of $Co_{20}Cr_{20}Fe_{20}Ni_{20}Mn_{20}$ (in at.%) were fabricated by the vacuum arc-melting method using a non-consumable tungsten electrode. High-purity metals were used and melted at least three times to promote chemical homogeneity. The as-cast alloy was homogenized at 1200 °C for 24 h under a vacuum environment and then cut into 6 × 6 × 3 mm and polished with 200#, 400#, 600#, and 800# sandpaper and diamond solution. Additionally, a Pt-labeled layer was deposited on the surface of the annealed sample using an auto-fine coater (JEOL JFC-1600) with a sputtering time of 300 s to determine the evolution of the diffusion of the elements during boriding.

2.2. Boronizing Process

Boronizing treatment was performed at 850, 900, and 950 °C for 3 h, 6 h, 9 h, 12 h, 15 h, and 18 h, respectively. The annealed CoCrFeNiMn samples were placed in the alumina crucible, then packed with high-purity boron powder. The alumina crucibles were tightly closed by a high-temperature sodium silicate and kaolin binder to make them airtight. After boriding was completed, the sealed crucible was extracted and air-cooled to room temperature, and the samples were ultrasonically cleaned using anhydrous ethanol to remove the boron powder from the surface.

2.3. Characterization

The cross-sectional microstructures were observed using a scanning electron microscope (SEM, EVO MA10) equipped with energy dispersive spectrometry (EDS). However, it was difficult to quantify the B atom by EDS, so the element composition of BL was tested on the electron probe microanalysis (EPMA, Shimadzu 1720, beam spot diameter 1 μm, acquisition time 10 s) equipped with a wavelength dispersive spectrometer (WDS). The phase structure was characterized by X-ray diffraction (XRD, Rigaku Ultimate IV) with Cu-Ka radiation. The friction and wear measurements were performed using a ball-on-plate tribometer (CFT-I Tribometer) under dry sliding conditions at 25 °C in an open-air atmosphere. The hardness and Young's modulus of the BL were measured using a nano-indenter.

3. Results

3.1. Microstructure Analyses

Figure 1 shows the surface and cross-sectional morphological images of the CoCrFeNiMn high-entropy alloy after boron penetration at 950 °C for 9 h. The results show that the surface of the high-entropy alloy after boriding is formed by many round rod-like boride assemblies, and there are discontinuous islands of borides, as shown in Figure 1a. According to the different contrast and phase numbers, the surface layer of boronized HEA can be divided into a single-phase layer (SGL), a transition layer (TL), a diffusion layer (DL), and a HEA substrate [32].

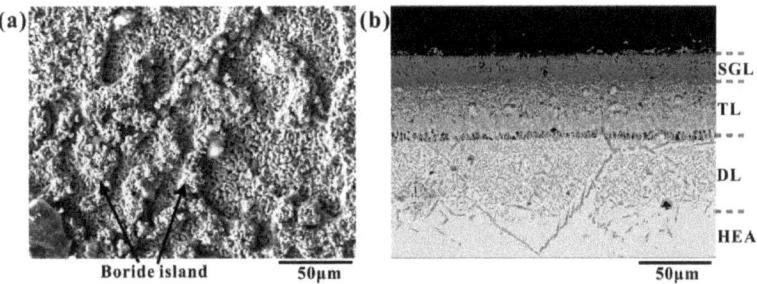

Figure 1. SEM images of boronized CoCrFeNiMn HEA at 950 °C for 9 h: (**a**) surface, (**b**) cross-section.

Figure 2 shows the cross-sectional SEM images of the CoCrFeNiMn HEA boronized at 950 °C at different times. As can be seen, the surface layer of the samples at different times is composed of the BL, the diffusion layer, and the substrate. The interface between the BL and the DL was smooth, which differs from the saw-tooth shape for the boronized pure iron and iron alloys. While the BL consists of an SGL and a two-phase TL, some discontinuous pores were also observed on the interface and the SGL; the interface between them is a saw-tooth shape. The thickness of the BL and total layer were measured, and the average thickness of the BL was 31.58, 41.7, 47.51, 51.09, 53.73, and 64.35 μm for the boronized alloys for 3 h, 6 h, 9 h, 12 h, 15 h, and 18 h, respectively.

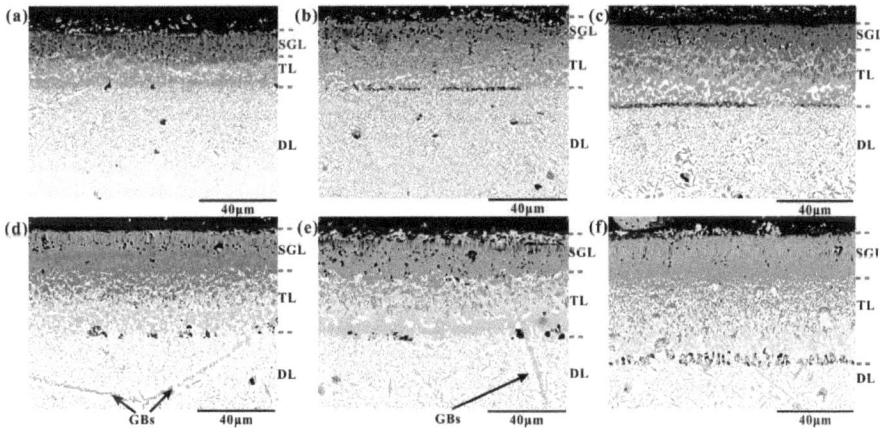

Figure 2. Cross-sectional SEM images of the CoCrFeNiMn HEA substrate with the boronizing times of (**a**) 3 h, (**b**) 6 h, (**c**) 9 h, (**d**) 12 h, (**e**) 15 h, and (**f**) 18 h.

3.2. Growth Kinetics Analyses

Previous studies have mainly investigated the growth kinetics of steel and its alloys. Andrijana Milinović et al. [33] investigated the growth kinetics of different steels, and the results showed that the frequency factor D_0 and diffusion activation energy Q of elemental boron in C15 steel are 3.17×10^{-4} m^2/s and 194.80 kJ/mol, respectively. Figure 3 shows the BL and the whole layer thickness as a function of time. It can be noticed that the thickness of the BL and the whole layer increases with the increase in boronizing time. With high-purity B powder as a boronizing agent, the influence of Si, C, and F in conventional boriding agents on the properties of BL can be excluded, which also simplifies the analysis of the diffusion behavior of different elements during the boronizing process. The growth rate was reduced significantly as the boronizing time increased.

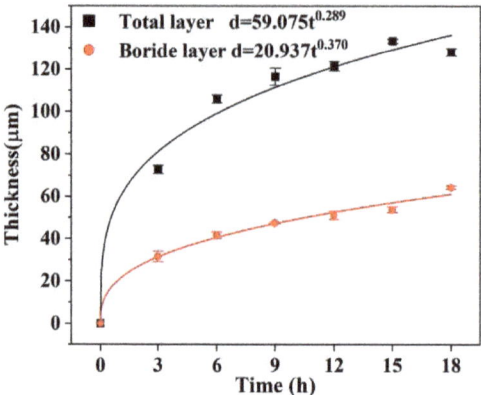

Figure 3. Variation of the boride layer thickness as a function of time for the boronized HEAs.

The diffusion rate is commonly expressed by the formula shown in Equation (1) [34]:

$$d = At^n \quad (1)$$

where d is the thickness of the BL, A is a time constant, t is the time, and n is the diffusion rate exponent. It is shown that the diffusion rate exponent n was calculated to be 0.370 by the BL thickness exponential fitting curve expression $d = 20.937t^{0.370}$ and the fitting curve is closer to parabolic. According to Wagner's hypotheses [35], this parabolic growth can imply that a diffusion process controls the boronizing rate.

The function between the thickness of the BL and time is in accordance with the parabolic law, and the expression is shown in Equation (2):

$$d^2 = Dt \quad (2)$$

where d is the thickness of the diffusion layer (m), D is the growth rate constant (m^2/s), and t is the diffusion time (s). Since it is a quadratic equation, the solution is obtained by finding its roots, combined with the Arrhenius formula:

$$D = D_0 \cdot e^{-Q/(RT)} \quad (3)$$

where D_0 is the frequency factor (m^2/s), Q is the diffusion activation energy (kJ/mol), T is the diffusion temperature (K), and R is the universal gas constant (kJ/(mol·K)). The frequency factor indicates the rate of molecular collisions in the reaction.

$$\ln D = \ln D_0 - \frac{Q}{R} \cdot \frac{1}{T} \quad (4)$$

Figure 4a, plotted according to Equation (3), shows the linear relationship between the thickness of the diffusion layer and the square root of time at different temperatures. According to Equation (4), the relationship between the growth rate constant's natural logarithm and the diffusion temperature's reciprocal can be represented by a line with slope Q/R, and $\ln D_0$ is the intersection of the line with the vertical coordinate. The value of the growth rate constant was obtained from the slope of the line; the fitted curve of the natural logarithm of the growth rate constant and the inverse of the temperature showed a linear relationship, as shown in Figure 4. The value of the diffusion activation energy of B in the CoCrFeNiMn HEA is determined by the slope of the line, while the natural logarithm of the frequency factor is determined by the intersection of the extrapolated line with the vertical coordinate. These values are given in Table 1, while the D_0 of HEA is lower than the D_0 of C15 due to the sluggish diffusion effect of HEA.

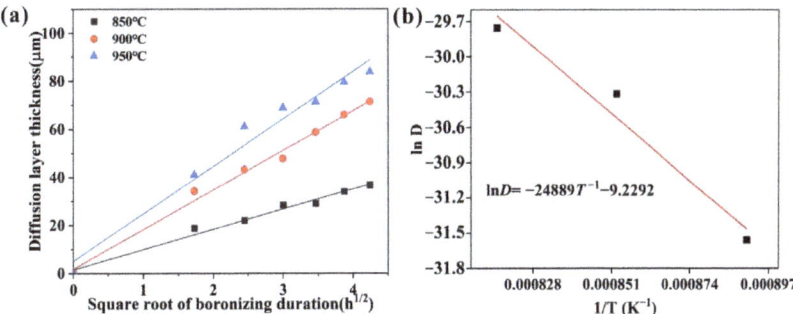

Figure 4. Growth kinetics of boride layers on HEA: (a) Boride layer thickness as a function of the square root of the boronizing duration; (b) natural logarithm of the growth rate constant as a function of the reciprocal boronizing temperature.

Table 1. Values of the frequency factor and the activation energy.

	Frequency Factor D_0, m^2/s	Activation Energy Q, kJ/mol
Boronized HEA	9.15×10^{-5}	206.93

3.3. Phase Structure and Chemical Composition

An XRD analysis was performed on the surface of the boronized HEA, as shown in Figure 5. The main diffraction peaks appear in the same position for boronized samples, and the BL consists of MB (M = Co, Cr, Fe, Ni, and Mn) and M_2B phases. While MB and M_2B are formed due to the replacement of Fe atoms in FeB and Fe_2B by Co, Cr, Ni, and Mn.

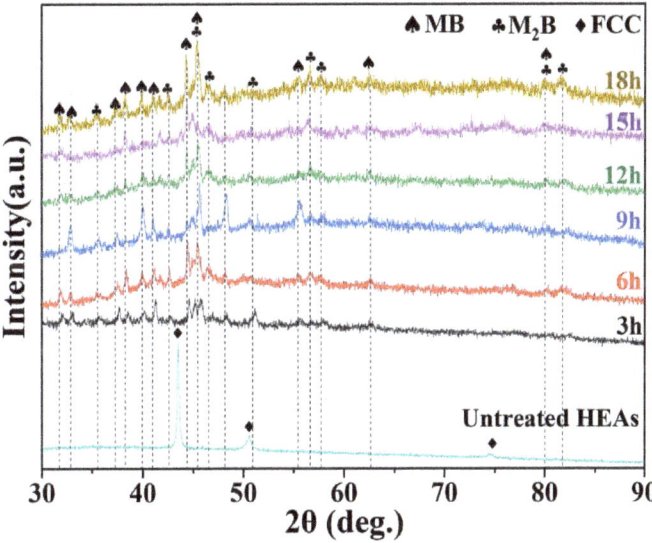

Figure 5. XRD patterns of boronized samples at 950 °C for different times.

The composition of each phase in boronized HEA from surface to substrate was tested by WDS, as shown in Figure 6a. The composition of the seven phases is listed in Table 2.

The BL has three different phases: the dark gray phase (spot 1), the gray phase (spot 3), and the white phase (spot 2). The element composition of BL was tested on the EPMA equipped with a WDS. The concentrations of B and M (M = Co, Cr, Fe, Ni, and Mn) at the outer layer are 45.49 at.% and 54.51 at.%, respectively. The atomic ratio B:M is close to 1:1. In addition, the XRD patterns of the boride surface show the existence of the MB, which has a similar microstructure to the CrB. Therefore, the single phase is identified as MB-type boride. According to the results of points 2–7, the phases from the surface to the substrate can be determined to be single-phase MB, Cr-rich MB + Ni-rich MB, Cr-rich M_2B + Ni-rich M_2B, Cr-poor FCC + Cr-rich M_2B, and FCC substrate. The EPMA elemental mapping and line analysis of the boronized CoCrFeNiMn HEA for 9 h are shown in Figure 6b–h. The concentration of B increases from surface to substrate, as shown in Figure 6b. B atoms mainly exist in the boride, and their solubility in an fcc solid solution is minimal. Based on the distribution of the Cr element, as shown in Figure 6c, Cr atoms have obvious segregation and are mainly concentrated in the DL and grain boundaries in the substrate close to the DL. The Cr element preferentially combines with the B element to form a Cr-rich boride. The distribution of Co, Fe, Ni, and Mn elements shows obvious delamination in the BL, and the (Co, Fe, Ni, Mn)-poor layer and the (Co, Fe, Ni, Mn)-rich layer are alternately arranged as shown in Figure 6d–g. Moreover, the content of these elements at the grain boundary was significantly lower than that in the grain. Figure 6h shows the line analysis from surface to substrate marked with arrows in Figure 6a, which further indicates the existence of the element segregation phenomenon.

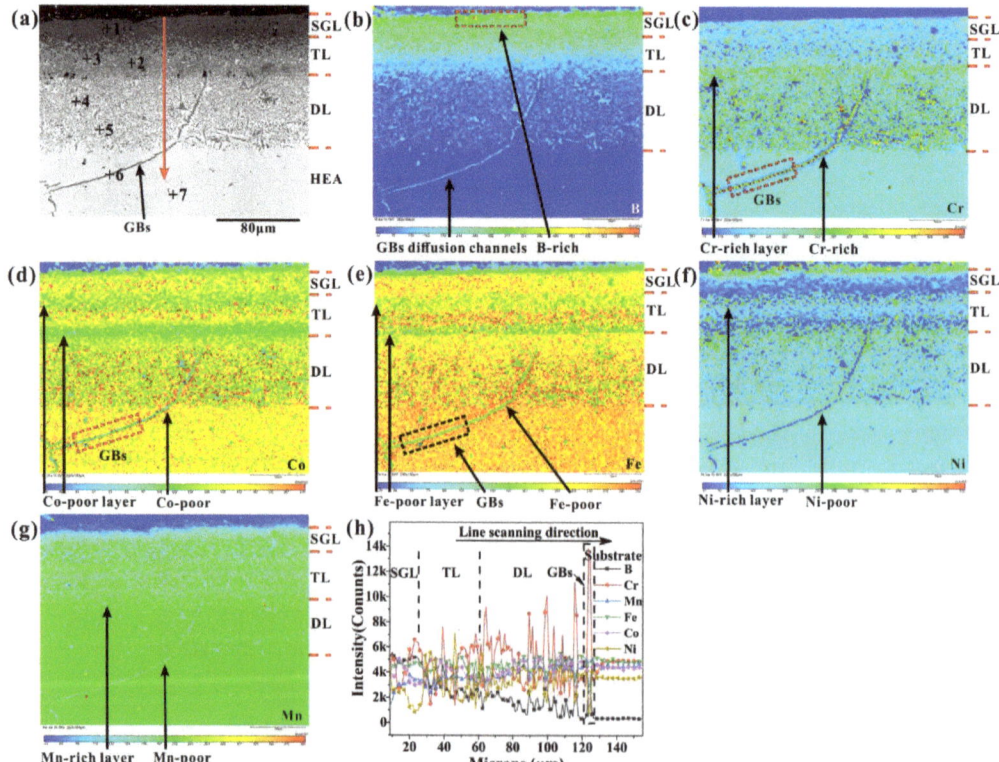

Figure 6. (a) Cross-sectional SEM images of boronized CoCrFeNiMn HEA at 950 °C for 9 h, (b–g) EPMA elemental mapping of B, Cr, Co, Fe, Ni, and Mn, respectively, (h) EPMA line analysis.

Table 2. EPMA elemental analysis of the boronized samples for 9h.

Position	Compositions						Phase
	B	Cr	Mn	Fe	Co	Ni	
1	45.5	13.5	11.4	13.6	11.2	4.8	Ni-poor
2	30.8	5.0	13.7	10.7	13.8	26.0	Ni-rich and Cr-poor M_2B-type boride
3	31.9	18.2	11.9	17.0	14.9	6.1	Ni-poor M_2B-type boride
4	29.8	28.8	12.2	12.5	10.2	6.5	Cr-rich and Ni-poor M_2B-type boride
5	1.9	9.4	21.4	24.0	23.5	19.8	Cr-poor FCC
6	27.9	32.3	11.9	9.8	9.4	8.7	Cr-rich and Ni-poor M_2B-type boride
7	1.6	20.5	20.5	21.9	19.4	17.1	Substrate

3.4. Boronizing Mechanism

Previously, Cengiz et al. [36] investigated the boronizing mechanisms of CoCrFeNi alloy and CoCrFeNiTi alloy. The diffusion of B and Si elements mainly occurred in the boronizing process, and the diffusion behavior of metal elements was not considered in their work. To investigate the diffusion behavior of elements during boronizing, the boronizing experiments were carried out using the Pt element labeling method. This method has been used in the past to study the oxidation behavior of steel [34,37]. The Pt layer shown in Figure 7a was sprayed on the surface of the annealed CoCrFeNiMn alloy before boronizing. Boronizing was performed at 950 °C for 9 h. The cross-sectional morphology of the boronized samples was obtained, as shown in Figure 7b,c. Comparing the interface morphology between the Pt-sprayed and unsprayed samples, it was observed that there was almost no difference between the samples except for the quantity and size of pores at the interface between the BL and the diffusion layer, indicating that the effect of the markers on boronizing was minimal. The Pt element is distributed between the boride and diffusion layers, and no Pt element is found in other locations, as shown in Figure 7d. It indicates that Pt did not diffuse during boronizing and that the interface moves outward from the position of the Pt mark after boronizing. The Pt-labeled experiment confirms the proposed mechanism, such as the "available space model" [38–44] in oxidation, which is the outward migration of metal atoms forming the BL and the inward migration of B forming the diffusion layer.

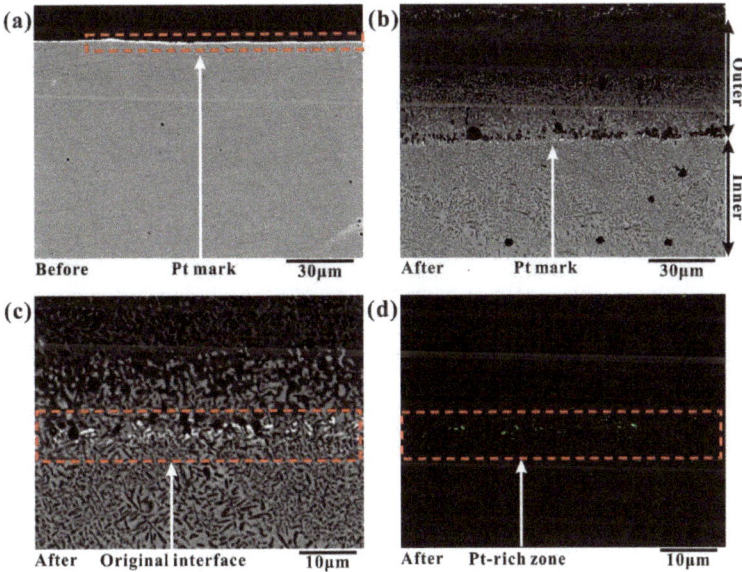

Figure 7. Cross-sectional SEM image of the Pt-marked CoCrFeNiMn HEA: (a) before boronizing, (b) after boronizing at 950 °C for 9 h, (c) the photograph of the original interface, (d) elemental mapping of Pt.

The schematic model of the boronizing mechanism is suggested based on the results above, as shown in Figure 8. The B atoms can be diffused easily into the surface of alloys due to their relatively small size (0.087 nm), and their diffusion is very fast at high temperatures. In the early stage, B atoms preferentially diffuse to the substrate along the grain boundary and then react easily with alloying elements, forming Cr-rich boride, while metal atoms diffuse outward and form M_2B-type boride. Meanwhile, due to the mutual diffusion of the metal atoms and B atoms, vacancies accumulate at the metal/boride interface, and metal creep cannot compensate for the volume of metal consumed by diffusion to the outer layers, resulting in the formation of pores at the metal/boride interface. The EPMA point analysis and WDS mapping results showed the existence of the element segregation phenomenon, and the content of Cr and Ni elements in boride changed most obviously. With increasing boronizing time, the thickness of the BL and DL increases, forming the microstructure as shown in Figure 8b. In addition, the metal atoms that diffuse to the surface are relatively smaller than those on the inner side and are unable to fully combine with the boriding agent to form borides. The retained B powder is removed from the SGL after grinding and polishing the samples, leaving the pores on the SGL, as shown in Figure 8c.

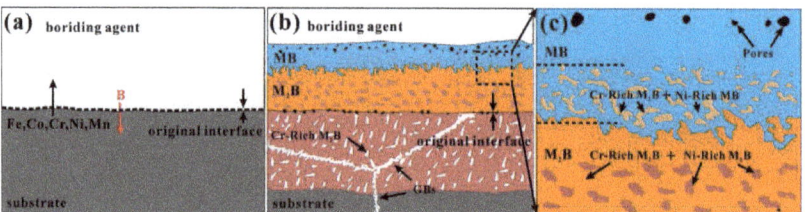

Figure 8. Schematic views of the boronizing mechanism of the CoCrFeNiMn HEA. (**a**) before boronizing, (**b**) after boronizing, (**c**) local magnification of (**b**).

3.5. Hardness and Wear Behavior

The effect of boronizing on the mechanical properties of CoCrFeNiMn alloys was investigated. The cross-sectional morphology of the boronized CoCrFeNiMn alloys after the nanoindentation test is shown in Figure 9a. 80 locations were tested according to the S-shaped path, and a series of load-displacement curves were obtained, as shown in Figure 9b. For the same indentation depth (400 nm), the 68th indentation band requires the lowest load, the 1st indentation point requires the highest load, and the 26th indentation point load is in between. The indentation morphology on the single-phase BL is shown in c. The microhardness and Young's modulus results at different locations were obtained, as shown in Table 3. It is indicated that the hardness of the BL is significantly higher than the DL and substrate, while the surface hardness of the HEA was markedly improved after boronizing. The variation curve of indentation hardness with depth was obtained from Table 3, as shown in Figure 9d. The average hardness of the substrate is 3.9 ± 0.1 GPa, and the average hardness of the BL is 23.8 ± 1.4 GPa, while the trend of modulus and hardness is similar. After the boriding treatment, a multilayer reinforced layer consisting of numerous high-hardness borides is formed on the surface of HEA, resulting in a significant improvement in the surface hardness and modulus of HEA.

The tribological behavior of the strengthening layer in dry conditions was investigated using reciprocating ball-on-plate tests. The wear tracks of unboronized and boronized alloys at 950 °C for 9 h under dry conditions were investigated using SEM, as shown in Figure 10. The width of friction tracks on the surface of the boronized CoCrFeNiMn HEA was significantly reduced, and the interior of the wear tracks was smoother compared with the unboronized samples, as shown in Figure 10a,b. The periodically localized fracture of the surface layer and the periodic accumulation and elimination of debris on the worn surface of CoCrFeNiMn HEA were observed, as shown in Figure 10c. The plastic deformation along the groove, some lamellar delamination, and pits could be seen. In addition, it is also found that the surface is attached to some wear debris, indicating that it is adhesive wear. The boronized HEA has no peeling pits, and more parallel grooves emerge. However, the small abrasive dust is significantly reduced, and a large number of parallel grooves are observed. The occurrence of parallel furrows is a representative characteristic of abrasive wear, as shown in Figure 10d.

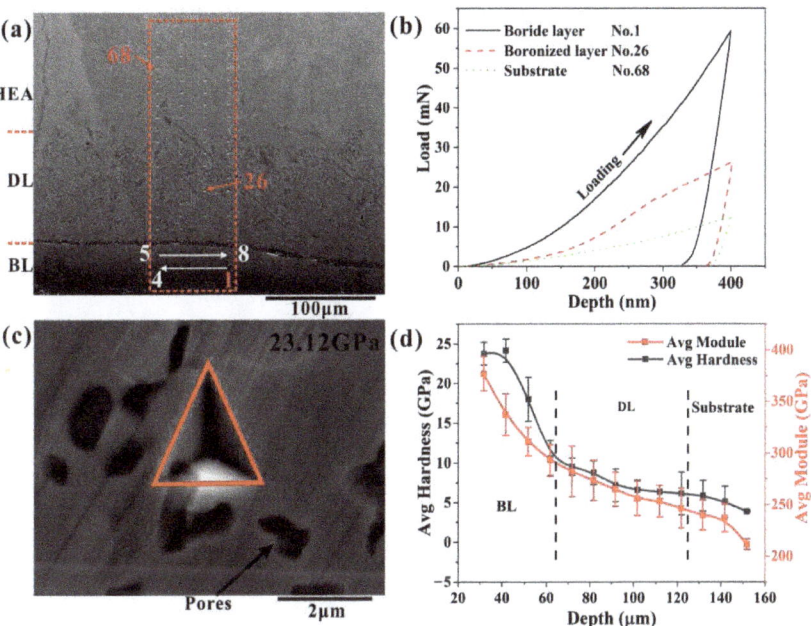

Figure 9. (a) Measurement position of the nanoindentation test, (b) load-displacement curves of the three representative nanoindentation points (Nos.1, 26, 68); (c) the indentation morphology on the single-phase layer; (d) average modulus and hardness vs. distance from the surface.

Table 3. Average modulus and hardness of the boronized sample for 18 h.

Group	Distance from the Surface (μm)	Serial Number	Group Avg Modulus (GPa)	Group Avg Hardness (GPa)
1	31.82	1–4	377.0 ± 16.6	23.8 ± 1.4
2	41.82	5–8	337.1 ± 20.0	24.1 ± 1.5
3	51.82	9–12	311.2 ± 13.7	18.0 ± 2.8
4	61.82	13–16	293.4 ± 14.9	10.6 ± 2.2
5	71.82	17–20	282.3 ± 24.5	9.5 ± 1.1
6	81.82	21–24	273.8 ± 19.5	8.8 ± 1.5
7	91.82	25–28	265.0 ± 16.4	7.2 ± 2.1
8	101.82	29–32	256.2 ± 16.6	6.6 ± 1.3
9	111.82	33–36	253.5 ± 15.2	6.4 ± 0.8
10	121.82	37–40	246.5 ± 19.3	6.2 ± 2.7
11	131.82	41–44	239.8 ± 14.8	5.9 ± 1.9
12	141.82	45–48	237.3 ± 14.0	5.1 ± 2.0
13	151.82	49–80	211.5 ± 5.2	3.9 ± 0.1

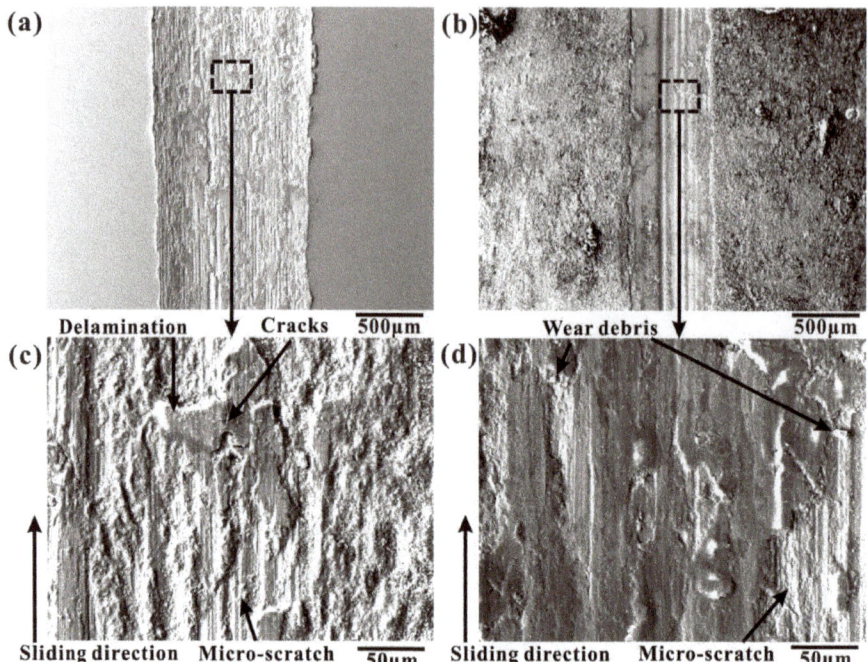

Figure 10. SEM images of the worn surface of the CoCrFeNiMn HEA: (**a,c**) as-cast CoCrFeNiMn HEA; (**b,d**) CoCrFeNiMn HEA boronized at 950 °C for 9 h.

Figure 11 shows the friction coefficient curves of the unboronized and boronized HEA. The friction coefficient of the boronized alloy (0.48~0.61) is lower than that of the unboronized alloy (0.86), and the friction coefficient shows a decreasing trend. Among these boronized samples, the friction coefficient of the 18h sample was the lowest at 0.48, while some broad waves with relatively large fluctuations are observed in the friction coefficient curve of the untreated alloy. Boronizing can effectively improve the wear resistance of HEA surfaces due to their higher surface hardness and larger thickness of the strengthened layer.

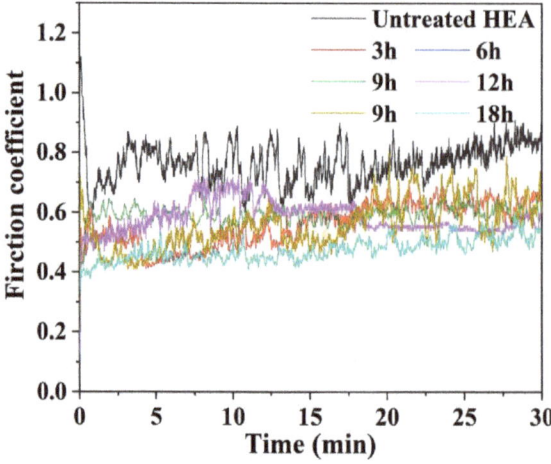

Figure 11. Friction coefficient plots of unboronized and boronized samples.

4. Conclusions

In this work, a reinforced layer on the surface of HEA was successfully synthesized by powder-pack boriding. The microstructure, microhardness, wear resistance, and boriding mechanism were investigated. The following conclusions can be summarized:

(1) The microstructure of the surface layer is mainly composed of MB-type boride and M_2B-type boride. The original interface of the surface layer is located at the interface between the BL and the diffusion layer, where element B diffuses inward and metal elements diffuse outward.

(2) The activation energy and frequency factor of the B element in CoCrFeMnNi HEA are 206.93 kJ/mol and 9.15×10^{-5} m^2/s, respectively. Increasing the boronizing duration and temperature resulted in an increase in the BL and diffusion layer thickness.

(3) The surface strengthening of CoCrFeMnNi HEA was achieved by the boriding treatment. Its surface microhardness has significantly increased from 3.9 ± 0.1 Gpa to 23.8 ± 1.4 Gpa. The surface layer shows a lower friction coefficient of 0.48 than that of the substrate (0.86). Depending on the wear mechanism, adhesive wear mainly occurs in unboronized samples, and abrasive wear is the main wear mechanism in boronized samples.

The findings of this study could provide insights into designing and developing high-hardness and high-wear resistance alloys through surface treatment, which can be used in the cast aluminum industry and allow HEA to be applied in special environments and increase service life.

Author Contributions: Conceptualization, X.O. and M.H.; data curation, M.H., X.Z. and Z.Z.; funding acquisition, F.Y. and X.W.; investigation, M.H., X.Z. and Z.Z.; methodology, X.O. and M.H.; project administration, F.Y., X.O. and X.W.; validation, M.H.; writing—original draft, M.H.; writing—review and editing, X.O. and M.H. All authors have read and agreed to the published version of the manuscript.

Funding: This work was supported by the National Natural Science Foundation of China (No. 52171017 and 51971189) and the Zhuzhou Joint Fund of the Hunan Natural Science Foundation, China (2019JJ60011).

Institutional Review Board Statement: Not applicable.

Informed Consent Statement: Not applicable.

Data Availability Statement: Data will be made available on request.

Conflicts of Interest: The authors declare that they have no known competing financial interests or personal relationships that could have appeared to influence the work reported in this paper.

References

1. Gludovatz, B.; Hohenwarter, A.; Catoor, D.; Chang, E.H.; George, E.P.; Ritchie, R.O. A fracture-resistant high-entropy alloy for cryogenic applications. *Science* **2014**, *345*, 1153–1158. [CrossRef] [PubMed]
2. Otto, F.; Dlouhý, A.; Somsen, C.; Bei, H.; Eggeler, G.; George, E.P. The influences of temperature and microstructure on the tensile properties of a CoCrFeMnNi high-entropy alloy. *Acta Mater.* **2013**, *61*, 5743–5755. [CrossRef]
3. Gali, A.; George, E.P. Tensile properties of high-and medium-entropy alloys. *Intermetallics* **2013**, *39*, 74–78. [CrossRef]
4. Haase, C.; Barrales-Mora, L.A. Influence of deformation and annealing twinning on the microstructure and texture evolution of face-centered cubic high-entropy alloys. *Acta Mater.* **2018**, *150*, 88–103. [CrossRef]
5. Zhang, T.; Xin, L.; Wu, F.; Zhao, R.; Xiang, J.; Chen, M.; Jiang, S.; Huang, Y.; Chen, S. Microstructure and mechanical properties of FexCoCrNiMn high-entropy alloys. *J. Mater. Sci. Technol.* **2019**, *35*, 2331–2335. [CrossRef]
6. Poletti, M.G.; Fiore, G.; Gili, F.; Mangherini, D.; Battezzati, L. Development of a new high entropy alloy for wear resistance: FeCoCrNiW$_{0.3}$ and FeCoCrNiW$_{0.3}$+ 5 at.% of C. *Mater. Des.* **2017**, *115*, 247–254. [CrossRef]
7. Stepanov, N.D.; Yurchenko, N.Y.; Tikhonovsky, M.A.; Salishchev, G.A. Effect of carbon content and annealing on structure and hardness of the CoCrFeNiMn-based high entropy alloys. *J. Alloys Compd.* **2016**, *687*, 59–71. [CrossRef]
8. Sza, B.; Yya, B.; Bza, B.; Zza, B.; Xiao, Y.A.; Zwa, B. Microstructure and wear behaviour of in-situ TiN-Al$_2$O$_3$ reinforced CoCrFeNiMn high-entropy alloys composite coatings fabricated by plasma cladding. *Mater. Lett.* **2020**, *272*, 127870.

9. Ye, W.; Xie, M.; Huang, Z.; Wang, H.; Zhou, Q.; Wang, L.; Chen, B.; Wang, H.; Liu, W. Microstructure and tribological properties of in-situ carbide/CoCrFeNiMn high entropy alloy composites synthesized by flake powder metallurgy. *Tribol. Int.* **2023**, *181*, 108295. [CrossRef]
10. He, F.; Wang, Z.; Wu, Q.; Niu, S.; Li, J.; Wang, J.; Liu, C. Solid solution island of the Co-Cr-Fe-Ni high entropy alloy system. *Scr. Mater.* **2017**, *131*, 42–46. [CrossRef]
11. Nutor, R.K.; Azeemullah, M.; Cao, Q.; Wang, X.; Zhang, D.; Jiang, J. Microstructure and properties of a Co-free $Fe_{50}Mn_{27}Ni_{10}Cr_{13}$ high entropy alloy. *J. Alloys Compd.* **2021**, *851*, 156842. [CrossRef]
12. Cui, Y.; Shen, J.; Manladan, S.M.; Geng, K.; Hu, S. Strengthening mechanism in two-phase FeCoCrNiMnAl high entropy alloy coating. *Appl. Surf. Sci.* **2020**, *530*, 147205. [CrossRef]
13. Tarakci, M.; Gencer, Y.; Calik, A. The pack-boronizing of pure vanadium under a controlled atmosphere. *Appl. Surf. Sci.* **2010**, *256*, 7612–7618. [CrossRef]
14. Dybkov, V.; Goncharuk, L.; Khoruzha, V.; Samelyuk, A.; Sidorko, V. Growth kinetics and abrasive wear resistance of boride layers on Fe–15Cr alloy. *Mater. Sci. Technol.* **2011**, *27*, 1502–1512. [CrossRef]
15. Liu, W.; He, J.; Huang, H.; Wang, H.; Lu, Z.; Liu, C. Effects of Nb additions on the microstructure and mechanical property of CoCrFeNi high-entropy alloys. *Intermetallics* **2015**, *60*, 1–8. [CrossRef]
16. Li, C.; Shen, B.; Li, G.; Yang, C. Effect of boronizing temperature and time on microstructure and abrasion wear resistance of $Cr_{12}Mn_2V_2$ high chromium cast iron. *Surf. Coat. Technol.* **2008**, *202*, 5882–5886. [CrossRef]
17. Cengiz, S.; Thuvander, M. The effect of Hf addition on the boronizing and siliciding behavior of CoCrFeNi high entropy alloys. *Materials* **2022**, *15*, 2282. [CrossRef]
18. Kulka, M. *Current trends in Boriding*; Springer International Publishing: Cham, Switzerland, 2019.
19. Campos-Silva, I.E.; Rodriguez-Castro, G.A. Boriding to improve the mechanical properties and corrosion resistance of steels. In *Thermochemical Surface Engineering of Steels*; Elsevier: Amsterdam, The Netherlands, 2015; pp. 651–702.
20. Contla-Pacheco, A.; Keddam, M.; Lartundo-Rojas, L.; Ortega-Avilés, M.; Mejía-Caballero, I.; Campos-Silva, I. Application of the heat balance integral method to the growth kinetics of nickel boride layers on an Inconel 718 superalloy. *Surf. Coat. Technol.* **2021**, *420*, 127355. [CrossRef]
21. Günen, A. Tribocorrosion behavior of boronized $Co_{1.19}Cr_{1.86}Fe_{1.30}Mn_{1.39}Ni_{1.05}Al_{0.17}B_{0.04}$ high entropy alloy. *Surf. Coat. Technol.* **2021**, *421*, 127426. [CrossRef]
22. Gómez-Vargas, O.; Solis-Romero, J.; Figueroa-López, U.; Ortiz-Domínguez, M.; Oseguera-Peña, J.; Neville, A. Boro-nitriding coating on pure iron by powder-pack boriding and nitriding processes. *Mater. Lett.* **2016**, *176*, 261–264. [CrossRef]
23. Joshi, A.A.; Hosmani, S.S.; Dumbre, J. Tribological performance of boronized, nitrided, and normalized AISI 4140 steel against hydrogenated diamond-like carbon-coated AISI D2 steel. *Tribol. Trans.* **2015**, *58*, 500–510. [CrossRef]
24. Hou, J.; Zhang, M.; Yang, H.; Qiao, J.; Wu, Y. Surface strengthening in $Al_{0.25}$CoCrFeNi high-entropy alloy by boronizing. *Mater. Lett.* **2019**, *238*, 258–260. [CrossRef]
25. Nakajo, H.; Nishimoto, A. Boronizing of CoCrFeMnNi High-Entropy Alloys Using Spark Plasma Sintering. *J. Manuf. Mater. Process.* **2022**, *6*, 29. [CrossRef]
26. Löbel, M.; Lindner, T.; Hunger, R.; Berger, R.; Lampke, T. Precipitation hardening of the HVOF sprayed single-phase high-entropy alloy CrFeCoNi. *Coatings* **2020**, *10*, 701. [CrossRef]
27. Löbel, M.; Lindner, T.; Clauß, S.; Pippig, R.; Dietrich, D.; Lampke, T. Microstructure and Wear Behavior of the High-Velocity-Oxygen-Fuel Sprayed and Spark Plasma Sintered High-Entropy Alloy AlCrFeCoNi. *Adv. Eng. Mater.* **2021**, *23*, 2001253. [CrossRef]
28. Zhu, Y.; Schwam, D.; Wallace, J.F.; Birceanu, S. Evaluation of soldering, washout and thermal fatigue resistance of advanced metal materials for aluminum die-casting dies. *Mater. Sci. Eng. A* **2004**, *379*, 420–431. [CrossRef]
29. Wang, Q.; Wang, W.J.; Liu, H.J.; Zeng, C.L. Corrosion behavior of zirconium diboride coated stainless steel in molten 6061 aluminum alloy. *Surf. Coat. Technol.* **2017**, *313*, 129–135. [CrossRef]
30. Long, Y.; Che, J.; Wu, Z.; Lin, H.-T.; Zhang, F. High entropy alloy borides prepared by powder metallurgy process and the enhanced fracture toughness by addition of yttrium. *Mater. Chem. Phys.* **2021**, *257*, 123715. [CrossRef]
31. Fernández-Valdés, D.; Meneses-Amador, A.; López-Liévano, A.; Ocampo-Ramírez, A. Sliding wear analysis in borided AISI 316L steels. *Mater. Lett.* **2021**, *285*, 129138. [CrossRef]
32. Lindner, T.; Löbel, M.; Sattler, B.; Lampke, T. Surface hardening of FCC phase high-entropy alloy system by powder-pack boriding. *Surf. Coat. Technol.* **2019**, *371*, 389–394. [CrossRef]
33. Milinovi, A.; Marui, V.; Konjati, P.; Beri, N. Effect of Carbon Content and Boronizing Parameters on Growth Kinetics of Boride Layers Obtained on Carbon Steels. *Materials* **2022**, *15*, 1858. [CrossRef] [PubMed]
34. Bischoff, J.; Motta, A.T.; Eichfeld, C.; Comstock, R.J.; Cao, G.; Allen, T.R. Corrosion of ferritic–martensitic steels in steam and supercritical water. *J. Nucl. Mater.* **2013**, *441*, 604–611. [CrossRef]
35. Wagner, C. The distribution of cations in metal oxide and metal sulphide solid solutions formed during the oxidation of alloys. *Corros. Sci.* **1969**, *9*, 91–109. [CrossRef]
36. Cengiz, S. Effect of refractory elements on boronizing properties of the CoCrFeNi high entropy alloy. *Int. J. Refract. Met. Hard Mater.* **2020**, *95*, 105418. [CrossRef]

37. Mrowec, S.; Werber, T.; Zastawnik, M. The mechanism of high temperature sulphur corrosion of nickel-chromium alloys. *Corros. Sci.* **1966**, *6*, 47–68. [CrossRef]
38. Atkinson, A. Surface and interface mass transport in ionic materials. *Solid State Ion.* **1988**, *28*, 1377–1387. [CrossRef]
39. Atkinson, A.; Smart, D. Transport of nickel and oxygen during the oxidation of nickel and dilute nickel/chromium alloy. *J. Electrochem. Soc.* **1988**, *135*, 2886. [CrossRef]
40. Brückman, A.; Emmerich, R.; Mrowec, S. Investigation of the high-temperature oxidation of Fe-Cr alloys by means of the isotope18O. *Oxid. Met.* **1972**, *5*, 137–147. [CrossRef]
41. Brückman, A.; Romanski, J. On the mechanism of sulphide scale formation on iron. *Corros. Sci.* **1965**, *5*, 185–191. [CrossRef]
42. Hales, R.; Hill, A.C. The role of metal lattice vacancies in the hightemperature oxidation of nickel. *Corros. Sci.* **1972**, *12*, 843–853. [CrossRef]
43. Robertson, J. The mechanism of high temperature aqueous corrosion of steel. *Corros. Sci.* **1989**, *29*, 1275–1291. [CrossRef]
44. Terachi, T.; Yamada, T.; Miyamoto, T.; Arioka, K.; Fukuya, K. Corrosion behavior of stainless steels in simulated PWR primary water—Effect of chromium content in alloys and dissolved hydrogen. *J. Nucl. Sci. Technol.* **2008**, *45*, 975–984. [CrossRef]

Disclaimer/Publisher's Note: The statements, opinions and data contained in all publications are solely those of the individual author(s) and contributor(s) and not of MDPI and/or the editor(s). MDPI and/or the editor(s) disclaim responsibility for any injury to people or property resulting from any ideas, methods, instructions or products referred to in the content.

Article

Influence of Al Addition on the Microstructure and Mechanical Properties of Mg-Zn-Sn-Mn-Ca Alloys

Shujuan Yan [1], Caihong Hou [2,*], Angui Zhang [1] and Fugang Qi [2,3,*]

[1] National Energy Group Ningxia Coal Industry Co., Ltd., Yinchuan 750001, China; 0411ysj@163.com (S.Y.); 15056070@chnenergy.com.cn (A.Z.)
[2] School of Materials Science and Engineering, Xiangtan University, Xiangtan 411105, China
[3] Hunan Bangzer Technology Co., Ltd., Xiangtan 411100, China
* Correspondence: houcaihong19@163.com (C.H.); qifugang@xtu.edu.cn (F.Q.); Tel.: +86-156-1673-2585 (F.Q.)

Abstract: The effects of Al addition on the microstructure and mechanical properties of Mg-Zn-Sn-Mn-Ca alloys are studied in this paper. It was found that the Mg-6Sn-4Zn-1Mn-0.2Ca-xAl (ZTM641-0.2Ca-xAl, x = 0, 0.5, 1, 2 wt.%; hereafter, all compositions are in weight percent unless stated otherwise) alloys have α-Mg, Mg_2Sn, Mg_7Zn_3, MgZn, α-Mn, CaMgSn, AlMn, $Mg_{32}(Al,Zn)_{49}$ phases. The grain is also refined when the Al element is added, and the angular-block AlMn phases are formed in the alloys. For the ZTM641-0.2Ca-xAl alloy, the higher Al content is beneficial to elongation, and the double-aged ZTM641-0.2Ca-2Al alloy has the highest elongation, which is 13.2%. The higher Al content enhances the high-temperature strength for the as-extruded ZTM641-0.2Ca alloy; overall, the as-extruded ZTM641-0.2Ca-2Al alloy has the best performance; that is, the tensile strength and yield strength of the ZTM641-0.2Ca-2Al alloy are 159 MPa and 132 MPa at 150 °C, and 103 MPa and 90 MPa at 200 °C, respectively.

Keywords: Mg-Zn-Sn-Mn-Ca alloy; Al; microstructure; mechanical properties

Citation: Yan, S.; Hou, C.; Zhang, A.; Qi, F. Influence of Al Addition on the Microstructure and Mechanical Properties of Mg-Zn-Sn-Mn-Ca Alloys. *Materials* **2023**, *16*, 3664. https://doi.org/10.3390/ma16103664

Academic Editor: Carmine Maletta

Received: 26 April 2023
Revised: 5 May 2023
Accepted: 9 May 2023
Published: 11 May 2023

Copyright: © 2023 by the authors. Licensee MDPI, Basel, Switzerland. This article is an open access article distributed under the terms and conditions of the Creative Commons Attribution (CC BY) license (https://creativecommons.org/licenses/by/4.0/).

1. Introduction

In this century, the rapid development of automobile, aerospace and other fields has brought a series of problems such as serious environmental pollution and excessive energy consumption. For reducing pollution, reducing energy consumption and improving the living environment, environmentally friendly alloy materials have gradually attracted attention [1]. As a light alloy, the Mg alloy has certain reproducibility and excellent properties such as good casting performance and formability, good shock absorption, good cutting machinability, and good electromagnetic shielding [2,3]. For the past few years, Mg alloys have gradually become the focus of the development and application of new materials [4–6]. However, the Mg alloy also has many deficiencies. Firstly, the Mg alloy is easily oxidized at room temperature, and the oxide layer is loose and porous, which greatly limits its application in industrial production. Secondly, the Mg alloy is more active and easy to burn, which makes it necessary to pay attention to the safe production in practical applications. Finally, although the specific strength of the Mg alloy is high, its absolute strength is low, and the high temperature strength is especially poor, which also limits its application. In recent years, many scholars have improved the properties by alloying, that is, by adding a trace or small amount of other elements to the Mg alloy during the cast process.

Due to the addition of different alloying elements, many kinds of alloy systems with different compositions and properties are formed. Among them, the Mg-Zn alloy is a good aging strengthening alloy due to the precipitation of the $MgZn_2$ phase [7,8]. The Mg-Sn alloys have a smaller composition span during the smelting process, so that the alloys have fewer defects and better performance [9]. However, it is often not enough to rely on the strengthening effect of a single strengthening phase. In addition, because the strengthening phase is single, the aging strengthening effect is not obvious in the subsequent heat treatment. Therefore, the Mg-Zn and Mg-Sn binary alloy are rarely applied to industrial production. Generally speaking, the alloy's properties are increased by adding other elements to the binary alloy. The Mg-Zn-Sn alloys are formed by combining the Mg-Zn alloy and the Mg-Sn alloy, and it has the advantages of both alloys. The development of Mg-Zn-Sn alloys have two major advantages: on the one hand is the phenomenon whereby the Mg alloy has poor strength and creep properties at high temperatures because the low melting point phase can be changed. On the other hand, these Mg-Zn-Sn alloys have a wider range of applications, because the price of Zn and Sn elements is lower, and the content on the earth is more, which is convenient for research and development. Therefore, the Mg-Zn-Sn alloys are promising high property deformed magnesium alloys. Based on the existing cognition of the strengthening mechanism of Mg-Zn-Sn alloys, these mechanical properties of Mg-Zn-Sn deformed alloys are further enhanced via adding some other elements or changing some heating treatments. The Mn element can remove harmful elements (such as Fe) in alloy melting, improve the corrosion resistance and casting performance of the alloy, and the Mn element also has a certain ability to refine grain [10].

The Ca element is a common alloying element. The Ca element will nucleate and precipitate on grain boundaries, which inhibit the growth of grains and play the role of grain refinement. At the same time, the Mg_2Ca phase with good thermal stability is formed when an appropriate amount of Ca is added to the Mg alloy, which can hinder the grain boundaries slip at higher temperature and improve the creep resistance of the alloy [11,12]. It was found that the Mg-Ca alloy has excellent corrosion resistance, high temperature creep performance, and high temperature mechanical properties [13]. Baghani et al. [14] found that when Ca > 2wt.% is added into the alloy, the CaMgSn and Mg_2Ca phases are generated, which correspondingly reduces the content of the Mg_2Sn phase. Through this transformation, the properties of the Mg-4Sn alloy are obviously improved. At the same time, many studies have shown that the Ca element should not be too high in the Mg alloy. If the Ca element is too high, the hot cracking tendency will be increased, and the performance will be worse. Therefore, 0.2% Ca element was selected to be added to the alloy in this paper. Al is also a common alloying element. It was found that the appropriate amount of Al will refine the grain and enhance the alloys' casting properties, and the Mg-Al alloy has good solution strengthening and ageing strengthening effects [15]. Jayaraj et al. [16] found that the Al element will effectively enhance the aging strengthening effect for the Mg-Ca alloy, and then effectively enhance the properties of the Mg-Ca alloy. Wang et al. [17] studied the microstructure and mechanical properties of an as-extruded and aged Mg-Zn-Al-Sn alloy, and found that the comprehensive properties of the Mg-4Zn-2Al-2Sn alloy reached the optimal level after aged treatment at 150 °C for 40 h. Wei et al. [18] researched the effect of different contents of Ca on the microstructure and properties of the Mg-4.5Zn-4.5Sn-2Al alloy, and found that the addition of an appropriate amount of Ca can refine the microstructure of the alloy; among them, the Mg-4.5Zn-4.5Sn-2Al-0.2Ca alloy has the best mechanical properties. Our research group has studied the Mg-Sn-Zn-Mn alloy, and found that the Mg-6Sn-4Zn-1Mn alloy has the best performance, and thus the Mg-6Sn-4Zn-1Mn-Al alloy and the Mg-6Sn-4Zn-1Mn-Ca alloy were chosen for study. However, the effect of the Al element on the Mg-6Sn-4Zn-1Mn-Ca alloy is still not fully understood. In order to study the effect of the Al element on the Mg-Zn-Sn-Mn-Ca alloy, and explore the potential properties of the Mg-Zn-Sn-Mn-Ca-xAl alloy, the different contents of Al are added to the Mg-6Zn-4Sn-1Mn-0.2Ca alloy, and the microstructure and properties of Mg-6Zn-4Sn-1Mn-0.2Ca-xAl alloy are investigated. It is expected that these studied alloys can be applied to the field of high-strength heat resistant deformation materials.

2. Experiment

The furnace temperature was set at 720 °C, and the mixed gas of CO_2 and SF_6 acted as the protective gas. Pure Mg ($\geq$99.9 wt.%), Mg-4.10 wt.% Mn, pure Al ($\geq$99.9 wt.%), pure Sn ($\geq$99.9 wt.%), pure Zn ($\geq$99.9 wt.%) and Mg-30 wt.% Ca were added successively. After the melting was completed, the ingot was poured into the mold to form. After that, the as-cast alloys were homogenized, that is, the as-cast alloy was placed in a heating furnace at 330 °C and keep at a constant temperature for 24 h. The homogenization treatment can eliminate the ingot's ingot composition and microstructure inhomogeneity, and can preheat the ingot for plastic deformation. The plastic deformation process adopted in this paper was extrusion treatment, because compared with deformation processes such as forging and rolling, the extrusion treatment process is subjected to compressive stress in three directions at the same time, which can make the plasticity of the alloy play out to the maximum. In this experiment, the extrusion equipment was a 500 t horizontal extrusion press, the extrusion process was forward extrusion, and the diameter × length of the extrusion cylinder was φ 80 mm×450 mm. The detailed extrusion parameters are shown in Table 1. After extrusion, solution treatment and aging treatment will further enhance the alloys' mechanical properties. This solution treatment is adopted in this paper on 440 °C/2 h, and the aging treatment includes single-stage aging and two-stage aging, respectively. The detailed process is shown in Table 2.

Table 1. Extrusion parameters for the studied alloys.

Test Materials	Billet Temperature (°C)	Extrusion Chamber Temperature (°C)	Mold Hole Diameter (mm)	Extrusion Speed (m/min)	Extrusion Ratio
Mg-6Zn-4Sn-1Mn-0.2Ca-xAl	360	350	16	2	25

Table 2. Heat treatment parameters for the studied alloys.

Type	Solution Treatment	Aging Treatment
T4	440 °C/2 h	—
T4+single aging		180 °C/12 h
T4+double aging		90 °C/24 h + 180 °C/8 h

In this paper, an U1tima IV X-ray diffractometer was chosen to identify the experimental alloys' phases. The specific test parameters are as follows: the scanning angle was 10~90°, and the scanning speed was 4°/min. The microstructure of the alloy was observed by optical microscopy, scanning electron microscopy and transmission electron microscopy. The optical microscope used was an Olympus BX53M. The model of the scanning electron microscope was a JSM-6360, and the scanning electron microscope was combined with an Energy Dispersive Spectrometer (EDS) for point measurement, line scanning, and surface scanning at the same time. The EDS used was an Oxford INCA Energy 350. Secondary electron scanning electron microscope (SE-SEM) and back scattered electron scanning electron microscope (BSE-SEM) were also used. A FEI Tecnai G2 F20 Transmission Electron Microscope (TEM) was used, and the alloy sample was prepared according to the needs of the Bright field image (Bright field, BF) and the high-angle annular dark-field scanning transmission electron microscope (HAADF-STEM).

In order to evaluate the comprehensive properties of the alloy, the extrusion state and the aging state of the alloy are selected to carry out a unidirectional tensile test at room temperature. The room temperature performance test followed the ASTM B557M-02 standard. The experimental instrument used was a Xin think CMT-5105 microcomputer controlled electronic universal testing machine. The unidirectional tensile test was carried out at uniform speed, and the tensile rate was 3 mm/s. Before stretching, according to the GB228-2002 standard, the stretching sample was made into the stretching key in Figure 1a. On the basis of the room temperature tensile test of the developed alloy, the extruding alloy was further selected. According to the GB228-2002 standard, the tensile bond was made, as shown in Figure 1b. After that, the high temperature performance tests were carried out at 150 °C and 200 °C, respectively, using an Instron 3369 electronic universal material testing machine. The high temperature performance test followed the ASTM E21→92(1998) e1 standard. We adopted uniform unidirectional drawing, held the heat for 10min before drawing, and the drawing speed was 2 mm/min.

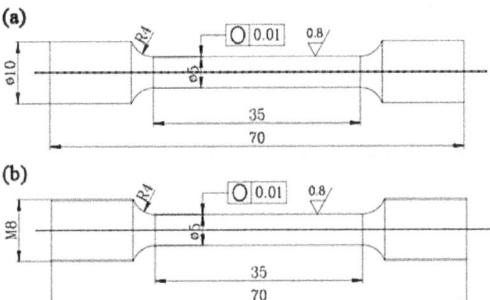

Figure 1. The schematic of the tensile sample at room and high temperature, (a) room temperature, (b) high temperature.

3. Results and Discussion

3.1. As-Cast Microstructure

One characteristic of the Mg alloy is its low density. In order to confirm whether the density of the studied alloy still meets the requirements of the light alloy, the density test was conducted on the as-cast ZTM641-0.2Ca-xAl (x = 0, 0.5, 1, 2) alloy, and the test results are shown in Table 3. It can be seen from the table that the density of the as-cast ZTM641-0.2Ca is 1.7666 g/cm^3, the density of ZTM641-0.2Ca-0.5Al is 1.8240 g/cm^3, and the density of ZTM641-0.2Ca-1Al is 1.8284 g/cm^3. The density of ZTM641-0.2Ca-2Al is 1.8359 g/cm^3. Generally speaking, the alloy density increases with the increase in Al content.

Table 3. The density of as-cast ZTM641-0.2Ca-xAl alloys.

Alloys	ZTM641-0.2Ca	ZTM641-0.2Ca-0.5Al	ZTM641-0.2Ca-1Al	ZTM641-0.2Ca-2Al
Density (g/cm^3)	1.7666	1.8240	1.8284	1.8359

Figure 2 is the XRD graph of the as-cast ZTM641-0.2Ca xAl (x − 0, 0.5, 1, 2) alloy. From the figure, we can see that the ZTM641-0.2Ca alloy mainly has α-Mg, Mg$_2$Sn, Mg$_7$Zn$_3$, MgZn, α-Mn and CaMgSn phases. As 0.5%Al is added to the alloy, the new diffraction peak of the AlMn phase appears. As the Al element continues to increase to 1%, the diffraction peak of the AlMn phase is significantly enhanced, which indicates that the number of AlMn phases in the ZTM641-0.2Ca-1Al alloy increases significantly. When the Al element continues to increase, a new diffraction peak of Mg$_{32}$(Al,Zn)$_{49}$ phases is found in the ZTM641-0.2Ca-2Al alloy, but the diffraction peak intensity of the AlMn phases is slightly weakened.

Figure 3 is the Optical graphs of as-cast ZTM641-0.2Ca-xAl (x = 0, 0.5, 1, 2) alloys. From the figure, we can see that the as-cast microstructure mainly consists of an Mg matrix, dendrites, and a dispersed-distributed second phase. Compared with Figure 3a–d, the secondary growth of the dendrite is more sufficient, and the dendrite distribution is more dense with the addition of Al from 0 to 2%. This is due to the formation of the component supercooling of the Al element at the liquid-solid interface. When the amount of the Al element increases, the composition undercooling is obvious, which accelerates the growth rate of the dendrite tip, and then leads to the spacing decrease in the secondary dendrite.

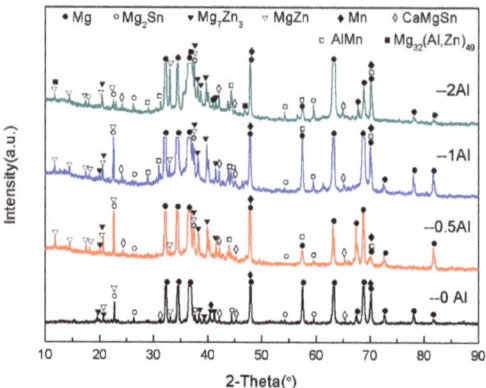

Figure 2. XRD patterns of as-cast ZTM641-0.2Ca-xAl (x = 0, 0.5, 1, 2) alloy.

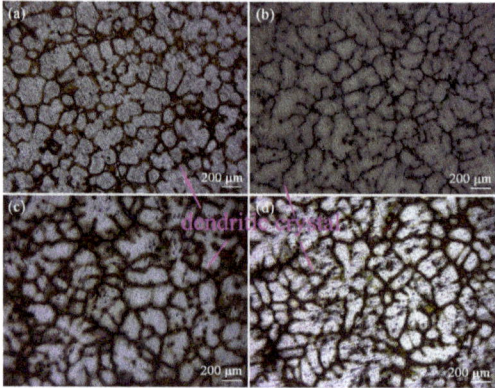

Figure 3. Optical micrographs of as-cast ZTM641-0.2Ca-xAl alloys: (**a**) x = 0; (**b**) x = 0.5; (**c**) x = 1; (**d**) x = 2.

Figure 4 shows the SEM graphs of the dendrites of the as-cast ZTM641-0.2Ca-xAl (x = 0.5, 1, 2) alloy. It can be seen that the dendrite of the ZTM641-0.2Ca-0.5Al alloy mainly has two phases, one being the Mg_2Sn phase and the other the Mg_7Zn_3 phase, in which the Mg_7Zn_3 phase shows a network structure. As the Al element increases to 1%, the type for the dendrite phases does not change, but the morphology of the Mg_7Zn_3 phase changes significantly, and its network structure is obviously refined, as depicted in Figure 4b. When the Al content continues to increase, the morphology and type of dendrite phase are changed for the ZTM641-0.2Ca-2Al alloy. It mainly has an Mg_2Sn phase, an Mg_7Zn_3 phase, and an $Mg_{32}(Al,Zn)_{49}$ phase. Meanwhile, the network structure of the Mg_7Zn_3 phase is further refined, as shown in Figure 4c. The gradual refinement of the network structure of the Mg_7Zn_3 phase may result from the aliquation of the Al element on the Mg-Zn eutectic compound. According to the EDS data for point A, B and C, The Mg_7Zn_3 phase contains different Al content, and the greater the Al element, the finer the network structure.

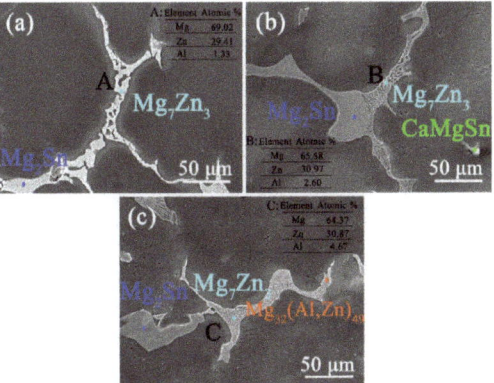

Figure 4. BSE-SEM micrographs of as-cast ZTM641-0.2Ca-xAl alloys: (**a**) x = 0.5; (**b**) x = 1; (**c**) x = 2.

3.2. As-Homogenized Microstructure

Figure 5 shows the SEM graphs for the as-homogenized ZTM641-0.2Ca-xAl (x = 0.5, 1, 2) alloys. It shows that the microstructure is made up of the Mg matrix, the dendritic phase, and the dispersed second phase. Compared to an as-cast microstructure, the as-homogenized dendritic phases are of discontinuous distribution. This indicates that a part of the dendrites has been dissolved in the matrix. Figure 5d is an enlargement of the green box in Figure 5c. As we can see in this graph that a large number of white fine points are distributed around the dendrite, showing a gray transition shape, this indicates that a part of the dendrite structure is melted by the heat treatment, while the white fine points are residual eutectic compounds.

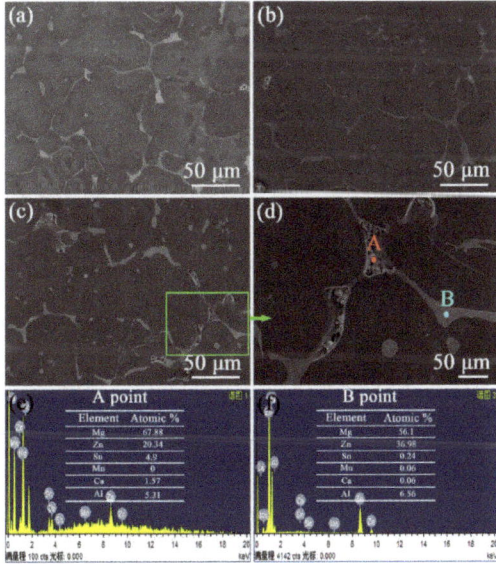

Figure 5. BSE-SEM micrographs of as-homogenized ZTM641-0.2Ca-xAl alloys: (**a**) x = 0.5, (**b**) x = 1, (**c**) x = 2, (**d**) Enlargement of circular rectangular part in (**c**); (**e**,**f**) Corresponding EDS results of points in (**d**).

In order to determine the composition of the dendrite structure, an EDS analysis was conducted, and the results are shown in Figure 5e,f. We can see that the composition of point A is mainly Mg, Zn, Sn, Ca and Al, while point B mainly contains Mg and Zn elements, mixed with a small amount of the Al element. Combined with the XRD results, it can be determined that point A and B are of the Mg_7Zn_3 phase.

3.3. As-Extruded Microstructure

Table 4 shows the density of the as-extruded ZTM641-0.2Ca-xAl (x = 0, 0.5, 1, 2) alloy. It can be seen from the table that the density of as-extruded ZTM641-0.2Ca-0.5Al is 1.8318 g/cm^3. The density of as-extruded ZTM641-0.2Ca-1Al and ZTM641-0.2Ca-2Al is 1.8469 g/cm^3 and 1.8641 g/cm^3, respectively. The density increases with the increase of the Al content, which is the same as that of the as-cast alloy. However, the density of the as-extruded alloy is slightly higher. This is because after the homogenization and extrusion treatment, the as-extruded composition is more uniform and the microstructure is more closely distributed, which can be confirmed in the subsequent microstructure analysis.

Table 4. The density of as-extruded ZTM641-0.2Ca-xAl alloys.

Alloys	ZTM641-0.2Ca	ZTM641-0.2Ca-0.5Al	ZTM641-0.2Ca-1Al	ZTM641-0.2Ca-2Al
Density (g/cm^3)	1.7889	1.8318	1.8469	1.8641

Figure 6 shows the XRD patterns for the as-extruded ZTM641-0.2Ca-xAl (x = 0, 0.5, 1, 2) alloy. According to the results, the as-extruded ZTM641-0.2Ca-xAl alloys mainly consist of the α-Mg, Mg_2Sn, Mg_7Zn_3, MgZn, α-Mn, CaMgSn, $MgZn_2$, AlMn, and $Mg_{32}(Al,Zn)_{49}$ phases. Unlike the as-cast alloys, these fine diffraction impurity peaks in the as-extruded alloys are significantly reduced after homogenization treatment and extrusion plastic deformation. At the same time, a new diffraction peak of the $MgZn_2$ phase appears in the as-extruded alloy, and the diffraction peak intensity of the $MgZn_2$ phases is significantly weakened after the addition of the Al element, which may be due to the combination of Al with the Mg and Zn elements to generate the $Mg_{32}(Al,Zn)_{49}$ phase, which consumes part of Zn element. The variation of diffraction peak intensity of the other phases is similar to that of an as-cast alloy.

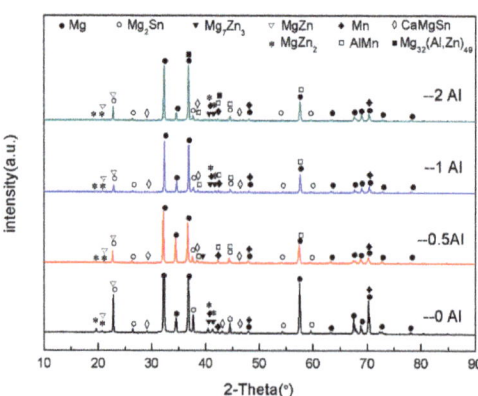

Figure 6. XRD patterns of as-extruded ZTM641-0.2Ca-xAl (x = 0, 0.5, 1, 2) alloy.

Figure 7a–d are the longitudinal metallographic for the as-extruded ZTM641-0.2Ca-xAl (x = 0, 0.5, 1, 2) alloys. It shows that dynamic recrystallization occurs after the hot extrusion treatment, and these bulk second phases are broken into small particles; these small particles are streamlined distribution. Compared with figure (a), (b), (c) and (d), it was determined that the recrystallized grain size of the ZTM641-0.2Ca-1Al alloy was the smallest. According to the research of Humphreys F J [19], the large second phase will cause lattice distortion, produce large-angle grain boundaries, promote the crystal grains nucleation, increase the number of crystal grains, and significantly reduce the crystal grain diameter. For studying these second phases of the as-extruded alloys, the as-extruded ZTM641-0.2Ca-2Al alloy was chosen for SEM observation. The observation results are shown in Figure 7e. It can be seen that the white second phase particles are distributed into the matrix with a streamlined distribution, and some large second phases are also observed. Through the EDS results in Figure 7f, as we can see that point A mainly contains three elements: Mg, Zn and Al. According to the XRD pattern in Figure 6, the second phase can be determined as the $Mg_{32}(Al,Zn)_{49}$ phase. Point B mainly contains the three elements of Mg, Al and Mn, and Al:Mn = 1:1. Combined with the XRD pattern, the second phase can be determined as the AlMn phase.

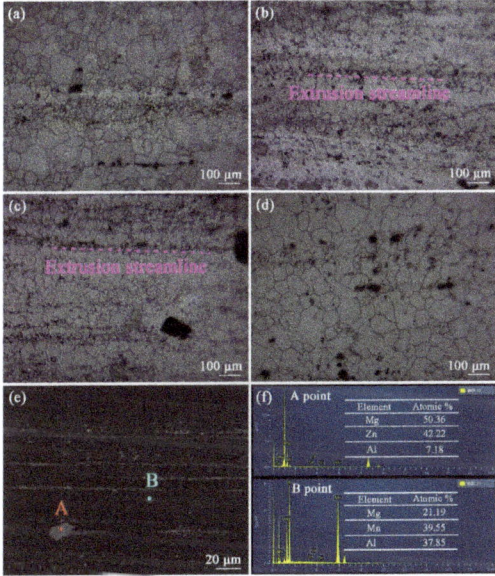

Figure 7. Optical micrographs of as-extruded ZTM641-0.2Ca-xAl alloys: (**a**) x = 0; (**b**) x = 0.5; (**c**) x = 1; (**d**) x = 2; (**e**) BSE-SEM micrographs and (**f**) EDS results of as-extruded ZTM641-0.2Ca-2Al alloys.

Figure 8 shows the TEM micrographs of the as-extruded ZTM641-0.2Ca-2Al alloy. As shown in Figure 8a, the microstructure of the as-extruded ZTM641-0.2Ca-2Al alloy contains some rod-like phases of about 1 μm in diameter, and the distribution is relatively concentrated. The aggregation of rod-like phases may be due to the large number of boundary surfaces of the rod-like phases, and the aggregation growth can reduce part of the interface. According to the minimum energy criterion, the decrease of interface energy can make the system tend to a more stable state. The EDS results and XRD patterns confirm that these rod-like phases are MgZn phases. The rod-like MgZn phase is also observed in Figure 8b. Through the EDS results, the CaMgSn phase is observed in the dislocation pile-up. This indicates that the Ca element and Sn element are enriched here. This is due to the lower energy required for the movement of atoms in dislocations, which makes it easier for Ca to combine with the Sn and Mg atoms. Therefore, the CaMgSn phase is nucleated and precipitated with dislocation and dislocation entanglement.

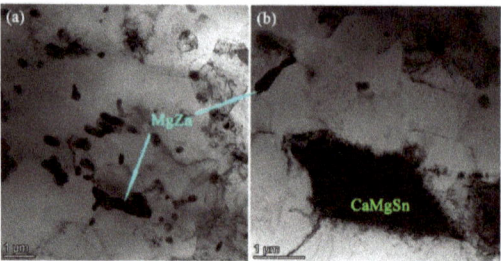

Figure 8. TEM micrographs of as-extruded ZTM641-0.2Ca-2Al alloy.

3.4. Solution-Treated and Aged Microstructures

The phases of the solid solution ZTM641-0.2Ca-xAl (x = 0, 0.5, 1, 2) alloy are analyzed by XRD patterns. The results are shown in Figure 9. Among them, the solid solution ZTM641-0.2Ca alloy mainly consists of α-Mg, Mg_2Sn, $MgZn_2$, α-Mn and CaMgSn phases. As 0.5%Al is added to the alloy, the ZTM641-0.2Ca-0.5Al alloy is α-Mg, Mg_2Sn, $MgZn_2$, α-Mn, CaMgSn and AlMn phases, and the Al element is mainly used to form the AlMn phase. As the Al content increases to 1%, the phase type does not change. As the Al element continues to increase to 2%, the phase of ZTM641-0.2Ca-2Al changes into the α-Mg, Mg_2Sn, $MgZn_2$, α-Mn, CaMgSn, AlMn and $Mg_{32}(Al,Zn)_{49}$ phases. Due to the further increase of the Al content, the Al element is mainly used to form the AlMn phase and the $Mg_{32}(Al,Zn)_{49}$ ternary phase. Compared with the as-extruded alloys, the diffraction peaks of the Mg_7Zn_3 and MgZn phases are not detected. This indicates that these two phases all melt into the matrix, and the Zn element is mainly combined with Mg and Al to form the $MgZn_2$ phase and the $Mg_{32}(Al,Zn)_{49}$ phase. Comparing the several curves, it is found that when the Al element increases, the diffraction peak intensity of the AlMn phase increases significantly.

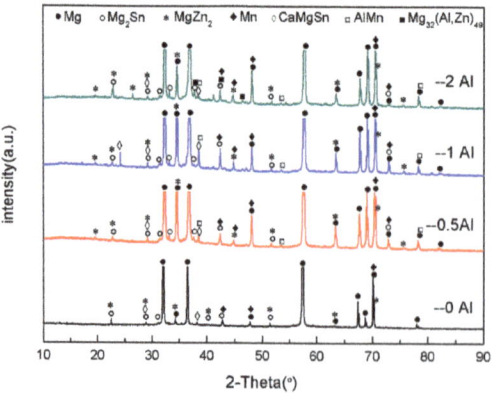

Figure 9. XRD patterns of solid solution ZTM641-0.2Ca-xAl (x = 0, 0.5, 1, 2) alloys.

Figure 10 shows the optical micrographs of the solid solution ZTM641-0.2Ca-xAl (x = 0, 0.5, 1, 2) alloy. According to the figure, the microstructure for the solid solution alloy includes the Mg matrix, grain, and second phases. After this solution treatment, the recrystallized grain grows further, and the average grain size decline with the Al element rises, although the change is not very obvious, which indicates that the Al element plays a certain role in grain refinement for the alloys. The grain refinement may be due to the combination of the Al element and the Mn element to form the AlMn phase. The AlMn phase preferably precipitates and concentrates at the front-end α-Mg phase, which pins the grain boundary and hinders the grain growth. For the ZTM641-0.2Ca alloy, the majority of second phases have been dissolved into the matrix, as shown in Figure 10a. When 0.5%Al is being added, some black second phase with a diameter of about 5 μm is observed in the microstructure, a large number of them are distributed on the grain boundary, and a few are distributed in the grain. As the Al element increases to 1%, the number of second phase particles rise, and the diameters of these particles rise to about 10 μm. All of them are distributed on the grain boundaries, and some of them are of an aggregated distribution. When the Al content continues to increase to 2%, the second phase particle continues to grow.

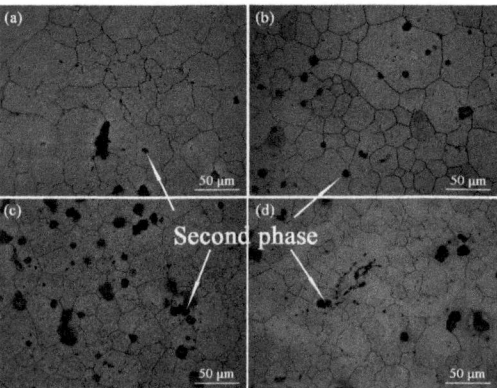

Figure 10. Optical micrographs of solid solution ZTM641-0.2Ca-xAl alloys: (**a**) x = 0; (**b**) x = 0.5; (**c**) x = 1; (**d**) x = 2.

To observe the microstructure and morphology for solid solution alloys from a clearer perspective, the solid solution ZTM641-0.2Ca-0.5Al alloy is selected for scanning electron microscope observation. The observation results are shown in Figure 11a. We can see that the solid solution structure has substrate and second phase particles, and the particle sizes are not unique, among which the particles with a small diameter are point-like. They are mainly located in the grain boundary, although a small portion of them are located in the grain interior. The second phase, which is larger in size, is angular and massive, mainly distributed on the grain boundary. Figure 10b is the red box's magnification in Figure 10a. We can clearly observe that these second phases of the bright white angular block are located at the intersection of several grain boundaries. The grain diameter is about 5 μm. In order to determine the specific composition of the second phase particles in the figure, the region shown in Figure 10b was scanned by the energy spectrum, as shown in Figure 11. It can be observed that Mg mainly exists in the matrix. Zn appears in the matrix and second phases. Sn elements are also segregated in the matrix and second phases. The Mn element appears to be segregated in the second phase and disperses in the matrix. The Ca elements are uniformly distributed in the matrix and second phases. This distribution of Al elements in the matrix is uniform, and there is segregation in the second phase particles. In general, these second phases of the angular block are AlMn phases.

After solid solution treatment, the ZTM641-0.2Ca-xAl (x = 0, 0.5, 1, 2) alloy is subjected to artificial aging treatment. Because there is no obvious difference in the microstructure between one-stage and two-stage aged alloys, we selected the two-stage aged alloy as that representing the microstructure analysis in this paper. Figure 12 is the optical micrograph of the two-stage aged ZTM641-0.2Ca-xAl (x = 0, 0.5, 1, 2) alloy. We can see that these second phase particles at the microstructure are less for the ZTM641-0.2Ca-0.5Al alloy. The second phase particles increased significantly for the ZTM641-0.2Ca-1Al alloy, most of which are located on the grain boundary, as shown in Figure 12c. When the Al element increases to 2%, the more second phases particles that appear in the microstructure, the more diffuse is the distribution, the majority of them are located in grain boundaries, and a few of them are distributed in grains. Compared with Figure 12a–d, it was found that the grain size declines when the Al element increases, and the grain size of the ZTM641-0.2Ca-2Al alloy is the smallest.

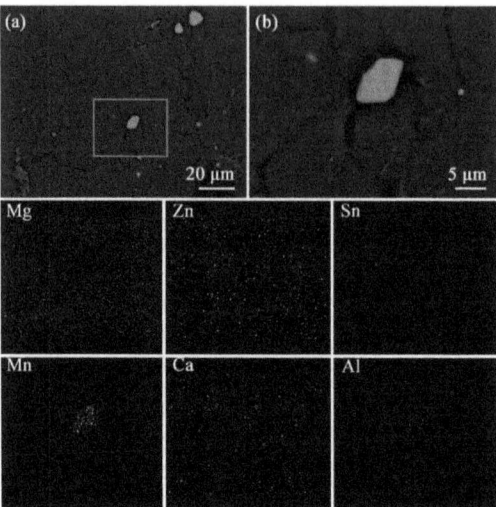

Figure 11. BSE-SEM micrographs and mapping of solid solution ZTM641-0.2Ca-0.5Al alloys: (**a**) BSE-SEM micrograph, (**b**) Enlargement of the rectangular part in (**a**).

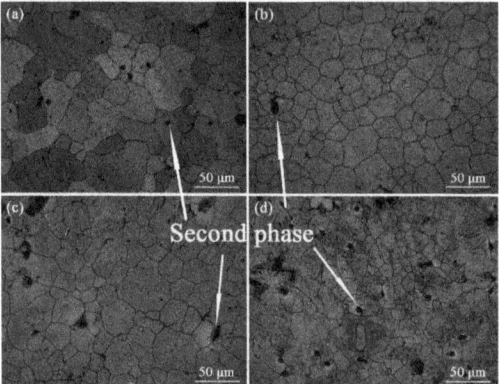

Figure 12. Optical micrographs of two-stage aged ZTM641-0.2Ca-xAl alloys: (**a**) x = 0; (**b**) x = 0.5; (**c**) x = 1; (**d**) x = 2.

Figure 13 shows the XRD patterns of the two-stage aged ZTM641-0.2Ca-xAl (x = 0, 0.5, 1, 2) alloy. From this figure, we can see that the phase composition of the two-stage aged alloy are mainly the α-Mg, Mg_2Sn, $MgZn_2$, α-Mn, CaMgSn, AlMn and $Mg_{32}(Al,Zn)_{49}$ phases. The addition of Al mainly forms the AlMn phase and the $Mg_{32}(Al,Zn)_{49}$ phase with the other elements in the alloy. The variation of other precipitated phases is similar to that of the solid solution alloy, but these diffraction peaks of Mg_2Sn and $MgZn_2$ phases for the two-stage aged alloy are obviously wide, which indicates that the particle size for these two precipitated phases is small.

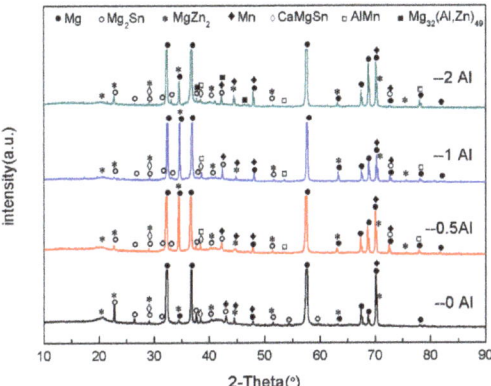

Figure 13. XRD patterns of two-stage aged ZTM641-0.2Ca-xAl (x = 0, 0.5, 1, 2) alloy.

To clearly observe the microstructure of the aged alloy, the two-stage aged ZTM641-0.2Ca-xAl (x = 0.5, 1, 2) alloy is observed in these SEM images. As shown in Figure 14, the microstructure of the alloys mainly consists of the Mg matrix and second phases. Among them, the ZTM641-0.2Ca-0.5Al alloy has a large second phase with a diameter of about 10 μm and a blocky morphology. In addition, most of these second phases are located on grain boundaries, while a few are located at the grain. For the ZTM641-0.2Ca-1Al alloy, the number of large second phase particles of these alloys increases, while the particles' morphology and distribution are similar to the ZTM641-0.2Ca-0.5Al alloy. For the ZTM641-0.2Ca-2Al alloy, the number of second phase particles of these alloys increases further, and the morphology is still angular and blocky. It can be seen in Figure (c) that these second phase particles are mainly located on the grain boundary.

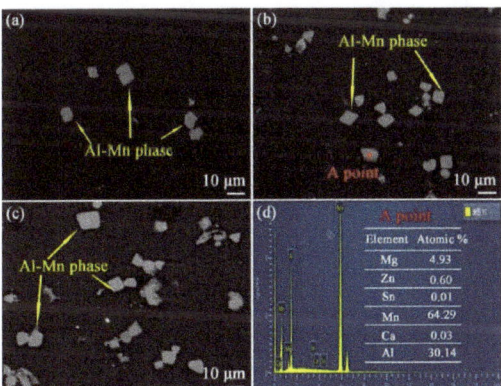

Figure 14. BSE-SEM micrographs of two-stage aged ZTM641-0.2Ca-xAl alloys: (**a**) x = 0.5, (**b**) x = 1, (**c**) x = 2; (**d**) The corresponding EDS result of the A point.

In order to determine the composition of these second phases, EDS testing was chosen, and these results are shown at Figure 14d. As shown in the graph, the elemental composition of the second phase in point A is mainly Mg, Al and Mn. Due to the second phase being located on the matrix, the Mg element will inevitably exist in the EDS results. Therefore, this second phase is the Al-Mn phase. Compared with Figures 10 and 12, the aged alloy has more second phase particles. This is due to most of the Al and Mn atoms being dissolved into the matrix during the solid solution process. When the alloy is subjected to an aging treatment, the Al and Mn atoms will continue to precipitate with the form of

Al-Mn phase. It is worthy of note that in the EDS result in Figure 14d, the atomic ratio of the precipitated phase is Al:Mn = 1:2, but the XRD graph in Figure 13 shows that the Al-Mn phase is the AlMn phase. These results conflict with each other. According to the previous literature [20], some researchers think that the Al-Mn phase is the Al_8Mn_5 phase. In this paper, this phase was tentatively named the Al-Mn phase after thorough consideration.

3.5. Mechanical Properties

Figure 15 shows the mechanical properties of the as-extruded ZTM641-0.2Ca-xAl (x = 0, 0.5, 1, 2) alloy at room temperature. As can be seen, the as-extruded ZTM641-0.2Ca alloy has the highest comprehensive mechanical property; that is, the UTS, YS and elongation are 336 MPa, 230 MPa and 14.5%. According to our previous research, the UTS, YS and the elongation of the Mg-6Sn-4Zn-1Mn alloy are 328 MPa, 255 MPa and 10.76%, respectively. When 0.5% Al was added, the properties of the alloy decreased slightly, with the UTS, YS and elongation at 329 MPa, 224 MPa and 11.9%, respectively. When the Al content increases to 1%, the UTS and YS decrease to 327 MPa and 221 MPa, but the elongation increases to 13%. When the Al content is 2%, the UTS is 324 MPa, the YS is 218 MPa, and the elongation is 14%. In general, the properties of the as-extruded ZTM641-0.2Ca-xAl (x = 0, 0.5, 1, 2) alloy decreased slightly with the increase of Al content, but the overall difference was not significant. The reason for this variation trend is that the Al element was added to the alloy and combined with the Mn element in the alloy to form the AlMn phase. According to the previous microstructure analysis, the higher Al content, the larger the second phase, and the worse binding force with the matrix. When the alloy is deformed, the coordinating deformation ability of the alloy is poor, resulting in a decrease in the strength of the alloy.

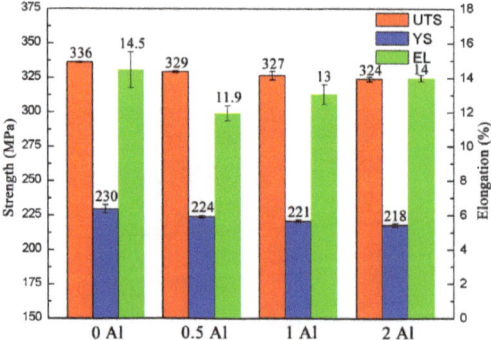

Figure 15. Mechanical properties of as-extruded ZTM641-0.2Ca-xAl alloys tested at room temperature.

Figure 16 shows the strength and plasticity of the aged ZTM641-0.2Ca-xAl (x = 0, 0.5, 1, 2) alloy at room temperature, and Figure (a) and Figure (b) correspond to the one-stage aging and two-stage aging, respectively. From Figure (a), we can see that the one-stage aged ZTM641-0.2Ca alloy has the highest strength, with the UTS and YS being 392 MPa and 365 MPa, respectively. According to our previous research [21], the UTS and YS of the one-stage aged ZTM641-0.2Al alloy are 382 MPa and 352 MPa, and the UTS and YS of one-stage aged ZTM641-0.5Al alloy are 350 MPa and 299 MPa. When 0.5% Al is added to the alloy, the strength and plasticity show a decreasing trend, while the UTS and YS are 365 MPa and 325 MPa, respectively. When the Al content increases to 1%, the UTS and YS continually decrease to 350 MPa and 304 MPa, respectively. In addition, the UTS and YS of the ZTM641-0.2Ca-2Al alloy are further reduced to 333 MPa and 269 MPa, respectively. On the other hand, with the gradual increase of the Al content, the elongation of the alloy gradually increases. When the Al content increases to 2%, the elongation increases significantly, to 13.2%. For the two-stage aged alloy, the variation of properties is consistent with that of the one-stage aged alloy. As shown in Figure 16b, the two-stage aged ZTM641-0.2Ca alloy also has the highest strength, with UTS and YS at 407 MPa and 392 MPa, respectively. According to our previous research, the UTS and YS of the two-stage aged ZTM641-0.2Al alloy are 384 MPa and 360 MPa. When 0.5%Al is added to the alloy, the alloy strength decreases slightly, and the UTS and YS are 369 MPa and 328 MPa, respectively. When the Al content continues to increase, the strength of the alloy also continues to decrease. When the Al content is 2%, the UTS and YS of the alloy decreases to 350 MPa and 305 MPa. When the Al atom content is 2%, the elongation of the alloy is the best, at 12%. The gradual improvement of the plasticity may be due to the more Al atoms that are dissolved into the matrix with the increase of Al content, which effectively reduces the stacking fault energy of the alloy. Due to the low stacking fault energy of the alloy, dislocation cells do not form easily, which effectively prevents the nucleation and growth of cracks, and improves the plasticity of the alloy [22]. As for the alloy strength gradually decreasing, this may be due to the large size of the Al-Mn phase in the alloy. On the one hand, the large size of the second phase reduces the coordinated deformation ability of the matrix. On the other hand, it can be clearly observed from Figure 14 that the large size of the Al-Mn phase is mainly distributed on the grain boundary. With the increase of Al content, the amount of the Al-Mn phase on the grain boundary also increased. Due to the size of the second phase particles on the grain boundary affecting the plastic deformation ability, when the size and number of the second phase particles on the grain boundary are larger, the cracks are more likely to appear in the alloy, and the corresponding strength is lower. Therefore, the strength of the ZTM641-0.2Ca-2Al alloy is the lowest.

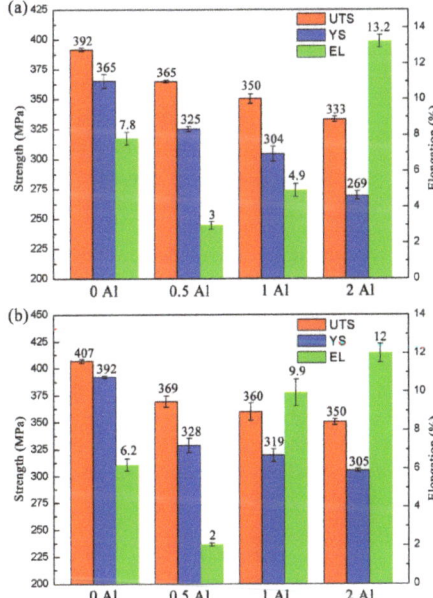

Figure 16. Mechanical properties of aged ZTM641-0.2Ca-xAl alloys tested at room temperature: (**a**) one-stage aging; (**b**) two-stage aging.

Figure 17a,b show the strength and plasticity of the as-extruded ZTM641-0.2Ca-xAl (x = 0, 0.5, 1, 2) alloy at 150 °C and 200 °C, respectively. It can be seen from Figure (a) that the UTS and YS of the ZTM641-0.2Ca-xAl alloy gradually increases with the addition of the Al element. Among them, the strength of the ZTM641-0.2Ca-2Al alloy i the highest, and the UTS and YS are 159 MPa and 132 MPa, respectively. However, the elongation of the ZTM641-0.2Ca-xAl alloy decreases gradually with the addition of Al; the elongation of ZTM641-0.2Ca-2Al alloy is only 5%. According to Figure (b), the variation trend of the mechanical properties at 200 °C is consistent with that at 150 °C. The UTS and YS of the ZTM641-0.2Ca-2Al alloy are the highest, at 103 MPa and 90 MPa. However, the elongation is lower, at 59.8%.

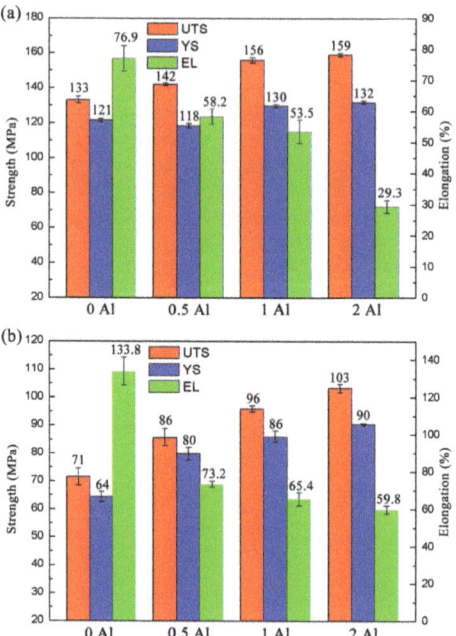

Figure 17. Mechanical properties of as-extruded ZTM641-0.2Ca-xAl alloys tested at high temperature: (**a**) 150 °C; (**b**) 200 °C.

According to the previous as-extruded microstructure, it can be seen that the Al element is added to the alloy and combines with the Mn element to generate the AlMn phase. The AlMn phase is a high-melting phase with a melting point of 670 °C. When the alloy is deformed under high temperature, the AlMn phase can pin the grain boundaries, hindering dislocation slip. According to metal theory [23], when the alloy is subjected to stress and deformation at high temperatures, if the size of second phase is large, the dislocation cannot cut through the second phase, but can only bypass the second phase. In this process, the dislocation must overcome the resistance caused by the bending dislocation tension. Specifically, the shear stress required by the dislocation bypass phase is shown in the following formula:

$$\tau = G \frac{b}{\lambda}$$

where G is the tangent modulus of elasticity, b is the Berkovian vector of dislocations, and λ is the distance between the dislocation and the precipitate. Based on the previous microstructure analysis, when the Al content increases from 0.5% to 1% and 2%, the number and size of the second phase also increases correspondingly, and the required shear stress also increases, which improves the strength. However, due to the AlMn phase being a brittle phase, the higher content of the AlMn phase will inevitably lead to the deterioration of the plasticity.

4. Conclusions

In this paper, the Al element was added to the ZTM641-0.2Ca alloy, and the microstructure and mechanical properties of the ZTM641-0.2Ca-xAl (x = 0, 0.5, 1, 2) alloy were explored by means of XRD, OM, SEM, TEM, EDS and a unidirectional tensile experiment. The conclusions obtained are summarized as follows:

(1) The ZTM641-0.2Ca-xAl (x = 0.5, 1, 2) alloy is mainly composed of the α-Mg, Mg_2Sn, Mg_7Zn_3, MgZn, $MgZn_2$, α-Mn, CaMgSn, AlMn and $Mg_{32}(Al,Zn)_{49}$ phases. When the Al element is added to the alloy, the AlMn and $Mg_{32}(Al,Zn)_{49}$ phases are formed.
(2) When the Al element is added to the ZTM641-0.2Ca alloy, the alloy grains are refined, and the angular bulk AlMn phase is formed, which is mainly distributed at the grain boundaries. With the increase in Al element content, the amount of the AlMn phase increases, but the morphology does not change.
(3) At room temperature, the higher content of Al can improve the elongation of the aged alloy, and the two-stage aged ZTM641-0.2Ca-2Al alloy has the highest elongation, at13.2%. At high temperatures, the Al element can improve the strength, and the strength of the as-extruded ZTM641-0.2Ca-2Al alloy at 150 °C is the highest, with a UTS and YS of 159 MPa and 132 MPa, respectively.

Author Contributions: Conceptualization, C.H. and F.Q.; methodology, S.Y. and A.Z.; investigation, A.Z.; resources, S.Y.; data curation, F.Q.; writing—original draft preparation, C.H. and S.Y.; writing—review and editing, C.H. and F.Q.; funding acquisition, S.Y. and F.Q. All authors have read and agreed to the published version of the manuscript.

Funding: This work was supported by the Ningdong base science and Technology Innovation Development Special Project (2019NDKJLX0008), National Natural Science Foundation of China (52271117), and High Technology Research and Development Program of the Hunan Province of China (2022GK4038).

Data Availability Statement: Data will be made available on request.

Conflicts of Interest: The authors declare that they have no conflict of interest.

References

1. Polmear, I.J. Recent Developments in Light Alloys. *Jpn. Inst. Met.* **1996**, *37*, 12–17. [CrossRef]
2. Eliezer, D.; Aghion, E.; Froes, F.H. Magnesium Science, Technology and Applications. *Adv. Perform. Mater.* **1998**, *5*, 201–212. [CrossRef]
3. Luo, A.A. Recent magnesium alloy development for elevated temperature applications. *Int. Mater. Rev.* **2004**, *49*, 13–30. [CrossRef]
4. Kulekci, M.K. Magnesium and its alloys applications in automotive industry. *Int. J. Adv. Manuf. Technol.* **2008**, *39*, 851–865. [CrossRef]
5. Pan, F.S.; Yang, M.B.; Chen, X.H. A Review on Casting Magnesium Alloys: Modification of Commercial Alloys and Development of New Alloys. *J. Mater. Sci. Technol.* **2016**, *32*, 1211–1221. [CrossRef]
6. Mo, N.; Tan, Q.; Bermingham, M.; Huang, Y.; Dieringa, H.; Hort, N.; Zhang, M.X. Current development of creep-resistant magnesium cast alloys: A review. *Mater. Des.* **2018**, *155*, 422–442. [CrossRef]
7. Blake, A.H.; Cáceres, C.H. Solid-solution hardening and softening in Mg–Zn alloys. *Mater. Sci. Eng. A* **2006**, *483*, 161–163. [CrossRef]
8. Zhao, C.; Chen, X.; Pan, F.; Wang, J.; Gao, S.; Tu, T.; Liu, C.; Yao, J.; Atrens, A. Strain hardening of as-extruded Mg-xZn (x = 1, 2, 3 and 4 wt%) alloys. *J. Mater. Sci. Technol.* **2019**, *35*, 142–150. [CrossRef]
9. Cheng, W.; Bai, Y.; Wang, L.; Wang, H.; Bian, L.; Yu, H. Strengthening Effect of Extruded Mg-8Sn-2Zn-2Al Alloy: Influence of Micro and Nano-Size Mg_2Sn Precipitates. *Materials* **2017**, *10*, 822. [CrossRef]
10. Hou, C.; Cao, H.; Qi, F.; Wang, Q.; Li, L.; Zhao, N.; Zhang, D.; Ouyang, X. Investigation on microstructures and mechanical properties of Mg–6Zn–0.5Ce–xMn (x = 0 and 1) wrought magnesium alloys. *J. Magnes. Alloy.* **2022**, *10*, 993–1003.
11. Chai, Y.; Jiang, B.; Song, J.; Wang, Q.; Gao, H.; Liu, B.; Huang, G.; Zhang, D.; Pan, F. Improvement of mechanical properties and reduction of yield asymmetry of extruded Mg-Sn-Zn alloy through Ca addition. *J. Alloy. Compd.* **2018**, *782*, 1076–1086. [CrossRef]
12. Pulido-Gonzalez, N.; Garcia-Rodriguez, S.; Torres, B.; Rams, J. Effect of Heat Treatment on the Dry Sliding Wear Behavior of the Mg-3Zn-0.4Ca Alloy for Biodegradable Implants. *Materials* **2023**, *16*, 661. [CrossRef]
13. Pan, H.; Yang, C.; Yang, Y.; Dai, Y.; Zhou, D.; Chai, L.; Huang, Q.; Yang, Q.; Liu, S.; Ren, Y.; et al. Ultra-fine grain size and exceptionally high strength in dilute Mg–Ca alloys achieved by conventional one-step extrusion. *Mater. Lett.* **2019**, *237*, 65–68. [CrossRef]
14. Baghani, A.; Khalilpour, H.; Miresmaeili, S.M. Microstructural evolution and creep properties of Mg-4Sn alloys by addition of calcium up to 4 wt.%. *Trans. Nonferrous Met. Soc. China* **2020**, *30*, 896–904. [CrossRef]
15. Cáceres, C.H.; Rovera, D.M. Solid solution strengthening in concentrated Mg–Al alloys. *J. Light Met.* **2001**, *1*, 151–156. [CrossRef]

16. Jayaraj, J.; Mendis, C.L.; Ohkubo, T.; Oh-Ishi, K.; Hono, K. Enhanced precipitation hardening of Mg–Ca alloy by Al addition. *Scr. Mater.* **2010**, *63*, 831–834. [CrossRef]
17. Wang, B.; Pan, F.; Chen, X.; Guo, W.; Mao, J. Microstructure and mechanical properties of as-extruded and as-aged Mg–Zn–Al–Sn alloys. *Mater. Sci. Eng. A* **2016**, *656*, 165–173. [CrossRef]
18. Wei, J.; Chen, J.; Yan, H.; Su, B.; Pan, X. Effects of minor Ca addition on microstructure and mechanical properties of the Mg–4.5Zn–4.5Sn–2Al-based alloy system. *J. Alloy. Compd.* **2013**, *548*, 52–59. [CrossRef]
19. Humphreys, F.J.; Hatherly, M. *Recrystallization and Related Annealing Phenomena*; Pergamon Press: Oxford, UK, 1996.
20. Yu, Z.W.; Tang, A.T.; Wang, Q.; Gao, Z.Y.; He, J.J.; She, J.; Song, K.; Pan, F.S. High strength and superior ductility of an ultra-fine grained magnesium–manganese alloy. *Mater. Sci. Eng. A* **2015**, *648*, 202–207. [CrossRef]
21. Hou, C.H.; Ye, Z.S.; Qi, F.G.; Lu, L.W.; She, J.; Wang, L.F.; Ouyang, X.P.; Zhao, N.; Chen, J. Microstructure and Mechanical Properties of As-Aged Mg-Zn-Sn-Mn-Al Alloys. *Materials* **2023**, *16*, 109.
22. Zhang, J.; Dou, Y.C.; Dong, H.B. Intrinsic ductility of Mg-based binary alloys: A first-principles study. *Scr. Mater.* **2014**, *89*, 13–16. [CrossRef]
23. Wang, F.L.; Bhattacharyya, J.J.; Agnew, S.R. Effect of precipitate shape and orientation on Orowan strengthening of non-basal slip modes in hexagonal crystals, application to magnesium alloys. *Mater. Sci. Eng. A* **2016**, *666*, 114–122. [CrossRef]

Disclaimer/Publisher's Note: The statements, opinions and data contained in all publications are solely those of the individual author(s) and contributor(s) and not of MDPI and/or the editor(s). MDPI and/or the editor(s) disclaim responsibility for any injury to people or property resulting from any ideas, methods, instructions or products referred to in the content.

Article

Influence of Retrogression Time on the Fatigue Crack Growth Behavior of a Modified AA7475 Aluminum Alloy

Xu Zheng [1,2], Yi Yang [3], Jianguo Tang [1,4], Baoshuai Han [5], Yanjin Xu [5,*], Yuansong Zeng [5] and Yong Zhang [1,4,*]

1. School of Materials Science and Engineering, Central South University, Changsha 410083, China
2. ALG Aluminium Inc., Guangxi Key Laboratory of Materials and Processes of Aluminum Alloys, Nanning 530031, China
3. Shanxi Aircraft Industry Corporation, Ltd., Hanzhong 723213, China
4. Key Laboratory of Non-Ferrous Metals Science and Engineering, Ministry of Education, Changsha 410083, China
5. AVIC Manufacturing Technology Institute, Beijing 100024, China
* Correspondence: xuyj020@avic.com (Y.X.); yong.zhang@csu.edu.cn (Y.Z.)

Abstract: This paper investigates the effect of retrogression time on the fatigue crack growth of a modified AA7475 aluminum alloy. Tests including tensile strength, fracture toughness, and fatigue limits were performed to understand the changes in properties with different retrogression procedures at 180 °C. The microstructure was characterized using scanning electron microscopy (SEM) and transmission electron microscopy (TEM). The findings indicated that as the retrogression time increased, the yield strength decreased from 508 MPa to 461 MPa, whereas the fracture toughness increased from 48 MPa$\sqrt{m}$ to 63.5 MPa$\sqrt{m}$. The highest fracture toughness of 63.5 MPa$\sqrt{m}$ was seen after 5 h of retrogression. The measured diameter of η′ precipitates increased from 6.13 nm at the retrogression 1 h condition to 6.50 nm at the retrogression 5 h condition. Prolonged retrogression also increased the chance of crack initiation, with slower crack growth rate in the long transverse direction compared to the longitudinal direction. An empirical relationship was established between fracture toughness and the volume fraction of age-hardening precipitates, with increasing number density of precipitates seen with increasing retrogression time.

Keywords: retrogression and re-ageing; Al-Zn-Mg alloys; fatigue crack growth; η′ precipitates; AA7475 aluminum alloy

Citation: Zheng, X.; Yang, Y.; Tang, J.; Han, B.; Xu, Y.; Zeng, Y.; Zhang, Y. Influence of Retrogression Time on the Fatigue Crack Growth Behavior of a Modified AA7475 Aluminum Alloy. *Materials* **2023**, *16*, 2733. https://doi.org/10.3390/ma16072733

Academic Editor: Andrey Belyakov

Received: 23 February 2023
Revised: 24 March 2023
Accepted: 24 March 2023
Published: 29 March 2023

Copyright: © 2023 by the authors. Licensee MDPI, Basel, Switzerland. This article is an open access article distributed under the terms and conditions of the Creative Commons Attribution (CC BY) license (https://creativecommons.org/licenses/by/4.0/).

1. Introduction

Aluminum alloys made of high-strength Al-Zn-Mg-Cu have been a popular choice for use in aerospace and other civilian applications since the 1940s [1]. These alloys achieve their highest strength through peak aging conditions; however, they are also susceptible to stress corrosion cracking (SCC). Moreover, the safe life design approach necessitates a high level of fracture toughness and a low fracture cracking growth rate. Therefore, the fatigue crack growth behavior under spectrum loading is increasingly being considered when choosing alloys for aircraft structures that are critical to fatigue.

The AA7475 aluminum alloy is a good combination of high strength and superior fatigue crack growth resistance when compared with AA7050 and AA7075 in comparable tempers [2–4]. The AA7475 alloy originates from AA7075 and AA7175 aluminum alloys. It has a higher zinc to magnesium ratio, meaning a greater zinc content and less magnesium content. The copper content range is comparable between the two alloys, with AA7075 having 1.2–2.0% weight compared to 1.2–1.9% weight in AA7475. AA7475 and AA7075 alloys are both categorized as Al-Zn-Mg-Cu alloys with added Mn and Cr; however, AA7475 alloy only has a restricted amount of Mn added (0.06 wt% maximum) [5]. Y Zhou et al. conducted a study on the distribution of Mg and Cr in the dendrite arms of direct-chilled AA7475 ingots. Their findings indicate that the effective partition coefficients of Mg and Cr are 0.650 and 1.392, respectively, and illustrate the heterogeneous distribution of Mg and Cr along the radial direction of the dendrite arms' cross-section [6]. Ahmed et al. have studied the homogenization treatment of Zr-containing AA7475 alloys. They give an optimum homogenization treatment so that dense and uniformly distributed Zr-bearing dispersoids can be obtained [7]. It is believed that the dispersoids particles are able to stabilize the grain structure and increase the recrystallization resistance by pinning grains and sub-grain boundaries [6,8]. The crack propagation resistance of Al-Zn-Mg-Cu alloys can be influenced by controlling grain structures. For instance, elongated pancake grains can effectively slow down crack growth when the crack is spreading at a right angle to the main axis of the grain. This results in transgranular propagation being preferred, which increases the energy dissipated during the damage process [9]. Tsai and Chuang studied the impact of grain size on the susceptibility to stress corrosion cracking (SCC) of the AA7475 alloy. Their findings indicated that homogeneous slip mode is a key factor in reducing the susceptivity of SCC. A finer grain size leads to a more homogeneous slip mode and smaller grain boundary precipitates, which makes the material more resistant to SCC [10]. The mechanism of hydrogen embrittlement in high-strength aluminum alloys containing various dispersoids was explored by M. Safyari et al. Their research demonstrates that the hydrogen level at the GBs/vacancies is considerably smaller in the Zr-added alloy than it is in the Cr-added alloy. They explained this as being caused by the elastic interaction between hydrogen and coherency stain [11].

The typical fracture toughness values for AA7475 alloy plate are around 40% higher when compared to those of AA7075 alloy in the same tempers [2]. According to a recent publication, it is possible to attain a fracture toughness (K_{IC}) value of 65.2 MPa$\sqrt{m}$ and a critical stress intensity factor (K_{ISCC}) of 37.9 MPa$\sqrt{m}$ for 7X75 alloy in the T73 temper [3]. It suggests that the crack propagation behavior can be controlled through the deliberate selection of alloy composition and thermal mechanical processing. AA7475 aluminum alloys are often available in T7351 and T7451 conditions. Its resistance to corrosion and ability to withstand fatigue are comparable to, and in some instances surpass, other high-strength aerospace alloys, such as AA7075, AA7050, and AA2024. This is also attributed to its lower Fe and Si content, as noted in various publications [4,12,13]. The Fe-containing phase in the Al-Zn-Mg-Cu series, such as Al_7Cu_2Fe or $FeAl_m$, are brittle intermetallic particles that can nucleate cavities and cracks during the services period [14,15]. Verma et al. studied fatigue crack propagation (FCP) of AA7475 alloy. It was noted that the fatigue crack initiated from a surface grain, and the crack extension was dominated by ductile striations [16]. K. Wen and colleagues investigated the FCP behavior of an Al-Zn-Mg-Cu alloy with high zinc content under various aging conditions. They discovered that the FCP resistance improved during the aging process, which they attributed largely to changes in matrix precipitates [17].

Islam and Wallace have studied the retrogression and re-aging response of alloy AA7475. The results indicate that stress-corrosion-crack growth rates can vary independently of yield strength, and that T6 strength levels can be achieved in materials with stress-corrosion resistance comparable to that of the T73 condition for alloy AA7475 [18]. Their studies have also shown that stress-corrosion crack growth rates in retrogressed and re-aged materials are comparable to those in T73 temper [19]. Poole et al. claimed that the peak stress occurred after a short aging time. The acceleration of the kinetics of over-aging could be due to the deformation during aging [20]. Ohnishi et al. studied the retrogression and re-aging (RRA) process for the AA7475 alloy with the aim of enhancing its fracture toughness and SCC properties while preserving its high T6 level strength. Their findings indicated that the preservation of high strength was due to the forming of fine dispersions of age-hardening precipitates. The improvement in SCC resistance was attributed to the coarsening of grain boundary precipitates [4].

It should be noted that Cina and Ranis developed RRA in 1974 to improve corrosion resistance to that of the T73 temper while maintaining the peak aged strength [21]. The original study involved briefly heating 7075-T6 samples (from a few seconds to a few minutes) in the temperature range of 200–280 °C (reversion process). Today, the reversion temperatures have decreased to 160–180 °C, with the reversion time remaining in the range of tens of seconds or hundreds of seconds. For instance, T. Marlaud suggests a 20-min reversion process at 185 °C for their Al-Zn-Mg-Cu alloy study [22]. Nicolas discovered that the dissolution occurs rapidly and should be completed within 100 s if the retrogression process is carried out within a temperature range of 240–300 °C [23,24]. Neither scenario would be suitable for industrial production since the large plates require a significant amount of time to heat up or cool down evenly throughout their entire thickness. Additionally, the retrogression temperature must not be too low to avoid insufficient dissolution of precipitates.

We designed industrial-suited RRA experiment procedures to explore the impact of retrogression time on other mechanical properties, such as fracture toughness and fatigue limits. Comprehensive testing was conducted to examine the effect of RRA on these properties, with particular focus on the fatigue crack growth behavior of the alloy AA7475.

2. Experimental

The material is a modified AA7475 (namely 7X75) 80 mm hot-rolled plate produced by Guangxi Alnan Aluminium Inc. in China. The measured composition is Al-1.5Cu-2.6Mg-6.0Zn-0.2Cr-0.03Mn (wt%). The heat treatment procedures for the studied alloys are shown in Table 1.

Table 1. Heat treatment procedures for studied materials.

Temper	Condition	Aging Process
T6	Peak-aged	24 h/121 °C
T73	Over-aged	6 h/121 °C + 30 h/163 °C
RRA	–	24 h/121 °C + (0.5–5) h/180 °C + 12 h/121 °C

The received plate was solution treated at a temperature of 473 °C for 4 h, then cooled rapidly in a Roller Hearth Furnace to minimize the transfer time. The plate was stretched immediately after quenching to minimize residual stress, then cut into 1000 mm × 1000 mm pieces for further aging processes. The T6 and T73 temper procedures are standard, without any modifications, and their results will serve as reference. However, the RRA process altered the retrogression time (which was changed from 0.5 to 5 h) while being performed at a temperature of 180 °C. The RRA process was conducted in a customer-designed aging furnace that has three isolated temperature zones, each set to designed temperatures. The furnace's roller transports the aluminum plates from one temperature zone to another at a controlled speed, allowing for controlled soaking time. The temperature deviation in each zone is ±1 °C.

To fully understand the changes in properties, mechanical properties (ultimate tensile strength (UTS) and yield strength (YS)) tests were carried out along with the longitudinal direction. Dog-bone-shaped tensile specimens with a gauge length of 50 mm and diameter of 10 mm were used. Tensile tests were performed at a strain rate of 10^{-3} s^{-1} at room temperature. Five parallel tests were conducted.

A 60 mm thick compact tension (CT) (L-T direction) was used to measure the fracture toughness. The thickness (B > 2.5(K_{IC}/σ_{YS})^2) was selected to ensure that a valid plane-strain fracture toughness (K_{IC}) was obtained. High-cycle fatigue (HCF) and Fatigue crack growth (FCG) tests were performed using a 110 kN MTS landmark hydraulic fatigue machine. Both L-T direction and T-L direction samples were tested. Hourglass-shaped specimens were utilized for high-cycle fatigue (HCF) testing, conforming to the size specified in ASTM standard E466-15. A 12.5-mm thick CT sample based on ASTM standard E647-15e1 was used for the FCG test with a load ratio of R = 0.06 at a maximum load of 7.0 kN and a frequency of 4.0 Hz. The crack length was measured using a Crack Opening Displacement (COD) gauge.

Microstructures were characterized using a ZEISS EVOMA10 scanning electron microscope (SEM) with an OXFORD Energy Dispersive Spectroscopy (EDS) and an Electron Backscattered Diffraction (EBSD) detector. The related EBSD data analysis was carried out using CHANNEL 5 software.

The FEI Tecnai G^2 F20 Transmission electron microscope (TEM) equipped with an OXFORD INCA EDS detector was used to characterize the precipitation under different conditions. The TEM samples were circular discs 3 mm in diameter, cut from an 80 μm thick foil obtained through manual thinning. They were further polished using twin-jet electropolishing in a solution of 80% methanol and 20% nitric acid at a temperature below -25 °C. The particle sizes were measured through TEM image analysis using Image J software 1.53t. More than one hundred particles were analyzed for each condition.

3. Results and Discussions

3.1. Mechanical Properties, Fracture Toughness and Other Properties of Studied Alloy

The impact of retrogression time on strength, fatigue limits, and fracture toughness is depicted in Figure 1, with reference lines for T6 and T73 temper results. As seen in Figure 1a,b, both the UTS and YS decrease with increasing retrogression times, but remain within the range of T6 (565 MPa) and T73 (507 MPa) temper. It is also evident that the changes in YS are more pronounced than those in UTS at 2 h, 3 h, and 5 h, with a continuous decrease in YS and little change in UTS. The YS values for T6 and T73 temper are indicated by dashed lines in Figure 1b as 510 MPa and 410 MPa, respectively. Figure 1c displays a linear decrease in the fatigue limits with increasing retrogression time. It is evident that the fatigue limit (FL) exceeds the T6 temper (180 MPa) when retrogression occurs for 0.5 and 1 h. However, after 2 h of retrogression at 180 °C, the fatigue limit (FL) is equal to that of the T6 temper. While the fatigue limit decreases with retrogression time, it still exceeds the value for the T73 condition (170 MPa).

As depicted in Figure 1d, the fracture toughness shows a sharp rise with increasing retrogression time, particularly for retrogression times of 0.5 h, 1 h, and 2 h. Beyond a retrogression time of 2 h, the change in fracture toughness is more modest. The fracture toughness values fall within the range of T73 (64.7 MPa$\sqrt{m}$) and T6 temper (47.3 MPa$\sqrt{m}$) and the highest K_{IC} value is 63.5 MPa$\sqrt{m}$; this surpasses the traditional AA7475 report's L-T value of 52 MPa$\sqrt{m}$ and T-L value of 42 MPa$\sqrt{m}$ [25]. Figure 1e displays exceptional K_{ISCC} values for samples subjected to RRA treatment; these are within the published range of 19–22 MPa$\sqrt{m}$ s [26,27]. The K_{ISCC} value also increases with longer retrogression times, with a significant increase observed between 1 h and 2 h. Beyond that, the K_{ISCC} value stabilizes irrespective of retrogression time changes.

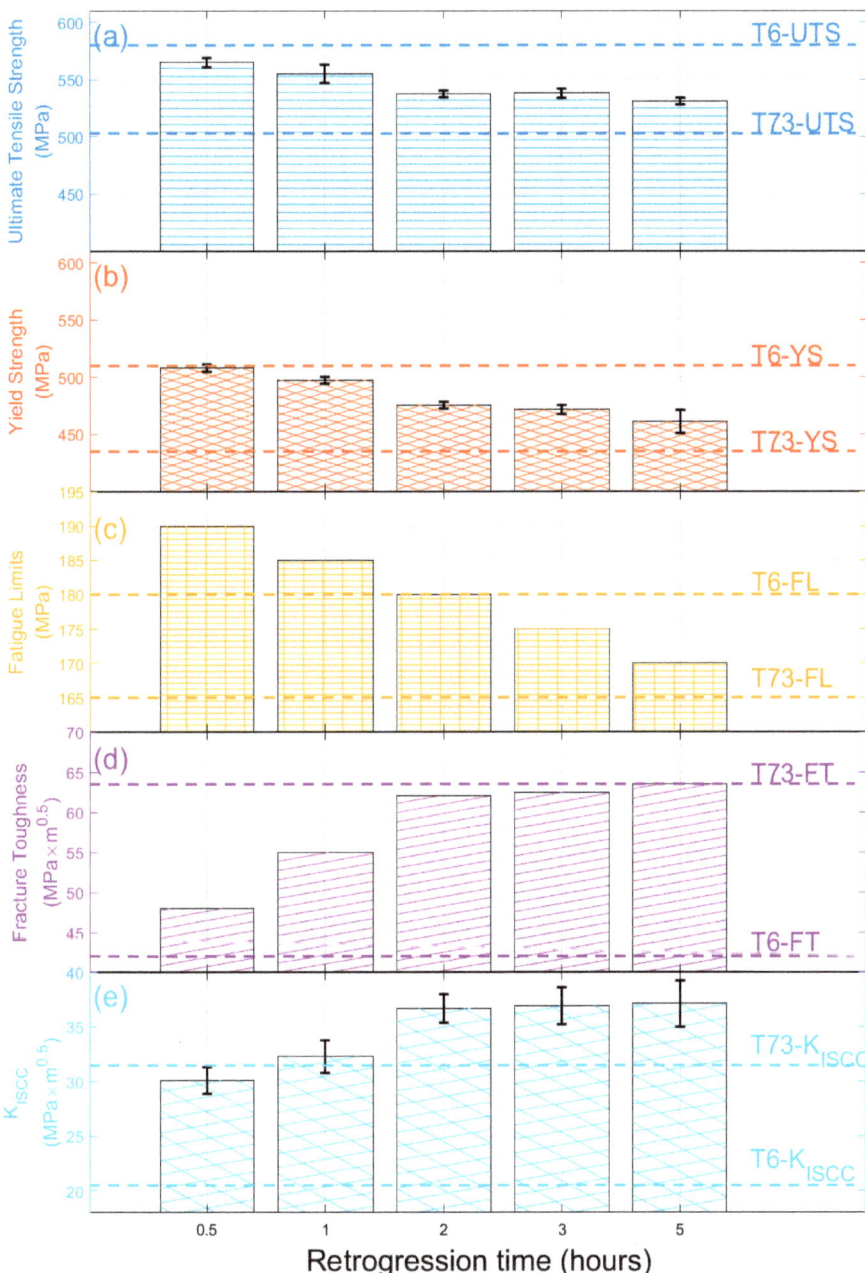

Figure 1. Effects of retrogression time at 180 °C on (**a**) ultimate tensile strength (UTS, MPa), (**b**) Yield strength (YS, MPa), (**c**) Fatigue endurance limits (FL, MPa), (**d**) Fracture toughness (FT, MPa$\sqrt{m}$), and (**e**) K_{ISCC} (MPa$\sqrt{m}$). Symbols "T6" and "T73" in the figure represented the benchmark of these conventional treatments.

3.2. Fatigue Crack Growth Study

Figure 2 illustrates the FCGR curves (da/dN vs. ΔK) for alloys experiencing different retrogression times in the L-T direction (Figure 2a) and T-L direction (Figure 2b). It is shown that the crack growth rate along the long transverse direction (L-T) is less than the longitudinal direction (T-L). This can be seen by the crack growth value of ~0.08 × 10^{-3} mm/cycle (Figure 2a) in the L-T direction when compared to ~1 × 10^{-3} mm/cycle (Figure 2b) in the T-L direction at ΔK = 20 MPa$\sqrt{m}$. The result agrees with the data shown in other publications [16]. The decreased ΔK$_{th}$ by about 2–3 units in the 2-h retrogression sample is also displayed in both figures. The da/dN value is more scattered when close to the ΔK$_{th}$ value. This is because constant force amplitude test procedures are usually well suited for fatigue crack growth rates above 10^{-6} mm/cycle. The crack growth rate is, on average, smaller than on atomic spacing per cycle, and such a crack can be considered dormant. The COD method can barely record the crack change in this ΔK region. When a precise ΔK$_{th}$ value is wanted, a K-decreasing test procedure is recommended in the standard [28]. We calculated the crack growth threshold by using the interpolation method. This involves taking the average of the five smallest da/dN values, as shown by the filled circles in Figure 2. The resulting mean ΔK value is estimated as ΔK$_{th}$, which represents the fatigue threshold. The figures demonstrate that the ΔK$_{th}$ value for RRA1h is higher than that for RRA2h in both directions, indicating that the alloy is more prone to cracking when subjected to longer retrogression treatment.

In the crack propagation state, which is characterized by higher ΔK values, the crack growth rate follows the Paris–Erdogan law [29].

$$\frac{da}{dN} = C \cdot \Delta K^m \tag{1}$$

In this law, C and m are considered material constants that depend on factors such as frequency, temperature, and stress ratio. The fitted constants are displayed in the figures. The results show that the parameter C has a magnitude of approximately 1 × 10^{-6} for both L-T and T-L directions, except for the RRA2h sample in the T-L direction, where C has a value of 1 × 10^{-7}. The coefficient m, which falls between 2.1–2.4, does not show significant variation. This m value is consistent with Verma's findings for the same series of aluminum alloys [5]. It is interesting to compare the fatigue crack propagation (FCP) behavior with other alloys. Figure 2a includes the FCGR for alloy AA7050-T7451 from MMPDS for comparison [30]. The MMPDS data were obtained at a higher stress ratio (R = 0.1) and, therefore, the FCGR is expected to be higher than at R = 0.06 for the same ΔK values. However, as shown in the figure, the FCGR for both alloys is very similar in the low ΔK value range. In the high ΔK value range, the studied 7X75 alloy exhibits significantly lower values, indicating that it is highly resistant to fatigue crack growth in the L-T direction after retrogression treatment.

Figure 2b displays the fracture surface of the samples tested in T-L directions, roughly divided into three stages of crack behavior. However, the boundaries of each stage require clarification; these are the initial crack stage (I), the steady crack propagation stage (II), and the catastrophic failure stage (III). The initial crack stage and steady crack propagation stage appear brighter as their fracture surface is relatively smooth, while the catastrophic failure stage appears darker due to its unstable crack propagation. Further fractographic analysis is presented in the following section.

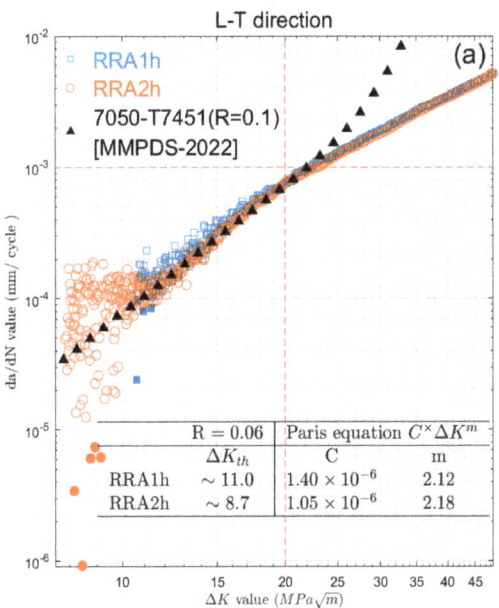
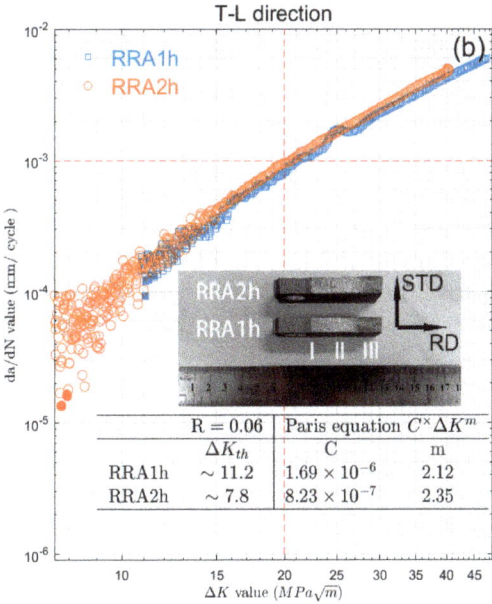

Figure 2. FCGR curves (da/dN vs. ΔK) for alloy retrogression at 180 °C for 1 h and 2 h for (**a**) L-T direction, the FCGR data of 7050-T7451 is from [30] and (**b**) T-L direction, the enclosed figure shows the fracture surface of tested samples.

3.3. Fractographic Characterizations

Figure 3 shows a detailed SEM characterization of the fractography of T-L samples at different retrogression times. The first column displays the fractography of RRA1h samples, and the second column shows the fractography of RRA2h samples.

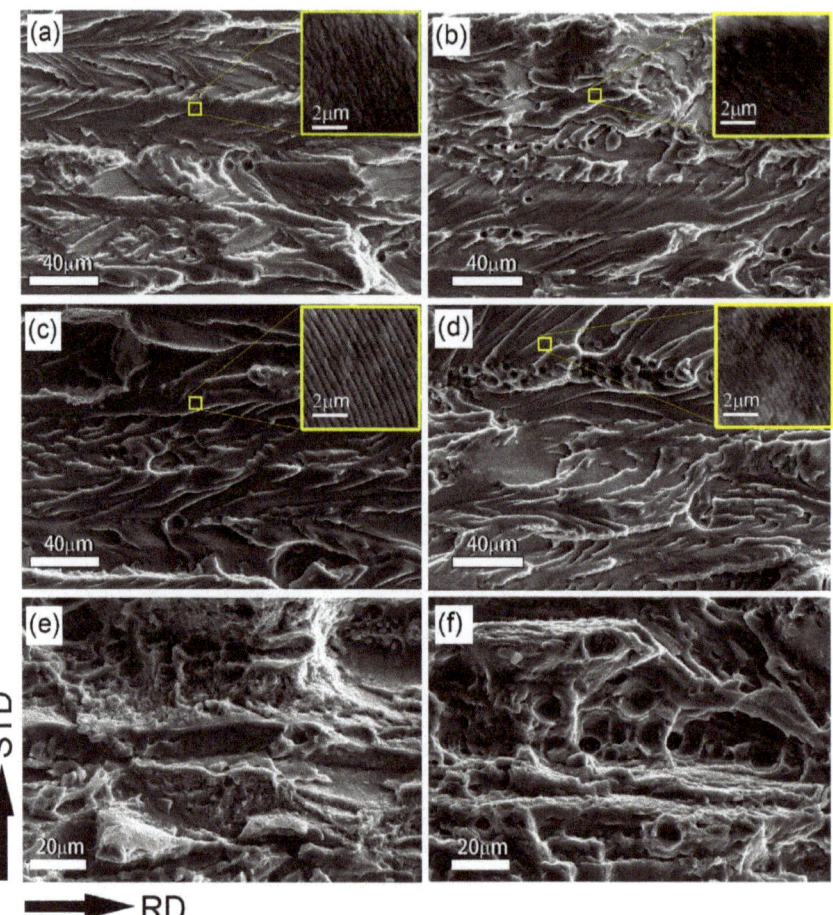

Figure 3. SEM characterization of fractography of T-L direction samples after retrogression at 180 °C for 1 h (first column) and 2 h (second column). Images (**a,b**) are taken from stage I-crack initial, (**c,d**) are taken from stage II-crack steadily propagation, and (**e,f**) are taken from stage III- catastrophic failure. The enclosed image in the top-right corner shows higher magnifications of the fatigue striations produced from amplitude loading.

The initial crack stage (Figure 3a,b) and steady crack propagation stage (Figure 3c,d) exhibit a typical cleavage fracture pattern, with straight cleavage cracks along the RD direction and some separations at certain angles. There is no evidence that the separations are linked to the crystallographic planes. Some prominent cracks exhibit the fiber grain structure, while the separations appear denser than in other conditions. RRA2h samples (Figure 3a) contain more micro-voids than RRA1h samples (Figure 3b); these result from plastic flow and will eventually enlarge during fracture. In stage II, the steady propagation of the main crack is less noticeable, and separated cracks at certain angles are the dominant behavior. Fracture voids are frequently observed in RRA2h conditions. The dimpled texture indicates better fracture toughness; this is consistent with the fracture toughness values in Figure 1.

Fatigue striations can be seen in both the crack initiation and propagation stages, with their morphology depicted in the top-right corner of images from Figure 3a–d. Striation spacings of approximately 430 nm were observed in the RRA1h condition, and approximately 315 nm in the RRA2h condition. Striations are traces left behind by the progressive expansion of a fatigue crack on the fracture surface. The location of the crack tip at the moment it was created is indicated by a striation. According to the observed striation spacings, FCGR is likely quicker in the RRA1h condition than the RRA2h condition. Thus, RRA1h samples should have lower fatigue limits than RRA2h samples. The fatigue limit, according to Figure 1c, is 180 MPa for the RRA2h condition and roughly 185 MPa for the RRA1h condition. In Figure 2b, the measured FCGR for the RRA1h condition is marginally lower than for the RRA2h condition. Hence, the striation spacing may thus necessitate a more thorough statistical analysis over the full fracture length. According to W.C. Connors, estimating the total number of striations would be more accurate if striation spacings were measured at a few different spots; though hundreds of striations are often present on fracture surfaces [31].

The formation of slip bands indicates a concentrated unidirectional slip on certain planes, causing a stress concentration. Typically, slip bands induce surface steps (e.g., roughness due persistent slip bands during fatigue) and a stress concentration which can be a crack nucleation site. Slip bands extend until impinged by a boundary, and the generated stress from dislocations pile-up against that boundary will either stop or transmit the operating slip depending on its (mis)orientation [32,33].

Figure 3e,f depict a catastrophic failure after the crack propagation stage. During this stage, fractography becomes more complicated. The stress intensity increases to the point where the crack grows rapidly, resulting in a complex fractography featuring both plastic and brittle fracture morphologies. Additionally, the fracture cavities are noticeably larger when compared to earlier stages.

3.4. Microstructure Characterizations and Discussions

The initial microstructure of a 1/4 thickness sample is displayed in Figure 4. It shows typical deformed grain structures with two planes that exhibit distinct grain structures. Figure 4a shows that the grains in the L-T plane have been squeezed due to rolling, with an aspect ratio (grain length in LTD/grain length in STD) of around 4.2. The T-L plane exhibits a pancake-like grain structure with a much higher aspect ratio of around 20.5. The EBSD image analysis reveals a low recrystallization fraction (less than 10%) due to the specific thermomechanical processing. This results in unrecrystallized grains that can dissipate crack growth energy and arrest crack paths [34]. The results in Figure 2 demonstrate that the FCGR is lower in the L-T direction compared to the T-L direction, leading to the conclusion that crack propagation is more favorable in pancake grains with high aspect ratio than in those with low aspect ratio. Many publications claim that the transgranular propagation along the pancake grains increases the energy dissipated during the damage process [9]. According to A. Garner, the high aspect ratio and large grain size produce little deflection or branching chances, which might lessen the local fracture tip stress intensity driving the crack propagation [35]. These findings have important ramifications for our knowledge of fracture propagation in Al-Zn-Mg-Cu alloys since the pancake grains were substantially larger and more complicated in shape and deformation structure in 3D when compared to what was previously believed from 2D.

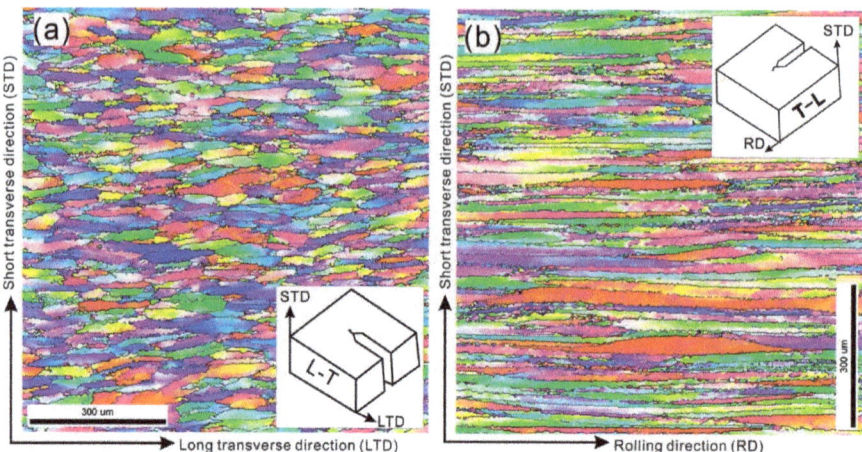

Figure 4. Inverse Pole Figures (IPF) show the grain structure of 7X75 aluminum alloy at 1/4 thicknesses of L-T direction (**a**) and T-L direction (**b**).

The RRA process does not alter the grain structures of the alloys being examined. Many studies have demonstrated that retrogression leads to a decrease in strength initially, followed by a subsequent increase up to a secondary peak. This initial drop is typically the result of the dissolving of age-hardening precipitates [22,36,37]. However, the decrease is usually only temporary during the short duration of retrogression and does not impact industrial manufacturing processes.

The evolution of age-hardening precipitates during retrogression has been analyzed using TEM and is depicted in Figure 5. The first column shows the precipitates within the grains. The size of the precipitates does not vary significantly with increasing retrogression time, but the number density or volume fraction has significantly increased. The statistics of precipitates within grains are presented at the top of the figures. More than a hundred particles were analyzed for each condition. The average diameter increased from 6.13 nm at the RRA1h condition to 6.50 nm at the RRA5h condition. They are slightly larger than those at the T6 condition (approximately 4.8 nm) and smaller than those at the T7 condition (around 7.6 nm) [24]. The statistics also show that the number density of the precipitates increased with increasing retrogression time.

The second column in Figure 5 shows the precipitates along the grain/subgrain boundaries. It can be seen in Figure 5b,d,f that the width of the PFZs ranges from 30 to 50 nm and remains relatively unchanged with increasing retrogression time. The size of grain boundary precipitates (GBPs) is also within a similar range of 20 to 60 nm, similar to PFZs. The GBPs are widely dispersed rather than continuous along the grain/subgrain boundaries. This is considered beneficial for both SCC properties and fatigue propagation as dislocations can move to the surface, initiating cracks. Alternatively, dislocations can move to the PFZs, causing strain localization and stress concentrations during cyclic loading [25,38].

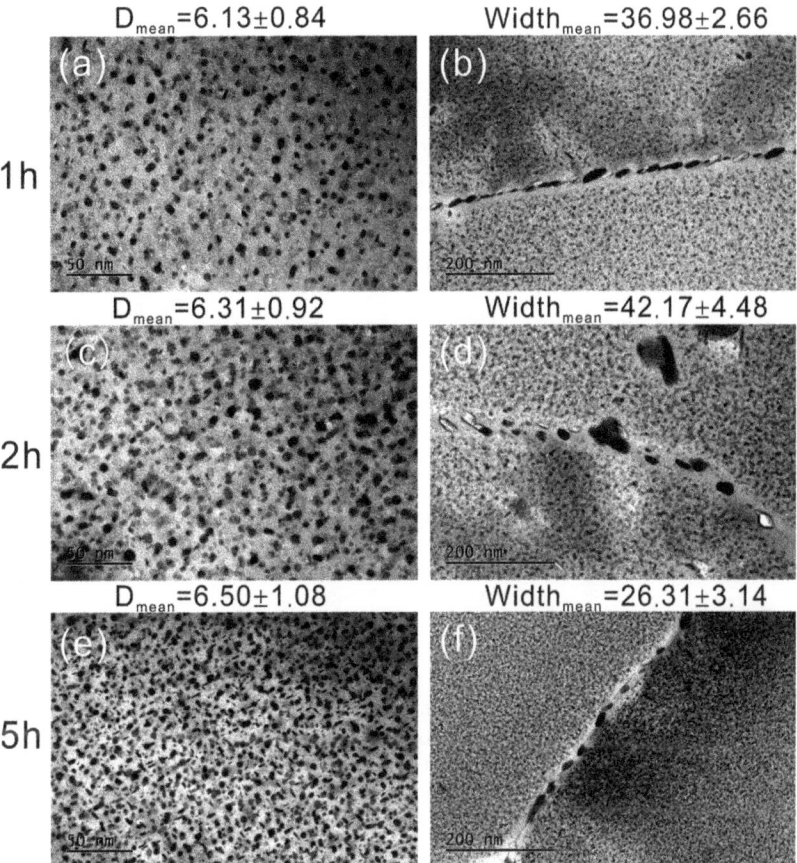

Figure 5. The age-hardening precipitates of RRA treated samples, (**a**,**b**) retrogression for 1 h, (**c**,**d**) retrogression for 2 h, and (**e**,**f**) retrogression for 5 h. The first column represents the precipitate within grains, the second column represents the precipitates on the grain/subgrain boundaries, and the third column shows the statistical analysis of the precipitates within grains.

The crack propagation behavior is thought to be influenced by multiple factors, including the grain structure size, the size or number density of precipitates within grains, precipitates along grain boundaries, and the width of PFZs. Different crack propagation mechanisms may apply during different stages, and other strengthening mechanisms must also be taken into account. As a result, creating a sophisticated function that describes the relationship between microstructure parameters and fracture toughness or fatigue endurance limits is challenging. However, some models can help explain the fatigue crack growth behavior of aluminum alloys. For instance, K. Wen's research primarily focused on the interaction between dislocations and age-hardening precipitates [6]. The study demonstrates that the rate of fatigue crack propagation (FCP) is proportional to the precipitate radius (*r*) and the reciprocal of the volume fraction (*fv*):

$$\frac{da}{dN} \propto r \cdot f_v^{-1/2} \qquad (2)$$

The volume fraction of age-hardening precipitates is a crucial factor that affects the FCP and fracture toughness of the conditions studied. In this research, we propose an empirical relationship between fracture toughness and the volume fraction (fv) of age-hardening precipitates.

$$f_v \approx \frac{C \cdot K_{IC}}{\sqrt{\sigma_y \cdot E \cdot d_p}} \tag{3}$$

where:

C is fitting constants; K_{IC} is the fracture toughness in MPa$\sqrt{m}$; σ_y is Yield strength; E is Young's modulus; d_p is the mean size of the particles; and f_v is the volume fraction of the given type of precipitate. The calculated volume fraction of age-hardening precipitates is shown in Table 2. The equation provides good results, and the increasing trend agrees with the microstructure characterizations.

Table 2. The calculated volume fraction of η′ precipitates.

Retrogression Time (Hours)	Young's Modulus (MPa)	Yield Strength (MPa)	K_{IC} (MPa$\sqrt{m}$)	Particle Size (nm)	Calculated Volume Fraction (%)
0.5	72,000	508	48	5.90	3.08
1	72,000	497.4	55	6.13	3.51
2	72,000	475.7	62	6.31	3.98
3	72,000	471.8	62.5	6.42	4.00
5	72,000	461	63.5	6.50	4.08

4. Conclusions

In the retrogression treatment process, the dissolution of fine and coherent precipitates that were previously formed during the initial aging treatment or natural aging occurs to some extent. The degree of dissolution is contingent on the temperature and duration of the retrogression treatment. This partial dissolution of precipitates leads to a more homogeneous distribution of solute atoms and vacancies, thereby increasing the driving force for the formation of new and smaller precipitates during subsequent aging. Consequently, the number density and volume fraction of η′ precipitates increase with increasing retrogression time, as supported by the findings presented in Table 2. As the size of age-hardening precipitates grows, their susceptibility to dislocation cutting decreases. This phenomenon, commonly known as the bypass mechanism, is well documented in literature [22,39,40]. It can result in the easy accumulation of dislocations in localized regions, and the FCGR for the RRA2h sample is correspondingly faster than that observed under other conditions.

In comparison to longer retrogression times, the size distribution of η′ precipitates is more varied when exposed to shorter retrogression times. As demonstrated by Figure 5a,c, both large and small precipitates can be observed. This dispersion of particles can result in a reduction of fatigue deformation, making it desirable to have a duplex precipitate structure in these types of alloys. Such a structure should consist of fine particles to provide high tensile properties and coarse particles to enhance fatigue strength.

Many other microstructures can influence FCGR. The presence of PFZs is well-known to be a soft region during cyclic stressing. While the η′ phase precipitates within the grains, the PFZs remain constant even with increasing retrogression time, as confirmed by the statistical analysis presented in Figure 5. The strength at grain boundaries and within individual grains therefore progressively balances, resulting in superior formability. These observations are consistent with the microstructural characterizations shown in Figure 3, where fracture voids are frequently observed in RRA2h conditions. Furthermore, the fracture toughness values presented in Figure 1 indicate that better fracture toughness can be achieved with prolonged retrogression treatments.

This paper studies the impact of retrogression time on the fatigue crack growth of a modified AA7475 aluminum alloy. To thoroughly examine the changes in properties with different RRA techniques, various tests, including tensile strength, fracture toughness, fatigue limits, etc., have been conducted. Results reveal that mechanical strength and fatigue limits decrease with increased retrogression time, while fracture toughness sharply increases, reaching its highest value of 63.5 MPa$\sqrt{\text{m}}$ after a 5-h retrogression treatment. Prolonged retrogression increases the likelihood of crack initiation, and the crack growth rate is slower in the long transverse direction compared to the longitudinal direction due to the high aspect ratio of pancake-like grain structure in the T-L plane. An empirical relationship between fracture toughness and the volume fraction of age-hardening precipitates is proposed, and the number density of these precipitates increases with increased retrogression time as revealed by both the equation and microstructural characterization.

Author Contributions: Data curation and original draft, X.Z.; Investigation and writing, review & editing, Y.Y.; Formal analysis and writing, review & editing, J.T.; Data curation and validation, B.H.; Resources and funding aquisition, review & editing, Y.X.; Supervision and validation, Y.Z. (Yuansong Zeng); Funding aquisition and supervision and writing, Y.Z. (Yong Zhang). All authors have read and agreed to the published version of the manuscript.

Funding: This research was funded by the National Key Research and Development Program of China (No. 2022YFB3403701) and the Guangxi Science and Technology Program (Guike AA22068075, GuikeAA22068075-1).

Institutional Review Board Statement: Not applicable.

Informed Consent Statement: Not applicable.

Data Availability Statement: We are unable to provide the data as it pertains to an ongoing project.

Acknowledgments: The authors thank ALG Aluminium Inc., China, for providing the materials.

Conflicts of Interest: The authors declare no conflict of interest.

References

1. Rometsch, P.A.; Zhang, Y.; Knight, S. Heat treatment of 7xxx series aluminium alloys—Some recent developments. *Trans. Nonferrous Met. Soc. China* **2014**, *24*, 2003–2017. [CrossRef]
2. Lezaack, M.B.; Hannard, F.; Simar, A. Understanding the ductility versus toughness and bendability decoupling of large elongated and fine grained Al 7475-T6 alloy. *Mater. Sci. Eng. A* **2022**, *839*, 142816. [CrossRef]
3. Han, B.; Zheng, X.; Wang, W.; Zhang, Y.; Xu, Y.; He, K.; Zeng, Y.; Zhang, X. Microstructures and properties of a high strength, toughness, and corrosion resistance Al-Zn-Mg-Cu alloy under an over-aging state. *Mater. Lett.* **2022**, *325*, 132674. [CrossRef]
4. Ohnishi, T.; Ibaraki, Y.; Ito, T. Improvement of Fracture Toughness in 7475 Aluminum Alloy by the RRA (Retrogression and Re-Aging) Process. *Mater. Trans. JIM* **1989**, *30*, 601–607. [CrossRef]
5. Association, T.A. *International Alloy Designations and Chemical Composition Limits for Wrought Aluminum and Wrought Aluminum Alloys*; The Aluminum Association, Inc.: Arlington, VA, USA, 2018.
6. Zhou, Y.-R.; Tian, N.; Liu, W.; Zeng, Y.; Wang, G.-D.; Han, S.-D.; Zhao, G.; Qin, G.-W. Mechanism of heterogeneous distribution of Cr-containing dispersoids in DC casting 7475 aluminum alloy. *Trans. Nonferrous Met. Soc. China* **2022**, *32*, 1416–1427. [CrossRef]
7. Ahmed, H.; Eivani, A.R.; Zhou, J.; Duszczyk, J. Thermomechanical Processing: Effect of Homogenization Treatment on the Sizes, Distributions and Compositions of Dispersoids in AA7475 Aluminum Alloy. In Proceedings of the 11th International Conference on Aluminium Alloys, Aachen, Germany, 22–26 September 2008; pp. 621–628.
8. Zhang, Y.; Rometsch, P.A.; Muddle, B.C. *Characterisation and Control of Al3Zr dispersoids in Al-Zn-Mg-Cu-Zr Alloys*; DGM: Bremen, Germany, 2011.
9. Ludtka, G.M.; Laughlin, D.E. The influence of microstructure and strength on the fracture mode and toughness of 7XXX series aluminum alloys. *Metall. Trans. A* **1982**, *13*, 411–425. [CrossRef]
10. Tsai, T.C.; Chuang, T.H. Role of grain size on the stress corrosion cracking of 7475 aluminum alloys. *Mater. Sci. Eng. A* **1997**, *225*, 135–144. [CrossRef]
11. Safyari, M.; Moshtaghi, M.; Hojo, T.; Akiyama, E. Mechanisms of hydrogen embrittlement in high-strength aluminum alloys containing coherent or incoherent dispersoids. *Corros. Sci.* **2022**, *194*, 109895. [CrossRef]
12. Chakraborty, P.; Tiwari, V. Dynamic fracture behaviour of AA7475-T7351 alloy at different strain rates and temperatures. *Eng. Fract. Mech.* **2023**, *279*, 109065. [CrossRef]
13. Yang, J.-G.; Ou, B.-L. Hot Ductility Behavior and HAZ Hot Cracking Susceptibility of 7475-T7351 Aluminum alloy. *Scand. J. Metall.* **2001**, *30*, 146–157. [CrossRef]
14. Xu, D.K.; Rometsch, P.A.; Birbilis, N. Improved solution treatment for an as-rolled Al-Zn-Mg-Cu alloy. Part I. Characterisation of constituent particles and overheating. *Mater. Sci. Eng. A* **2012**, *534*, 234–243. [CrossRef]
15. Allen, C.M.; O'Reilly, K.A.Q.; Cantor, B.; Evans, P.V. Intermetallic phase selection in 1XXX Al alloys. *Prog. Mater. Sci.* **1998**, *43*, 89–170. [CrossRef]

16. Verma, B.B.; Atkinson, J.D.; Kumar, M. Study of fatigue behaviour of 7475 aluminium alloy. *Bull. Mater. Sci.* **2001**, *24*, 231–236. [CrossRef]
17. Wen, K.; Xiong, B.; Zhang, Y.; Li, Z.; Li, X.; Huang, S.; Yan, L.; Yan, H.; Liu, H. Over-aging influenced matrix precipitate characteristics improve fatigue crack propagation in a high Zn-containing Al-Zn-Mg-Cu alloy. *Mater. Sci. Eng. A* **2018**, *716*, 42–54. [CrossRef]
18. Islam, M.U.; Wallace, W. Retrogression and reaging response of 7475 aluminium alloy. *Met. Technol.* **1983**, *10*, 386–392. [CrossRef]
19. Islam, M.U.; Wallace, W. Stresscorrosion-rack grovvth behaviour of 7475 T6 retrogressed and reaged aluminium alloy. *Met. Technol.* **1984**, *11*, 320–322. [CrossRef]
20. Poole, W.J.; Shercliff, H.R.; Castillo, T. Process Model for Two Step Age Hardening of 7475 Aluminium Alloy. *Mater. Sci. Technol.* **1997**, *13*, 897–903. [CrossRef]
21. Cina, B.M.; Gan, R. Reducing the Susceptibility of Alloys, Particularly Aluminium Alloys, to Stress Corrosion Cracking. U.S. Patent Application No. 3,856,584A, 24 December 1974.
22. Marlaud, T.; Deschamps, A.; Bley, F.; Lefebvre, W.; Baroux, B. Evolution of precipitate microstructures during the retrogression and re-ageing heat treatment of an Al–Zn–Mg–Cu alloy. *Acta Mater.* **2010**, *58*, 4814–4826. [CrossRef]
23. Nicolas, M.; Deschamps, A. Characterisation and Modelling of Precipitate Evolution in an Al-Zn-Mg Alloy during Non-isothermal Heat Treatment. *Acta Mater.* **2003**, *51*, 6077–6094. [CrossRef]
24. Nicolas, M. *Precipitation Evolution in an Al-Zn-Mg Alloy during Non-Isothermal Heat Treatments and in the Heat-Affected Zone of Welded Joints*; Grenoble Institute of Technology: Grenoble, France, 2002.
25. Bucci, R.J. Selecting aluminum alloys to resist failure by fracture mechanisms. *Eng. Fract. Mech.* **1996**, *12*, 407–441. [CrossRef]
26. McNaughton, D.; Worsfold, M.; Robinson, M.J. Corrosion product force measurements in the study of exfoliation and stress corrosion cracking in high strength aluminium alloys. *Corros. Sci.* **2003**, *45*, 2377–2389. [CrossRef]
27. Brown, B.F.; Beachem, C.D. A study of the stress factor in corrosion cracking by use of the pre-cracked cantilever beam specimen. *Corros. Sci.* **1965**, *5*, 745–750. [CrossRef]
28. ASTM. *Standard Test Method for Measurement of Fatigue Crack Growth Rates*; ASTM: West Conshohocken, PA, USA, 2008; Volume E674-08.
29. Paris, P.; Erdogan, F. A Critical Analysis of Crack Propagation Laws. *J. Basic Eng.* **1963**, *85*, 528–533. [CrossRef]
30. Battelle Memorial Institute. *Metallic Materials Properties Development and Standardization (MMPDS) Handbook*; Battelle Memorial Institute: Columbus, OH, USA, 2022.
31. Connors, W.C. Fatigue striation spacing analysis. *Mater. Charact.* **1994**, *33*, 245–253. [CrossRef]
32. Koko, A.; Elmukashfi, E.; Becker, T.H.; Karamched, P.S.; Wilkinson, A.J.; Marrow, T.J. In situ characterisation of the strain fields of intragranular slip bands in ferrite by high-resolution electron backscatter diffraction. *Acta Mater.* **2022**, *239*, 118284. [CrossRef]
33. Benjamin Britton, T.; Wilkinson, A.J. Stress fields and geometrically necessary dislocation density distributions near the head of a blocked slip band. *Acta Mater.* **2012**, *60*, 5773–5782. [CrossRef]
34. Marquis, F.D.S. Microstructural design of 7×50 aluminium alloys for fracture and fatigue. In *Nano and Microstructrual Design of Advanced Materials*; Elsevier: Amsterdam, The Netherlands, 2003.
35. Garner, A.; Donoghue, J.; Geurts, R.; Al Aboura, Y.; Winiarski, B.; Prangnell, P.B.; Burnett, T.L. Large-scale serial sectioning of environmentally assisted cracks in 7xxx Al alloys using femtosecond laser-PFIB. *Mater. Charact.* **2022**, *188*, 111890. [CrossRef]
36. Viana, F.; Pinto, A.M.P.; Santos, H.M.C.; Lopes, A.B. Retrogression and Re-ageing of 7075 Aluminium Alloy: Microstructural Characterization. *J. Mater. Process. Technol.* **1999**, *92-93*, 54–59. [CrossRef]
37. Grosvenor, A.R. *Microstructural Evolution during Retrogression and Reaging Treatment of Aluminium Alloy 7075*; Department of Materials Engineering, Monash University: Melbourne, Australia, 2008.
38. Zhang, Q.; Zhu, Y.; Gao, X.; Wu, Y.; Hutchinson, C. Training high-strength aluminum alloys to withstand fatigue. *Nat. Commun.* **2020**, *11*, 5198. [CrossRef]
39. Deschamps, A.; Hutchinson, C.R. Precipitation kinetics in metallic alloys: Experiments and modeling. *Acta Mater.* **2021**, *220*, 117338. [CrossRef]
40. Lee, S.-H.; Jung, J.-G.; Baik, S.-I.; Seidman, D.N.; Kim, M.-S.; Lee, Y.-K.; Euh, K. Precipitation strengthening in naturally aged Al–Zn–Mg–Cu alloy. *Mater. Sci. Eng. A* **2021**, *803*, 140719. [CrossRef]

Disclaimer/Publisher's Note: The statements, opinions and data contained in all publications are solely those of the individual author(s) and contributor(s) and not of MDPI and/or the editor(s). MDPI and/or the editor(s) disclaim responsibility for any injury to people or property resulting from any ideas, methods, instructions or products referred to in the content.

Article

Experimental and Numerical Studies on Hot Compressive Deformation Behavior of a Cu–Ni–Sn–Mn–Zn Alloy

Yufang Zhang [1], Zhu Xiao [2,3,*], Xiangpeng Meng [1,4], Lairong Xiao [2], Yongjun Pei [4] and Xueping Gan [1,*]

1. State Key Laboratory for Powder Metallurgy, Central South University, Changsha 410083, China
2. School of Materials Science and Engineering, Central South University, Changsha 410083, China
3. Key Laboratory of Non-Ferrous Metal Materials Science and Engineering, Ministry of Education, Changsha 410083, China
4. Ningbo Boway Alloy Material Co., Ltd., Ningbo 315135, China
* Correspondence: xiaozhumse@163.com (Z.X.); ganxueping@csu.edu.cn (X.G.)

Citation: Zhang, Y.; Xiao, Z.; Meng, X.; Xiao, L.; Pei, Y.; Gan, X. Experimental and Numerical Studies on Hot Compressive Deformation Behavior of a Cu–Ni–Sn–Mn–Zn Alloy. Materials 2023, 16, 1445. https://doi.org/10.3390/ma16041445

Academic Editor: Gábor Harsányi

Received: 15 November 2022
Revised: 9 January 2023
Accepted: 9 January 2023
Published: 9 February 2023
Corrected: 8 December 2023

Copyright: © 2023 by the authors. Licensee MDPI, Basel, Switzerland. This article is an open access article distributed under the terms and conditions of the Creative Commons Attribution (CC BY) license (https://creativecommons.org/licenses/by/4.0/).

Abstract: Cu–9Ni–6Sn alloys have received widespread attention due to their good mechanical properties and resistance to stress relaxation in the electronic and electrical industries. The hot compression deformation behaviors of the Cu–9Ni–6Sn–0.3Mn–0.2Zn alloy were investigated using the Gleeble-3500 thermal simulator at a temperature range of 700–900 °C and a strain rate range of 0.001–1 s^{-1}. The microstructural evolution of the Cu–9Ni–6Sn alloy during hot compression was studied by means of an optical microscope and a scanning electron microscope. The constitutive equation of hot compression of the alloy was constructed by peak flow stress, and the corresponding 3D hot processing maps were plotted. The results showed that the peak flow stress decreased with the increase in the compression temperature and the decrease in the strain rate. The hot deformation activation energy was calculated as 243.67 kJ/mol by the Arrhenius equation, and the optimum deformation parameters for the alloy were 740–760 °C and 840–900 °C with a strain rate of 0.001~0.01 s^{-1}. According to Deform-3D finite element simulation results, the distribution of the equivalent strain field in the hot deformation samples was inhomogeneous. The alloy was more sensitive to the deformation rate than to the temperature. The simulation results can provide a guideline for the optimization of the microstructure and hot deformation parameters of the Cu–9Ni–6Sn–0.3Mn–0.2Zn alloy.

Keywords: Cu–9Ni–6Sn; hot deformation; processing map; microstucture; finite element analysis

1. Introduction

Ultra-high-strength elastic copper alloys have been widely used in the electronic and electrical industries [1]. Beryllium copper is the most widely used elastic copper alloy [2]; however, it has a poor stress relaxation resistance performance at high temperatures. Cu-Ni-Sn alloys have received widespread attention due to their excellent overall mechanical properties and excellent resistance to stress relaxation [3]. The typical Cu–Ni–Sn alloys, including Cu–15Ni–8Sn [4], Cu–9Ni–6Sn [5], Cu–20Ni–5Sn [6], etc., have excellent overall mechanical properties, with hardness of 300~380 HV, electrical conductivities of 5~15% IACS, tensile strengths of 800~1100 MPa, and elongations of 3.8~5%. Among them, the Cu–9Ni–6Sn alloy (wt.%, designed as C72700 in an ASTM B740-84 standard) has a good balance of conductivity (of >12% IACS) and comprehensive mechanical properties (a hardness of >300 HV, a tensile strength of >800 MPa, and an elongation of >5%) [7].

Due to differences in the melting points and surface tensions of the alloying elements in the Cu–Ni–Sn alloy, segregation of the alloying elements is easy to occur during casting, which directly affects the subsequent processing performance of the materials. The microsegregation and some of the structural defects of the alloy ingot could be improved by the hot deformation process, and a suitable hot working process should be necessary for the Cu–9Ni–6Sn alloys to obtain good comprehensive performance [8]. The physical simulation test equipment for hot compression (the "Gleeble" series physical simulation test machine) is used to study the hot deformation behavior of the tested samples, and the hot working conditions of the alloy can be predicted according to the constitutive equation and the hot working diagram established in the dynamic material model (DDM) of hot compression simulations [9]. The deformation behavior and structural evolution of the alloys in the hot compression experiments provide a theoretical basis for the hot deformation process, thus, shortening the development period of the alloys [10]. This method has been successfully applied to various alloys, such as Ni [11], Ti [12], Al [13], Mg [14,15], and Cu [16]. In recent years, studies on the hot deformation behavior of Cu-Ni-Sn alloys have also been reported. Jiang et al. designed a Cu–20Ni–5Sn–0.25Zn–0.22Mn alloy, and a typical thermal processing map of the alloy was obtained. The hot deformation activation energy of the alloy was calculated to be 295.1 KJ/mol, and the optimum processing parameters were determined to be 760~880 °C/0.1~0.001 s^{-1} [17]. Zhao et al. performed hot deformation experiments on Cu–15Ni–8Sn alloys at deformation temperatures between 825 °C and 925 °C and strain rates of 0.0001–0.6 s^{-1}. It was found that Si and Ti promoted the dynamic recrystallization nucleation of the Cu–15Ni–8Sn alloys, and $Ni_{16}Si_7Ti_6$ particles precipitated, inhibiting the growth of recrystallized grains [18]. Niu compared three models of deformation (the Arrhenius model, the Johnson–Cook model, and the Zerilli–Armstrong model) and found that the Arrhenius model had the highest accuracy with a calculated hot deformation activation energy of 310 kJ/mol for the Cu–15Ni–8Sn alloy [19]. According to the current study, the hot deformation activation energy of copper alloys with different contents of alloying elements varies greatly, with a few obvious rules to follow. Therefore, it is necessary to study the hot deformation behavior of Cu–9Ni–6Sn alloys. Numerical simulation technology is an effective method to optimize the conditions during hot deformation [20]. At the same time, it can help researchers comprehensively understand the microstructure evolution of an alloy during deformation processing [21]. In numerical simulation methods, finite element (FE) is considered an important method to predict thermal mechanical processing. The Deform-2D software, which was used by Jia et al., simulated the hot compression and recrystallization behavior of a Ni alloy, and the comparison of the simulation results with the actual manufacturing results verified the accuracy of the model.

So far, there have been a few reports on the microstructure control and optimization of heat deformation conditions for Cu–9Ni–6Sn alloys by the combination of finite element simulations and hot compression simulation experiments. In this paper, hot compression experiments were performed on the Cu–9Ni–6Sn–0.3Mn–0.2Zn alloy at a strain rate range of 0.001~1 s^{-1} and a temperature range of 700~900 °C. Constitutive equations and thermal processing diagrams of the Cu–9Ni–6Sn alloy were established to fill the gap in the field of high temperature hot deformation of this system alloy. The microstructure evolution of typical regions during the compression process was studied by means of optical microscopy (OM) and scanning electron microscopy (EBSD). The thermal compression process of the alloy was simulated using the Deform-3D finite element software, and the accuracy of the model was verified by comparing the simulation and experimental results.

2. Materials and Methods

2.1. Materials

The raw materials used in this experiment were obtained by heating them to pure copper, pure nickel, pure tin, pure manganese, and pure zinc in an atmospheric smelting furnace. The composition of the alloy was measured by an inductively coupled plasma optical emission spectrometer (ICP-OES), as shown in Table 1. Figure 1a shows the dendritic microstructure of the casted ingot (as shown in Figure 1a). After being homogenized at 920 °C for 2 h, the dendritic microstructure disappeared and equiaxed grains with an average diameter of about 264.1 μm formed (Figure 1b). Figure 1c shows the XRD results of the homogenized alloy, where the phase composition can be determined [22]. According to the comparison with the Cu PDF card (#99-034), the sample mainly has a diffraction peak of pure copper, where the (200) peak has the largest strength.

Table 1. Composition of Cu–9Ni–6Sn–0.2Mn–0.1Zn (wt.%).

	Ni	Sn	Mn	Zn	Si	P	Cu
Composition	9.06	5.47	0.30	0.18	0.019	0.0016	Bal.

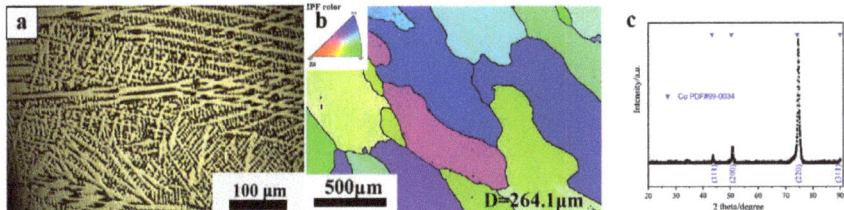

Figure 1. Microstructure of the Cu–9Ni–6Sn alloy before the test: (**a**) optical microstructure of the casted alloy; (**b**) IPF map of the homogenized alloy from an ND direction; (**c**) XRD results of the homogenized alloy.

2.2. Hot Compression

The homogenized alloys were machined into cylindrical specimens with a size of φ6 × 9 mm (a sample size ratio of 1:1.5) according to [23]. The cylindrical specimens were then subjected to isothermal thermal compression experiments on a Gleeble-3500 thermal simulator. According to the common experimental conditions of typical elastic copper alloys for hot deformation experiments [17,19], the testing temperatures were selected as 700 °C, 750 °C, 800 °C, 850 °C, and 900 °C, and the strain rates were selected as $0.001\ \text{s}^{-1}$, $0.01\ \text{s}^{-1}$, $0.1\ \text{s}^{-1}$, and $1\ \text{s}^{-1}$, respectively. The alloys were heated to the target temperatures at a heating rate of 5 °C/s before the compression and then held for 3 min to ensure a uniform temperature distribution. Figure 2 shows a schematic diagram of the technological process of the whole experiment. The hot compressed specimens were sliced along the central axis and mechanically or electrolytically polished. Subsequent processing of the sample in the central deformation region is generally performed; thus, samples from this part were selected for microstructure characterization. The microstructure and texture of the hot compressed samples were observed by metallographic microscopy (OM) and the JEOL 7200F field-emission scanning electron microscope (SEM) equipped with the dedicated software. The results were processed and analyzed using the HKL Channel 5 software.

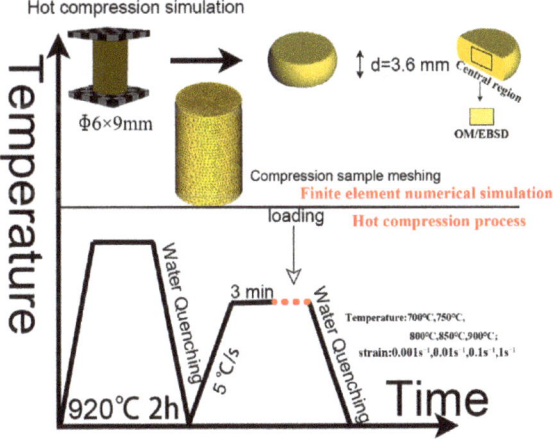

Figure 2. Schematic diagram of the technological process of the whole experiment.

2.3. Finite Element Simulation and Verification

2.3.1. Related Condition Settings

The 3D geometric model of the experiment was drawn by the SolidWorks software. As shown in Figure 2, the upper and lower indenters had a size of 110 mm × 10 mm × 1 mm, and the compressed specimens had a size of φ6 × 9 mm. The finite element simulation of the thermal compression process of the cylindrical specimens under different deformation conditions was carried out by the Deform-3D software. The mesh division method was chosen in the form of a tetrahedral cell mesh, and the compressed workpiece was divided into 32,000 mesh cells. The hot compression of the Cu–Ni–Sn alloy is a typical metal plastic forming process; thus, the effect of the elastic deformation on the deformation process was ignored. The simulation type was set to a rigid plastic finite element model with the LaGrange incremental simulation mode. The convergence conditions were as follows: 108 steps, and each step was set to move down by 0.05 mm· s^{-1} step. The speed of each step was 0.0054 mm·s^{-1}, 0.054 mm·s^{-1}, 0.54 mm·s^{-1}, and 5.4 mm·s^{-1} (corresponding to the strain rates of 0.001 s^{-1}, 0.01 s^{-1}, 0.1 s^{-1}, and 1 s^{-1}, respectively).

2.3.2. Theoretical Basis of Finite Element Numerical Simulation

The rigid visco-plastic finite element method was used in this paper to simulate and analyze the Cu–9Ni–6Sn alloy in the hot compression process, which is applicable to the plastic deformation process of thermally processed and strain rate sensitive materials [24]. The rigid visco-plastic finite element method was based on the following assumptions [24]:

(1) Ignore the effect of elastic deformation and bulk force of the material;

(2) Material deformation obeys the Levy–Mises flow theory;

(3) The material is homogeneous and incompressible, and its volume remains constant throughout the deformation process;

(4) The loading condition gives the boundary between the rigid and plastic regions;

When plastic deformation of the rigid visco-plastic materials occurs, the plastic mechanics can be expressed by the following basic equations:

(1) Balanced equations:

$$\sigma_{ij,j} = 0 \tag{1}$$

where σ_{ij} is the Cauchy stress component in the workpiece;

(2) Velocity–strain rate relationship equation:

$$\dot{\varepsilon}_{ij} = \frac{1}{2}\left(u_{i,j} + u_{j,i}\right) \tag{2}$$

(3) Levy–Mises stress–strain rate relationship equation:

$$\dot{\varepsilon}_{ij} = \dot{\lambda}\sigma_{ij}' \tag{3}$$

$$\dot{\lambda} = \frac{3}{2}\frac{\dot{\bar{\varepsilon}}}{\bar{\sigma}} \tag{4}$$

where $\dot{\bar{\varepsilon}} = \sqrt{\frac{2}{3}\dot{\varepsilon}_{ij}\dot{\varepsilon}_{ij}}$ is the rate of equivalent effect strain, and $\bar{\sigma} = \sqrt{\frac{3}{2}\sigma_{ij}'\sigma_{ij}'}$ is the equivalent effect stress;

(4) Mises yield guidelines:

$$\frac{1}{2}\sigma_{ij}'\sigma_{ij}' = k^2 \tag{5}$$

where $k = \frac{\bar{\sigma}}{\sqrt{3}}$ is for ideal rigid plastic materials, and k is the constant.

(5) Volumetric incompressible conditions:

$$\dot{\varepsilon}_v = \dot{\varepsilon}_{ij}\delta_{ij} = 0 \tag{6}$$

(6) Boundary conditions:

The boundary conditions were divided into force surface boundary conditions and velocity boundary conditions, respectively:

$$\sigma_{ij}n_{ij} = F_i \quad \text{on the stress } S_F, \tag{7}$$

$$\dot{u}_i = \bar{u}_i \quad \text{on the speed } S_U, \tag{8}$$

(7) Constitutive relationship (the Arrhenius model) [25]:

$$\dot{\varepsilon} = A[\sinh(\alpha\sigma)]^n \exp(Q/RT) \tag{9}$$

where $\dot{\varepsilon}$ is the strain rate, σ is the ture stress, Q is the heat of the deformation activation energy, R is the ideal gas constant 8.314 J/(mol·K), T is the hot deformation temperature, and A, α, and n are the constants, respectively.

2.4. Hot Deformation Intrinsic Model

In recent years, scholars have built on empirical observations and proposed to describe the dependence of alloy rheological stress on the strain rate, strain, and temperature with respect to stress by building an intrinsic model, e.g., the Johnson–Cook model [26,27], the Khan–Liang–Farrokh model [28], the Molinari–Ravichandran model [29], the Voce–Kocks model [30], and the Arrhenius model. Among them, the Arrhenius model and its improved model are relatively widely used. Sellars and McTegart first proposed the Arrhenius model in 1966 [31]. The Arrhenius model is expressed as follows:

$$\dot{\varepsilon} = A(\sin\alpha\sigma)^n \exp(-Q/RT) \tag{10}$$

In order to broaden the temperature range for the application of the Arrhenius model, the Zener–Hollomon parameter was used to further modify the model [25,32], which is as follows:

$$Z = A[\sin(\alpha\sigma)]^n = \dot{\varepsilon}\exp(Q/RT) \tag{11}$$

3. Results & Discussion

3.1. Effect of Deformation Conditions on the True Stress–Strain Curve

Figure 3 shows the true stress–strain curves of the Cu–9Ni–6Sn alloy, which is hot compressed at different temperatures and strain rates. In the initial stage of strain, the flow stress increased significantly with the increasing strain. The value of stress fluctuated in a small range after reaching the peak stress. It was found that the flow stress curves of 0.001 s^{-1} and 1 s^{-1} at 700 °C continued to increase with the increasing strain. The typical dynamic recovery (DRV) behavior was observed at 750 °C and 0.1 s^{-1} (Figure 3c), while the typical dynamic recrystallization (DRX) behavior was observed at 900 °C and 1 s^{-1} (Figure 3d). It showed a sawtooth shape in the middle part of the true stress–strain curve under the condition of 800 °C/0.001 s^{-1}, which could be associated with the incidence of dynamic recrystallization [33,34].

Figure 4 indicates the peak stresses of the samples at different conditions. The peak stress increased with the increasing strain rate and decreased with the increasing deformation temperature. When the strain rate was 1 s^{-1}, the peak stress decreased from 263 MPa at 700 °C to 115 MPa (at 900 °C). As the strain rate increased from 0.001 s^{-1} to 1 s^{-1}, the peak stress of the samples deformed at 700 °C increased from 82 MPa to 263 MPa.

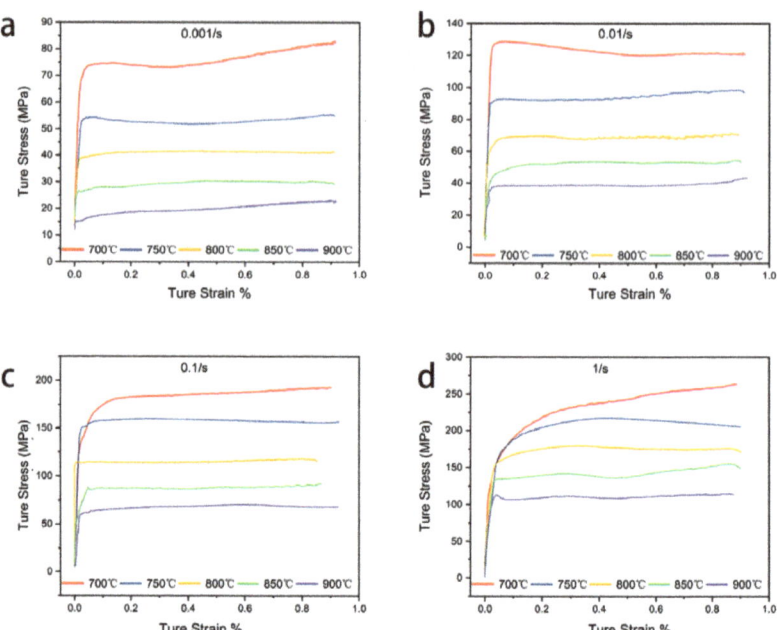

Figure 3. True stress–strain curves of the Cu–9Ni–6Sn alloy at different deformation temperatures and strain rates: (**a**) 0.001 s^{-1}; (**b**) 0.01 s^{-1}; (**c**) 1 s^{-1}; (**d**) 1 s^{-1}.

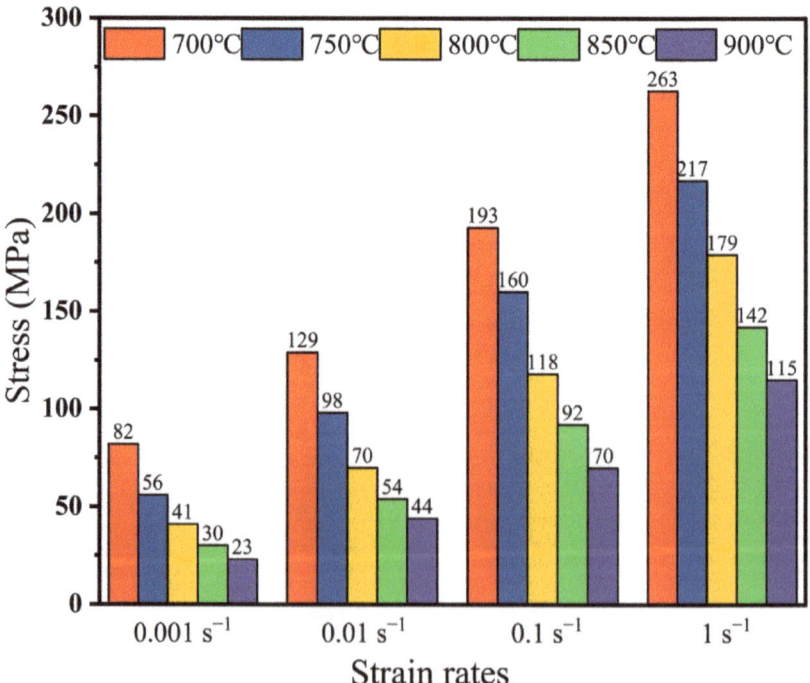

Figure 4. Peak stress of the Cu−9Ni−6Sn alloy under different deformation conditions.

In the initial stage of deformation, a large number of dislocations appeared in the matrix. The entanglement between the dislocations or the interaction between the dislocations and grain boundaries lead to a rapid increase in the strength of the alloy; thus, the flow stress of the alloy increased sharply. After reaching the peak of flow stress, the deformation of the alloy was completed in a short time at the high strain rate. Therefore, the dislocation density and flow stress values in the matrix were also quite high. At the low strain rate, the alloy underwent significant reduction and recrystallization because the hot deformation temperature was higher than the recrystallization temperature [35].

3.2. Flow Stress Constitutive Equation

The Arrhenius kinetic equation, which is summarized by Sellars et al., describes the thermal processing properties of a material through the characteristic parameters in the true stress–strain curve. The relationship between these characteristic parameters, such as the peak flow stress, strain rate, and temperature of processing, could be described as a unified constitutive equation [26] as follows:

$$\dot{\varepsilon} = A[\sinh(\alpha\sigma_p)]^n \exp\left(-\frac{Q}{RT}\right) = [\sinh(\alpha\sigma_p)]^n \exp\left(\ln A - \frac{Q}{RT}\right) \quad (12)$$

where σ_p is the peak stress.

Under low stress conditions (or $\alpha\sigma < 0.8$) and high stress conditions (or $\alpha\sigma > 1.2$), the constitutive equation can be written using Equations (13) and (14), respectively, which are as follows:

$$\dot{\varepsilon} = A_1 \sigma_p^{n_1} \exp\left(-\frac{Q}{RT}\right) \quad (13)$$

$$\dot{\varepsilon} = A_2 \exp(b\sigma_p) \exp\left(-\frac{Q}{RT}\right) \quad (14)$$

where A_1, A_2, n_1, and b are the constants, and $\alpha = b/n_1$.

Taking logarithms for each side of Equations (1)–(3):

$$\ln \dot{\varepsilon} = \ln A + n \ln[\sinh(\alpha\sigma_p)] - \frac{Q}{RT} \quad (15)$$

$$\ln \dot{\varepsilon} = \ln A_1 + n \ln \sigma_p - \frac{Q}{RT} \quad (16)$$

$$\ln \dot{\varepsilon} = \ln A_1 + b\sigma_p - \frac{Q}{RT} \quad (17)$$

The linear relationship of $\ln \dot{\varepsilon} \sim \ln[\sinh(\alpha\sigma_p)]$, $\ln \dot{\varepsilon} \sim \ln \sigma_p$, $\ln \dot{\varepsilon} \sim \sigma_p$ was obtained. The three relationships were plotted separately, as shown in Figure 5. As the thermal compression temperature increased, the slopes of the fitted curves all showed an increasing trend, except for the slope of $\ln \dot{\varepsilon} \sim \ln \sigma_p$ the fitted curve at 800 °C, which showed a sudden decrease. By linear fitting, it can be calculated that b and n_1 are equal to 0.0534 and 3.36, respectively.

Under the same temperature during the thermal processing, the hot deformation energy reflects the sensitivity of the deformation resistance to the strain rate, and it also directly represents the ease of deformation. For a given temperature and strain rate, the thermal activation energy Q, defined in [32] is as follows:

$$Q = R\left\{\frac{\partial \ln \dot{\varepsilon}^\circ}{\partial \ln[\sinh(\alpha\sigma_p)]}\right\}\left\{\frac{\partial \ln[\sinh(\alpha\sigma_p)]}{\partial\left(\frac{1}{T}\right)}\right\} = RNS \quad (18)$$

where R is the ideal gas constant, N is the average value of the slope obtained by fitting $\ln \dot{\varepsilon} \sim \ln[\sinh(\alpha\sigma_p)]$ at a given temperature, and S is the average value of the slope obtained by fitting $\ln[\sinh(\alpha\sigma_p)] \sim 1000/T$ at a given strain rate (as shown in Figure 6a).

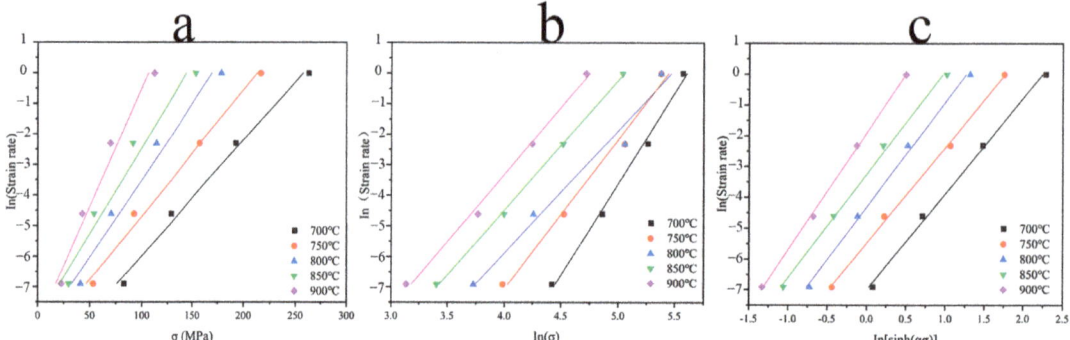

Figure 5. Relationship between peak stress and strain rate: (**a**) $\ln \dot{\varepsilon} \sim \sigma_p$ curve; (**b**) $\ln \dot{\varepsilon} \sim \ln \sigma_p$ curve; (**c**) $\ln \dot{\varepsilon} \sim \ln[\sinh(\alpha \sigma_p)]$ curve.

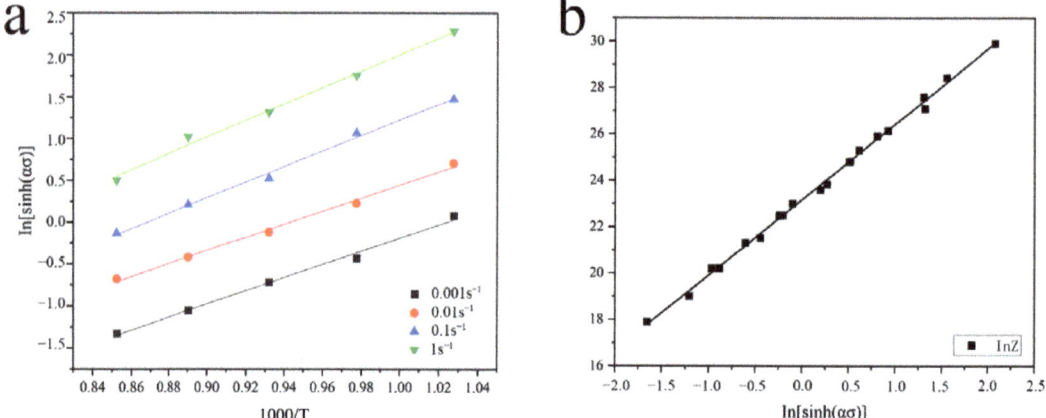

Figure 6. The linear relationship of (**a**) $\ln[\sinh(\alpha \sigma_p)] \sim \ln \frac{1000}{T}$ and (**b**) $\ln Z \sim \ln[\sinh(\alpha \sigma_p)]$.

The obtained hot deformation energy of the alloy (Q) was 243.67 kJmol^{-1}. The relationship between the strain rate and the processing temperature during thermal processing can be determined by temperature-compensated strain rate parameters, which Zener–Hollomon described as follows:

$$Z = \varepsilon \exp\left(\frac{Q}{RT}\right) = A[\sinh(\alpha \sigma_p)]^{n2} \qquad (19)$$

Taking the logarithm of both sides:

$$\ln Z = \ln A + n2 \ln[\sinh(\alpha \sigma_p)] \qquad (20)$$

Figure 6b shows the linear relationship obtained by fitting the line of $\ln Z \sim \ln[\sinh(\alpha \sigma_p)]$, where the intercept on the y-axis is ln$A$, from which the value of the parameter A can be obtained. The value of the slope n_2 (3.36) is averaged with the slope of the linear relationship between $\ln \dot{\varepsilon} \sim \ln[\sinh(\alpha \sigma_p)]$ to obtain the final value of n. The calculated results for each parameter of the constitutive equation are shown in Table 2.

Table 2. Calculation results of each parameter of the constitutive equation.

Parameter	
b	0.0534
α	0.0113
n	3.35
Q/kJ·mol^{-1}	243.67
lnA	22.93

Substituting the averaged parameter values into Equation (14), the constitutive equations of the alloy during hot deformation can be obtained as follows:

$$\dot{\varepsilon} = [\sinh(0.0113\sigma)]^{3.35} \exp\left(22.93 - \frac{243670}{RT}\right) \quad (21)$$

As the parameters of the flow stress curve are calculated by several linear fittings, the flow stress curve obtained from the parameters often has some errors compared with the actual measured curve. In order to evaluate the accuracy of the flow stress instantonal equation, the peak stresses under various test conditions were calculated according to the equation and compared with the experimental values. The relative error between the measured and calculated values was calculated using Equation (22), and the results are shown in Figure 7.

$$(\sigma_c - \sigma_m)/\sigma_m \times 100\% \quad (22)$$

where σ_c is the peak stress calculated from Equation (21), and σ_m is the actual measured peak stress. According to the calculation results, the minimum and maximum relative errors were 0.04% and −10.8%, respectively, and the average relative error was 4.4%. It was demonstrated that the calculated Arrhenius equation has a small error range and can accurately predict the variation of peak stress in the Cu–9Ni–6Sn alloy.

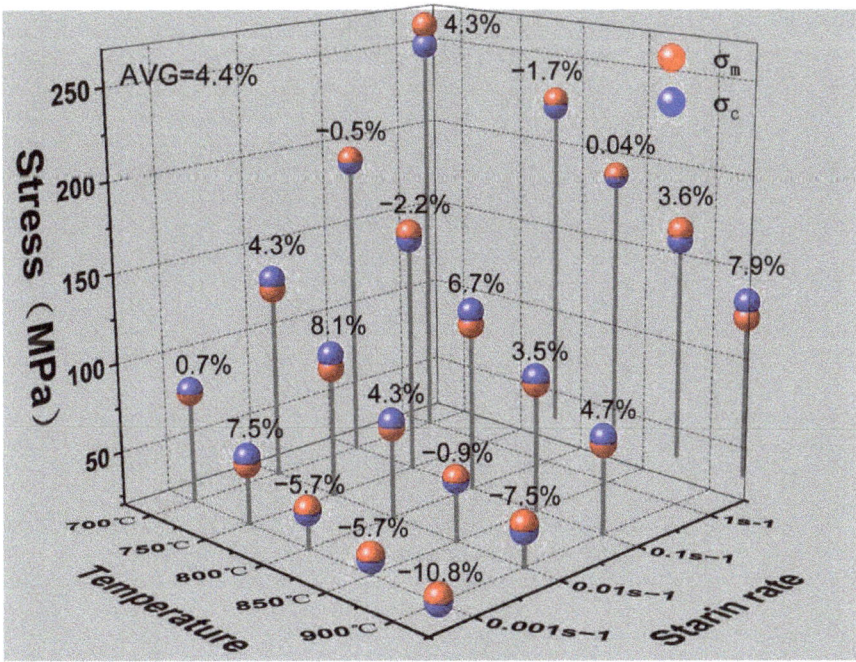

Figure 7. Correlation between the experiment and measured values of the peak stress.

3.3. Hot Deformation Processing Map

According to the dynamic material modeling (DMM) theory, the hot deformation constitutive equation of the Cu–9Ni–6Sn alloy can be expressed [36] as follows:

$$\sigma = K\dot{\varepsilon}^m \quad (23)$$

The strain rate sensitivity coefficient of the alloy was obtained by taking the partial derivative of Equation (23) as follows:

$$m = \frac{\partial \ln \sigma}{\partial \ln \dot{\varepsilon}} \quad (24)$$

where $\dot{\varepsilon}$ is the strain rate, σ is the flow stress, m is the strain rate sensitivity factor, and K is the constant. According to the dissipative structure theory, the power consumed by the alloy can be expressed [37] as follows:

$$P = \sigma\dot{\varepsilon} = G + J = \int_0^{\dot{\varepsilon}} \sigma d\dot{\varepsilon} + \int_0^{\sigma} \dot{\varepsilon} d\sigma \quad (25)$$

where G is the dissipation quantity, and J is the dissipation coefficient. The strain rate sensitivity factor m is defined as follows:

$$m = \frac{dJ}{dG} = \frac{\dot{\varepsilon} d\sigma}{\sigma d\dot{\varepsilon}} = \left[\frac{\partial(\log \sigma)}{\partial(\log \dot{\varepsilon})}\right]_{\tau, T} \approx \frac{\Delta \log \sigma}{\Delta \log \dot{\varepsilon}} \quad (26)$$

When the strain rate and the hot deformation temperature are constant, the dissipation coefficients of the alloy can be expressed as follows:

$$J = \int_0^{\sigma} \dot{\varepsilon} d\sigma = \frac{m}{m+1} \sigma\dot{\varepsilon} \quad (27)$$

At this time, the system has a steady-state flow stress, and the value of m is between 0 and 1, indicating that the dissipation state of the system is nonlinear dissipation. In addition, when the value of m is 0, the system does not dissipate energy. When the value of m is 1, the system dissipation state is ideal dissipation, and the dissipation coefficient of the alloy is at its maximum value (J_{max}) at this time, which is as follows:

$$J_{max} = \frac{\sigma\dot{\varepsilon}}{2} \quad (28)$$

The energy dissipation efficiency factor of the alloy (η) at a certain strain rate and hot deformation temperature can be obtained as follows:

$$\eta = \frac{J}{J_{max}} = \frac{2m}{m+1} \quad (29)$$

Ziegler's continuous instability condition can be expressed as follows:

$$\frac{\partial D}{\partial \dot{\varepsilon}} < \frac{D}{\dot{\varepsilon}} \quad (30)$$

where D is the dissipation function. Under the conditions of the dynamic material model, D is equal to the dissipation coefficient J.

Therefore, the flow instability criterion of the alloy can be expressed [38] as follows:

$$\xi(\dot{\varepsilon}) = \frac{\partial \ln\left(\frac{m}{m+1}\right)}{\partial \ln \dot{\varepsilon}} + m < 0 \quad (31)$$

Based on the true stress–strain curves obtained from the hot compression deformation experiments of the Cu–9Ni–6Sn alloy, the steady-state flow stress values of the true strain of the alloy can be obtained from 0 to 1 at different strain rates and deformation temperature conditions. The flow values and strain rates at each temperature can be fitted with a cubic polynomial, and the strain rate sensitivity and energy dissipation efficiency coefficients of the alloy can be derived from Equations (26) and (29). The flow instability intervals were obtained from the hot deformation temperature and strain rate, and the flow instability intervals of the alloy were superimposed with an energy dissipation diagram to obtain a thermal processing diagram of the sample under the steady-state flow conditions [39], as shown in Figure 8. The alloy instability region was located in the upper left corner of the thermal processing diagram at 700~820 °C/0.01~1 s^{-1}. The optimum deformation parameters of the Cu–9Ni–6Sn alloy should be between 740~760 °C with a strain rate of 0.01~0.001 s^{-1} and 840~900 °C with a strain rate of 0.001 s^{-1}.

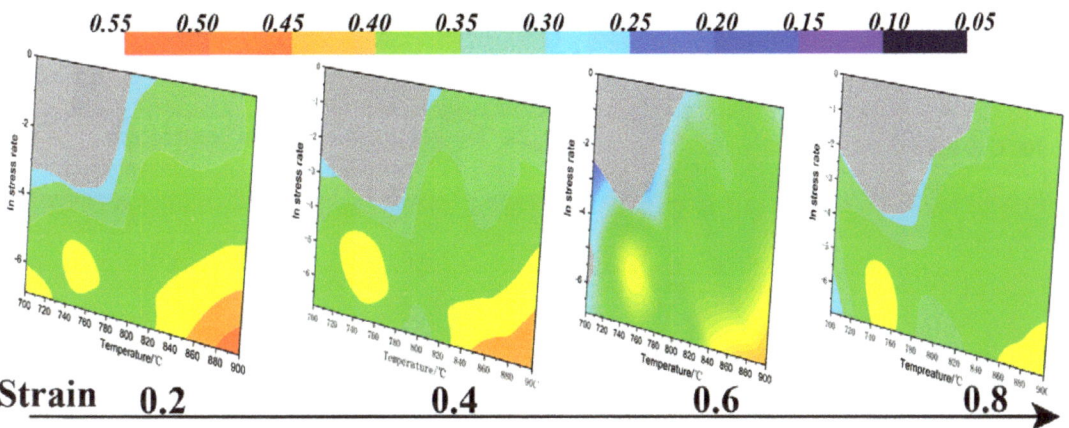

Figure 8. Hot process maps of the Cu−9Ni−6Sn alloy at different true strains.

3.4. Microstructure Evolution

The microstructure of the hot compression alloy in the central region of the sample under different compression conditions is shown in Figure 9. The 0.9 strain thermal processing map can be divided into the following three regions: unsafe deformation region "A", safe deformation region "B", and suitable processing region "C". The OM observations of the samples were performed in the three typical regions. The samples in the destabilization region were characterized by obvious microcracks on the grain boundaries. Very few dynamically recrystallized grains developed near the large angular grain boundaries of the original grains under the low temperatures or the high strain rates. Increasing the deformation temperature or decreasing the strain rate (region "B" or "C") resulted in a significant increase in the number of dynamically recrystallized grains. Increasing the temperature or decreasing the strain rate provided more favorable conditions for atomic diffusion, dislocation sliding, and grain boundary migration in the alloy [36].

proportional to the geometric dislocation density, suggesting that the geometric dislocation density and the stored energy were low in the optimum processing region.

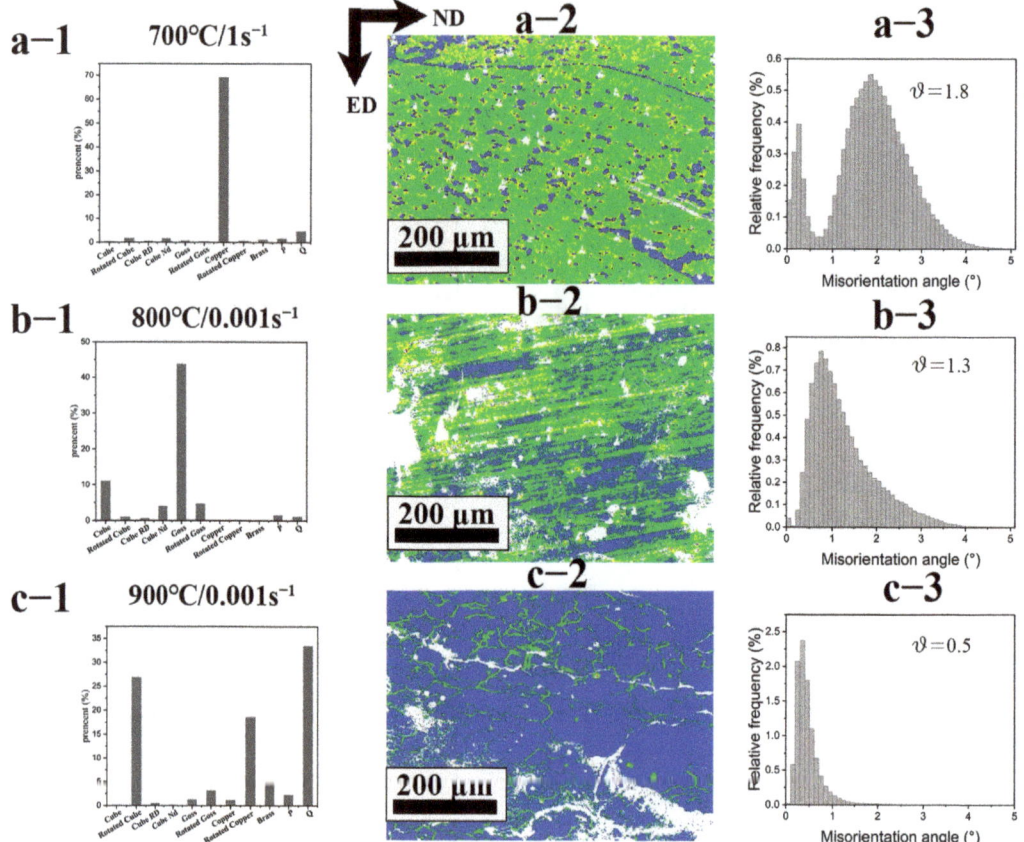

Figure 11. (**a1–c1**) show the distribution of several typical recrystallization structures at 700 °C/1 s^{-1}, 800 °C/0.001 s^{-1}, and 900 °C/0.001 s^{-1}, respectively; (**a2–c2**) show the local average misorientation (LAM) in the central region at 700 °C/1 s^{-1}, 800 °C/0.001 s^{-1}, and 900 °C/0.001 s^{-1}, respectively; (**a3–c3**) are histograms of the distribution of LAM at 700 °C/1 s^{-1}, 800 °C/0.001 s^{-1}, and 900 °C/0.001 s^{-1}, respectively.

Table 3. Several typical FCC metal recrystallization textures [40].

Designation	Miller Indices {hkl} <uvw>
Cube	{001} <100>
Rotated Cube	{001} <110>
Cube$_{ND}$	{001} <110>
Cube$_{RD}$	{013} <110>
Goss	{011} <100>/{110} <001>
Rotated Goss	{110} <110>
C(Copper)	{112} <111>
Bass	{011} <211>/{110} <112>
Rotated Bass	{110} <111>
P	{011} <122>
Q	{013} <122>

3.6. Finite Model Simulation Results and Validation

3.6.1. Simulation of Effective Strain Distribution under Different Deformation

Figure 12 shows the simulated distribution of the effective strain field of the 1/2 cylindrical Cu–Ni–Sn alloy specimen in the hot compression tests under different deformation conditions, where the x-axis indicates the effective strain and the y-axis indicates the percentage of different equivalent strains in the cross section. It shows the inhomogeneity of the effective strain distribution under different deformation conditions. The effective strains were symmetrically distributed along the radial centerline and axis of the specimens, showing a distribution phenomenon of dispersion from the center to the periphery. In contrast to Figure 12a–d, the maximum effective strain in the center of the specimen profile increased with the increasing strain rate at any given deformation temperature. In comparison to Figure 12e,f, the maximum effective strain also increased with the increasing deformation temperature at any given strain rate. The effective strain field distributions in Figure 12c (Figure 8, region "A") and Figure 11f (Figure 8, region "C") in the suitable processing region are very uniform.

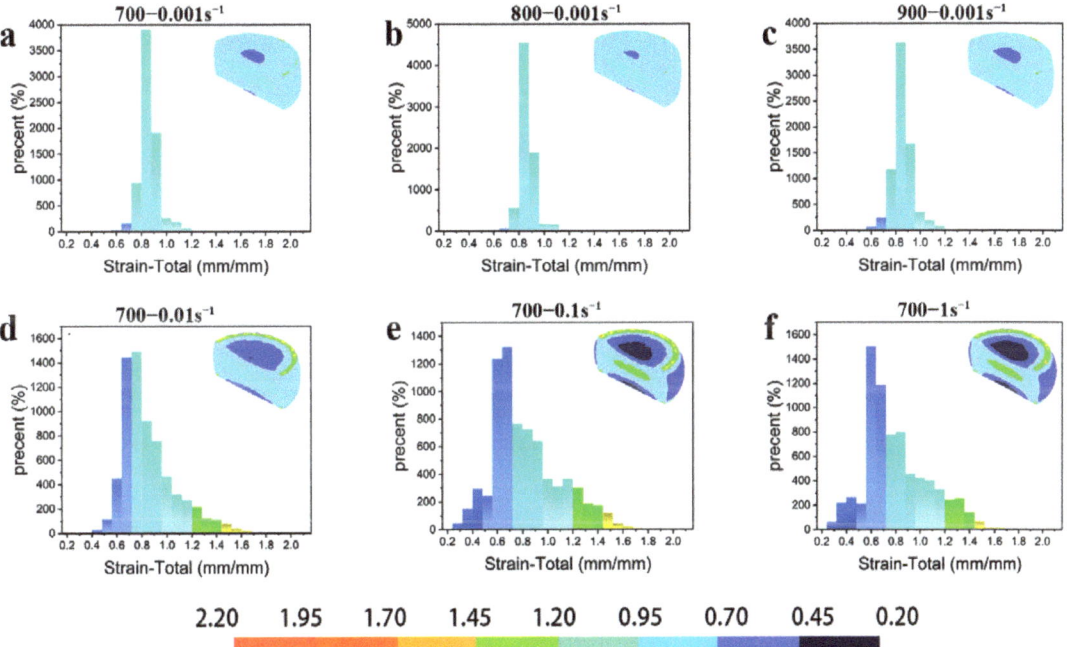

Figure 12. Simulation and prediction of the effective strain (100% of the true strain) distribution on half of the Cu–9Ni–6Sn under different deformation conditions: (**a**) 700–0.001 s^{-1}, (**b**) 800–0.001 s^{-1}, (**c**) 900–0.001s^{-1}, (**d**) 700–0.01s^{-1}, (**e**) 700–0.1s^{-1}, (**f**) 700–1s^{-1}, respectively.

3.6.2. Simulation of Effective Strain Distribution under Different True Strain

During the hot compression deformation, the internal structure and stress distribution of the alloy can change at any time. In order to explore the internal distribution rule of the samples in the three typical areas and analyze the equivalent strain changes of the alloy in the process of hot compression deformation, the simulation results at 700 °C/1 s^{-1}, 800 °C/0.001 s^{-1}, and 900 °C/0.001 s^{-1} deformation conditions were used to analyze the equivalent strain change rule of the Cu–9Ni–6Sn alloy during hot compression deformation. According to Figure 13, the drum shape of the sample is quite obvious, especially at 700 °C/1 s^{-1}. The distribution of the equivalent strain in the sample was extremely uneven. The maximum value of the equivalent strain was mainly concentrated at the edges of the upper and lower end surfaces and the core of the sample, while the value of the equivalent strain at the center of the end face of the sample was small. The equivalent strain at the waist of the sample varied greatly, and this area was less constrained; it is known as the free deformation area [40]. During the compression process, the equivalent strain value of the sample gradually increased, with the true strain increasing from 30% to 100% and the maximum equivalent strain increasing from 0.347 to 4.03 for the sample deformed at 700 °C/1 s^{-1} (Figure 11a), from 0.287 to 1.85 for the sample deformed at 800 °C/0.001 s^{-1} (Figure 11b), and from 0.285 to 2.11 for the sample deformed at 900 °C/0.001 s^{-1} (Figure 11c), respectively. The difference in the effect of the equivalent force values between the central region and the edge part of the sample in the unsafe deformation zone "A" (700 °C/1 s^{-1}) was the largest throughout the compression.

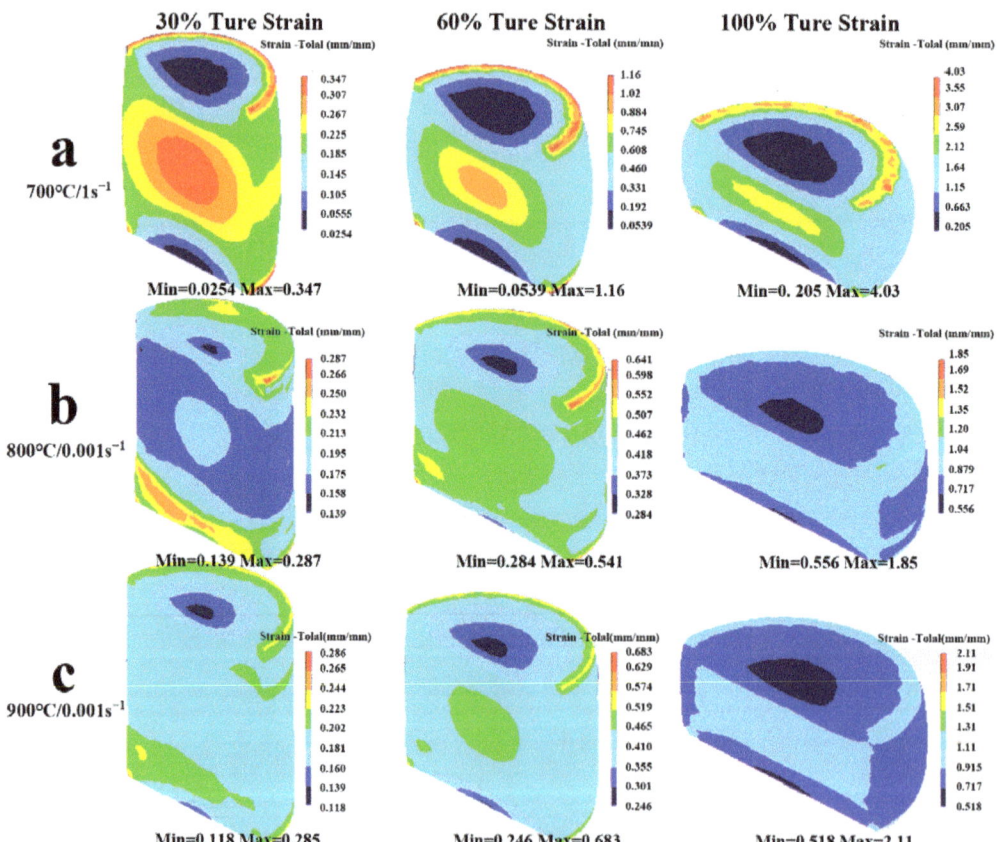

Figure 13. Simulation and prediction of the effective strain distribution of the Cu−9Ni−6Sn alloy under 30%, 60%, and 90% of the true strain at (**a**) 700 °C/1 s^{-1}, (**b**) 800 °C/0.001 s^{-1}, and (**c**) 900 °C/0.001 s^{-1}.

3.6.3. The Change Pattern of Strain Value in Different Regions

In order to further investigate the trend of the equivalent variation with time in the different locations of the two typical regions, the core large deformation region (P1) and the edge free deformation region (P2) [41] were selected for point tracking analysis (as shown in Figure 14a), and the results are shown in Figure 14b,c. All of the equivalent effect variation values of the different deformation regions increased with time, and the growth rate of the free variation region was lower than that of the core. The difference between P1 and P2 under the condition of 900 °C/0.001 s^{-1} was smaller, indicating that the deformation was more uniform. In contrast, the final equivalent variation of the core (deformed at 700 °C/1 s^{-1}) was about 1.2, and the free deformation was about 0.6.

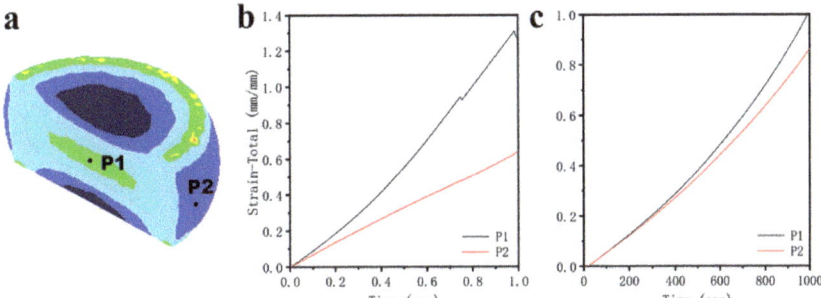

Figure 14. (a) Equivalent variation distribution at the end of compression; 700 °C/1 s^{-1} and (b) 900 °C/0.001 s^{-1}; (c) Changes in the strain values at P1 and P2 during compression, respectively.

4. Conclusions

In this work, the hot compression deformation behaviors of the Cu–9Ni–6Sn–0.3Mn–0.2Zn alloy were investigated, and the microstructure evolution and finite element simulation results during hot deformation were discussed. The main conclusions were as follows:

(1) The flow stress of the Cu–9Ni–6Sn–0.5Mn–0.2Zn alloy during hot compression experiments is sensitive to the process parameters, with the peak stress decreasing with the increasing deformation temperature and the decreasing strain rate.

(2) The thermal activation energy of the Cu–9Ni–6Sn–0.5Mn–0.2Zn alloy was 243.67 kJ mol^{-1}, and the hot deformation constitutive equation was determined as follows:

$$\dot{\varepsilon} = [\sinh(0.0113\sigma)]^{3.35} \exp\left(22.93 - \frac{243670}{RT}\right)$$

(3) Thermal processing diagrams were plotted for the alloys with the rheological instability zone temperatures ranging from 700 °C to 820 °C and the strain rates ranging from 1 s^{-1} to 0.01 s^{-1}. The optimum hot deformation parameters for the alloys were 740~760 °C with strain rates of 0.01~0.001 s^{-1} and 840~900 °C with strain rates of 0.001 s^{-1}.

(4) According to the Deform-3D finite element simulation results, the distribution of the equivalent strain field in the hot-pressed specimens were inhomogeneous. The alloy was more sensitive to the processing rate compared to the processing temperature. The simulation results can provide a guideline for optimizing the microstructure and hot deformation parameters of Cu–9Ni–6Sn alloys.

Author Contributions: Formal analysis, X.M., Y.P. and L.X.; Writing—original draft, Y.Z.; Writing—review & editing, Z.X. and X.G.; Visualization, Y.P.; Supervision, Z.X. All authors have read and agreed to the published version of the manuscript.

Funding: This research was funded by the National Key Research and Development Program of China (2021YFB3700700), the Key Technology Research Program of Ningbo, China (No. 2021Z051), and the grants from the Project of State Key Laboratory of Powder Metallurgy, Central South University, Changsha, China.

Institutional Review Board Statement: Not applicable.

Informed Consent Statement: Not applicable.

Data Availability Statement: The data used to support the findings of this study are included within the article.

Conflicts of Interest: The authors declare no conflict of interest.

References

1. Cribb, W.R.; Grensing, F.C. Property Analysis of a High Strength Cold Worked Cu-Ni-Sn Spinodal Alloy. In Proceedings of the Aeromat 20 Conference and Exposition American Society for Metals, Denver, CO, USA, 20 October 2009.
2. Zhang, H.; Jiang, Y.; Xie, J.; Li, Y.; Yue, L. Precipitation behavior, microstructure and properties of aged Cu-1.7 wt% Be alloy. *J. Alloys Compd.* **2018**, *773*, 1121–1130.
3. Yang, M.; Hu, Y.L.; Li, X.N.; Li, Z.M.; Zheng, Y.H.; Li, N.J.; Dong, C. Compositional interpretation of high elasticity Cu–Ni–Sn alloys using cluster-plus-glue-atom model. *J. Mater. Res. Technol.* **2022**, *17*, 1246–1258. [CrossRef]
4. Jiang, Y.; Li, Z.; Xiao, Z.; Xing, Y.; Zhang, Y.; Fang, M. Microstructure and Properties of a Cu-Ni-Sn Alloy Treated by Two-Stage Thermomechanical Processing. *JOM* **2019**, *71*, 2734–2741. [CrossRef]
5. Wang, L.; Tang, D.; Kong, C.; Yu, H. Crack-free Cu9Ni6Sn strips via twin-roll casting and subsequent asymmetric cryorolling. *Materialia* **2022**, *21*, 101283. [CrossRef]
6. Cai, W.; Yang, S.L.; Chen, L.; Lu, M.M.; Liao, Y.M. Effect of Alterative and Homogenization on the Microstructure of Cu-20Ni-5Sn Alloy. *Mater. Sci. Forum* **2013**, *749*, 421–424. [CrossRef]
7. Xu, S.; Li, Y.; Zhang, M.; Song, T.; Ding, H. The effects of Nb additions on the microstructure evolution in Cu–9Ni–6Sn alloy. *Intermetallics* **2022**, *143*, 107497. [CrossRef]
8. Yu, Q.X.; Li, X.N.; Wei, K.R.; Li, Z.M.; Zheng, Y.H.; Li, N.J.; Cheng, X.T.; Wang, C.Y.; Wang, Q.; Dong, C. Cu–Ni–Sn–Si alloys designed by cluster-plus-glue-atom model. *Mater. Des.* **2019**, *167*, 107641. [CrossRef]
9. Jonas, J.J.; Quelennec, X.; Jiang, L.; Martin, É. The Avrami kinetics of dynamic recrystallization. *Acta Mater.* **2009**, *57*, 2748–2756. [CrossRef]
10. Lin, Y.C.; Li, L.-T.; Fu, Y.-X.; Jiang, Y.-Q. Hot compressive deformation behavior of 7075 Al alloy under elevated temperature. *J. Mater. Sci.* **2011**, *47*, 1306–1318. [CrossRef]
11. He, G.; Liu, F.; Si, J.; Yang, C.; Jiang, L. Characterization of hot compression behavior of a new HIPed nickel-based P/M superalloy using processing maps. *Mater. Des.* **2015**, *87*, 256–265. [CrossRef]
12. Li, C.; Huang, L.; Zhao, M.; Guo, S.; Li, J. Hot deformation behavior and mechanism of a new metastable β titanium alloy Ti–6Cr–5Mo–5V–4Al in single phase region. *Mater. Sci. Eng. A* **2021**, *814*, 141231. [CrossRef]
13. Deng, L.; Zhang, H.-D.; Li, G.-A.; Tang, X.-F.; Yi, P.-S.; Liu, Z.; Wang, X.-Y.; Jin, J.-S. Processing map and hot deformation behavior of squeeze cast 6082 aluminum alloy. *Trans. Nonferrous Met. Soc. China* **2022**, *32*, 2150–2163. [CrossRef]
14. Zhang, X.-P.; Wang, H.-X.; Bian, L.-P.; Zhang, S.-X.; Zhuang, Y.-P.; Cheng, W.-l.; Liang, W. Microstructure evolution and mechanical properties of Mg-9Al-1Si-1SiC composites processed by multi-pass equal-channel angular pressing at various temperatures. *Int. J. Miner. Metall. Mater.* **2021**, *28*, 1966–1975. [CrossRef]
15. Liu, W.; Jiang, B.; Xiang, H.; Ye, Q.; Xia, S.; Chen, S.; Song, J.; Ma, Y.; Yang, M. High-temperature mechanical properties of as-extruded AZ80 magnesium alloy at different strain rates. *Int. J. Miner. Metall. Mater.* **2022**, *29*, 1373–1379. [CrossRef]
16. Yong, P.; Zhu, X.; Yanlin, J.; Rui, Z.; Jiang, Y.; Wenting, Q.; Zhou, L. Hot deformation behavior of a CuAlMn shape memory alloy. *J. Alloys Compd.* **2020**, *845*, 156161. [CrossRef]
17. Jiang, Y.; Wang, X.; Li, Z.; Xiao, Z.; Sheng, X.; Jiang, H.; Cai, G.; Zhang, X. Microstructure Evolution and Hot Deformation Behavior of a CuNiSn Alloy. *Processes* **2021**, *9*, 451. [CrossRef]
18. Zhao, C.; Wang, Z.; Pan, D.-q.; Li, D.-x.; Luo, Z.-q.; Zhang, D.-t.; Yang, C.; Zhang, W.-w. Effect of Si and Ti on dynamic recrystallization of high-performance Cu−15Ni−8Sn alloy during hot deformation. *Trans. Nonferrous Met. Soc. China* **2019**, *29*, 2556–2565. [CrossRef]
19. Niu, D.; Zhao, C.; Li, D.; Wang, Z.; Luo, Z.; Zhang, W. Constitutive Modeling of the Flow Stress Behavior for the Hot Deformation of Cu-15Ni-8Sn Alloys. *Front. Mater.* **2020**, *7*, 577867. [CrossRef]
20. Hussaini, S.M.; Singh, S.K.; Gupta, A.K. Experimental and numerical investigation of formability for austenitic stainless steel 316 at elevated temperatures. *J. Mater. Res. Technol.* **2014**, *3*, 17–24. [CrossRef]
21. Wu, H.; Liu, M.; Wang, Y.; Huang, Z.; Tan, G.; Yang, L. Experimental study and numerical simulation of dynamic recrystallization for a FGH96 superalloy during isothermal compression. *J. Mater. Res. Technol.* **2020**, *9*, 5090–5104. [CrossRef]
22. Ghandvar, H.; Jabbar, K.A.; Idris, M.H.; Ahmad, N.; Jahare, M.H.; Rahimian Koloor, S.S.; Petrů, M. Influence of barium addition on the formation of primary Mg$_2$Si crystals from Al–Mg–Si melts. *J. Mater. Res. Technol.* **2021**, *11*, 448–465. [CrossRef]
23. Shi, L.; Yang, H.; Guo, L.G.; Zhang, J. Constitutive modeling of deformation in high temperature of a forging 6005A aluminum alloy. *Mater. Des.* **2014**, *54*, 576–581. [CrossRef]
24. Lee, C.H.; Kobayashi, S. New Solutions to Rigid-Plastic Deformation Problems Using a Matrix Method. *J. Eng. Ind.* **1972**, *95*, 865. [CrossRef]
25. Xia, Y.F.; Long, S.; Wang, T.-Y.; Zhao, J. A Study at the Workability of Ultra-High Strength Steel Sheet by Processing Maps on the Basis of DMM. *High Temp. Mater. Process.* **2017**, *36*, 657–667. [CrossRef]
26. Jonas, J.J.; Sellars, C.M.; Tegart, W.M. Strength and structure under hot-working conditions. *Metall. Rev.* **1969**, *14*, 1–24. [CrossRef]
27. Grujicic, M.; Pandurangan, B.; Yen, C.F.; Cheeseman, B.A. Modifications in the AA5083 Johnson-Cook Material Model for Use in Friction Stir Welding Computational Analyses. *J. Mater. Eng. Perform.* **2012**, *21*, 2207–2217. [CrossRef]
28. Farrokh, B.; Khan, A.S. Grain size, strain rate, and temperature dependence of flow stress in ultra-fine grained and nanocrystalline Cu and Al: Synthesis, experiment, and constitutive modeling. *Int. J. Plast.* **2009**, *25*, 715–732. [CrossRef]

29. Molinari, A.; Ravichandran, G. Constitutive modeling of high-strain-rate deformation in metals based on the evolution of an effective microstructural length. *Mech. Mater.* **2005**, *37*, 737–752. [CrossRef]
30. Kocks, U. Laws for work-hardening and low-temperature creep. *J. Eng. Mater. Technol.* **1976**, *98*, 76–85. [CrossRef]
31. Sellars, C.M.; McTegart, W. On the mechanism of hot deformation. *Acta Met.* **1966**, *14*, 1136–1138.
32. Zener, C.; Hollomon, J.H. Effect of Strain Rate Upon Plastic Flow of Steel. *J. Appl. Phys.* **1944**, *15*, 22–32. [CrossRef]
33. Feng, D.; Zhang, X.M.; Liu, S.D.; Deng, Y.L. Constitutive equation and hot deformation behavior of homogenized Al–7.68Zn–2.12Mg–1.98Cu–0.12Zr alloy during compression at elevated temperature. *Mater. Sci. Eng. A* **2014**, *608*, 63–72. [CrossRef]
34. Hu, Y.L.; Lin, X.; Li, Y.L.; Zhang, S.Y.; Gao, X.H.; Liu, F.G.; Li, X.; Huang, W.D. Plastic deformation behavior and dynamic recrystallization of Inconel 625 superalloy fabricated by directed energy deposition. *Mater. Des.* **2020**, *186*, 108359. [CrossRef]
35. Liu, D.; Lu, Z.; Yu, J.; Shi, C.; Xiao, H.; Liu, W.; Jiang, S. Hot deformation behavior and microstructure evolution of NiAl-9HfO2 composite. *Intermetallics* **2021**, *139*, 107344. [CrossRef]
36. Wang, X.; Li, Z.; Xiao, Z.; Qiu, W.-T. Microstructure evolution and hot deformation behavior of Cu−3Ti−0.1Zr alloy with ultra-high strength. *Trans. Nonferrous Met. Soc. China* **2020**, *30*, 2737–2748. [CrossRef]
37. Prasad, Y.; Gegel, H.; Doraivelu, S.; Malas, J.; Morgan, J.; Lark, K.; Barker, D. Modeling of dynamic material behavior in hot deformation: Forging of Ti-6242. *Metall. Trans. A* **1984**, *15*, 1883–1892. [CrossRef]
38. Zhu, A.-Y.; Chen, J.-l.; Li, Z.; Luo, L.-Y.; Lei, Q.; Zhang, L.; Zhang, W. Hot deformation behavior of novel imitation-gold copper alloy. *Trans. Nonferrous Met. Soc. China* **2013**, *23*, 1349–1355. [CrossRef]
39. Prasad, Y. Performance. Processing maps: A status report. *J. Mater. Eng. Perform.* **2003**, *12*, 638–645. [CrossRef]
40. Lapeire, L.; Sidor, J.; Verleysen, P.; Verbeken, K.; De Graeve, I.; Terryn, H.; Kestens, L.A.I. Texture comparison between room temperature rolled and cryogenically rolled pure copper. *Acta Mater.* **2015**, *95*, 224–235. [CrossRef]
41. Chamanfar, A.; Valberg, H.S.; Templin, B.; Plumeri, J.E.; Misiolek, W.Z. Development and validation of a finite-element model for isothermal forging of a nickel-base superalloy. *Materialia* **2019**, *6*, 100319. [CrossRef]

Disclaimer/Publisher's Note: The statements, opinions and data contained in all publications are solely those of the individual author(s) and contributor(s) and not of MDPI and/or the editor(s). MDPI and/or the editor(s) disclaim responsibility for any injury to people or property resulting from any ideas, methods, instructions or products referred to in the content.

Article

Fabrication of Cu/Al/Cu Laminated Composites Reinforced with Graphene by Hot Pressing and Evaluation of Their Electrical Conductivity

Hang Zheng [1,2], Ruixiang Zhang [1], Qin Xu [2], Xiangqing Kong [1], Wanting Sun [1], Ying Fu [1,*], Muhong Wu [1,3] and Kaihui Liu [1,3]

[1] Songshan Lake Material Laboratory, Dongguan 523808, China
[2] College of Mechanical and Electrical Engineering, Henan University of Technology, Zhengzhou 450001, China
[3] State Key Laboratory for Mesoscopic Physics, Peking University, Beijing 100871, China
* Correspondence: fuying@sslab.org.cn

Abstract: Metal laminated composites are widely used in industrial and commercial applications due to their excellent overall performance. In this study, the copper/graphene-aluminum-copper/graphene (Cu/Gr-Al-Cu/Gr) laminated composites were prepared by ingenious hot pressing design. Raman, optical microscope (OM), scanning electron microscope (SEM), van der Pauw (vdP), and X-Ray Diffractometer (XRD) were used to investigate the graphene status, interface bonding, diffusion layer thickness, electrical conductivity, Miller indices and secondary phases, respectively. The results demonstrate that the Cu-Al interfaces in the Cu/Gr-Al-Cu/Gr composites were free of pores, cracks and other defects and bonded well. The number of graphene layers was varied by regulating the thickness of the Cu/Gr layer, with the Cu/Gr foils fabricated by chemical vapor deposition (CVD). The electrical conductivity of the composite was significantly improved by the induced high-quality interfaces Cu/Gr structure. The increased number of graphene layers is beneficial for enhancing the electrical conductivity of the Cu/Gr-Al-Cu/Gr composite, and the highest conductivity improved by 20.5% compared to that of raw Al.

Citation: Zheng, H.; Zhang, R.; Xu, Q.; Kong, X.; Sun, W.; Fu, Y.; Wu, M.; Liu, K. Fabrication of Cu/Al/Cu Laminated Composites Reinforced with Graphene by Hot Pressing and Evaluation of Their Electrical Conductivity. *Materials* **2023**, *16*, 622. https://doi.org/10.3390/ma16020622

Academic Editor: Irina V. Antonova

Received: 2 December 2022
Revised: 6 January 2023
Accepted: 6 January 2023
Published: 9 January 2023

Copyright: © 2023 by the authors. Licensee MDPI, Basel, Switzerland. This article is an open access article distributed under the terms and conditions of the Creative Commons Attribution (CC BY) license (https://creativecommons.org/licenses/by/4.0/).

Keywords: electrical conductivity; graphene; copper and aluminum composites; hot pressing; interface

1. Introduction

Metal laminated composites are comprised of at least two physical or chemical components. Their superior performance is given by the achievement of enhanced combinations of properties, which could hardly be achieved for single-phase materials. The composite materials have excellent properties due to the composite matrix, which is difficult to achieve for single-phase materials [1]. Different application scenarios have given rise to different metal composites, for example, Al/Mg [2], Cu/Ni [3], TiO_2/Cu [4], Al/Ni [5], Al/Sn [6], Cu/Zr [7], Al/Ti/Mg [8], Cu/Al [9–12], Cu/Al/Cu [13], etc. As a typical metal laminated composite, Cu/Al composite has been widely used in electric power, heat transmission, rail transit, and other fields due to the characteristics of high electrical conductivity and high thermal conductivity of Cu and the light weight and low cost of Al, which have attracted extensive attention [14]. Rimma et al. [15] achieved the combination of Cu powder and Al powder by four reciprocal extrusion passes at 400 °C. The electrical conductivity of Cu/Al composite increases with the increase of Cu content. Kocich et al. [16] used rotary swaging technology to produce Cu/Al clad composite wires with a diameter of 5 mm at 250 °C, ensuring a high strength while possessing good electrical conductivity. Han et al. [17] investigated the effect of Al/Cu diffusion bonding on the evolution of the interface at an isothermal temperature of 550 °C. The results showed that under the protection of vacuum and argon, the bonding time was increased from 15 to 25 min, and the intermetallic interactions such as Cu_9Al_4, CuAl, and $CuAl_2$ were generated in the interface region. Hu et al. [18] found that the thickness of the secondary phases in the Cu/Al composite increased from 25 μm to 300 μm, and the electrical conductivity decreased from an initial 5.29×10^5 S/cm to 3.83×10^5 S/cm. In recent years, with the development of the electronic information industry, automobile lightweight technology, the national defense industry and the transformation of consumer demand to high-end, high-performance and lightweight, higher requirements have been put forward for the performance of Cu/Al composites.

With the rapid development of nanotechnology, nanomaterials, as a new type of admixture, have a high specific surface area and high activity, providing new opportunities for the development of metal laminated composites. Indeed, numerous studies have demonstrated that introducing some emerging carbon nanomaterials, such as carbon nanotubes [19], graphene-oxide [20] and graphene [21], etc., into the metal composites endows metal composites with properties that cannot be obtained in their various components. Taking graphene as an example, it is reported that graphene is a hexagonal two-dimensional lattice nanomaterial (GNPs) composed of carbon atoms, and the thickness of a single layer is only 0.34 nm. Graphene has good functional properties with thermal conductivity up to 5000 $W \cdot m^{-1} \cdot K^{-1}$, and experimental carrier mobility up to 350,000 $cm^2/(V \cdot s)$, making it the thinnest, strongest, toughest, and best heat and conductive nanomaterial ever discovered [22,23]. Graphene has been known as an excellent reinforcement of metal matrix composites due to its excellent comprehensive properties. At present, a series of advancements have been achieved in the research of graphene-reinforced Cu matrix composites. Yu et al. [24] prepared copper graphene composites by electrodeposition; the graphene concentration in the electrolyte increased from 0 to 0.1 g/L, leading to the sample conductivity increasing from 88.3% IACS to 91.3% IACS. Dong et al. [25] found that graphene doping in $W_{70}Cu_{30}$ from 0 to 0.5 wt % increased the electrical conductivity from 42% IACS to 46% IACS. Cao et al. [26] prepared a laminar structure of Gr/Cu composites with an electrical conductivity of 93.8 to 97.1% IACS, which showed that the introduction of layered graphene can obtain excellent electrical conductivity due to the layered structure through the 2D catalytic growth of GR maintaining the electrical conductivity and thus the desired carrier transport conditions to maximize its performance. Chen et al. [27] prepared bulk Cu/Gr nanocomposites by an accumulative roll-compositing process in which graphite foil was sandwiched in a copper strip and subjected to high-cycle accumulative roll-bonding treatment, followed by hot rolling. The composites had superior strength, ductility and electrical conductivity. The electrical conductivity was above 70% IACS.

Previous research has suggested that adding graphene as reinforcement phase to Cu substrates can significantly improve the electrical conductivity of composites. However, the existing studies on graphene-reinforced Cu matrix composites mostly focus on the single Cu metal. There are insufficient studies on graphene-reinforced Cu/Al composites, and the related interface and electrical coupling response still need to be further elucidated. In addition, in order to obtain new materials with lightweight and high conductivity, graphene reinforced Cu/Al/Cu composites are worthy of attention. To address this issue, a lightweight, inexpensive, and highly conductive laminated composite material is developed in this paper. The innovative proposal is graphene-reinforced Cu in composites with Al. The graphene states, interface bonding, diffusion layer thickness, electrical conductivity, Miller indices and phase analysis of the novel graphene-reinforced Cu/Al/Cu laminated composites are investigated by using Raman, optical microscope (OM), scanning electron microscope (SEM), van der Pauw (vdP), and X-Ray Diffractometer (XRD). The contribution of graphene to the electrical conductivity of Cu/Al/Cu laminated composite in this process is demonstrated.

2. Materials and Methods

2.1. Materials

The experimental samples of commercial Cu foil (25 μm thick, 99.8%) were purchased from the Sichuan Oriental Stars Trading Co., Ltd. (Chengdu, China). The Al sheets (3 mm thick, 99.9%, 40 mm diameter) were purchased from Kierui Metal Materials Co. (Xingtai, China)

2.2. Methods

2.2.1. Methods of Sample Preparation

The experimental samples were prepared as shown in Figure 1. The commercial Cu foil was placed on a quartz substrate, loaded into a chemical vapor deposition (CVD) furnace, and annealed at 1030 °C in an atmosphere of 500 sccm argon and 10 sccm hydrogens to obtain single crystal Cu(111). Methane (1–5 sccm) was then passed through. The growth was continued with argon and hydrogen until the end of cooling [21]. Using a tool, the Cu(111)/Gr foil was cut into round foils of 40 mm in diameter. The corresponding amount of Cu(111)/Gr foils were sintered in a vacuum hot pressing furnace at 10 °C/min, 900 °C and 50 MPa for 1 h to obtain a round Cu(111)/Gr block of 0.5 mm and 0.6 mm thickness. The oxide film was removed from the surface of the industrial Al sheet with sandpaper, and the surface stains of the Cu and Al sheets were removed by ultrasound. The samples were placed into the mold in the order of Cu(111)/Gr, Al, and Cu(111)/Gr for vacuum hot pressing sintering with a heating rate of 10 °C/min, a hot pressing temperature of 530 °C, a pressure of 10 MPa, and a hot pressing sintering time of 1 h. The final Cu/Gr-Al-Cu/Gr laminated composite specimens were obtained.

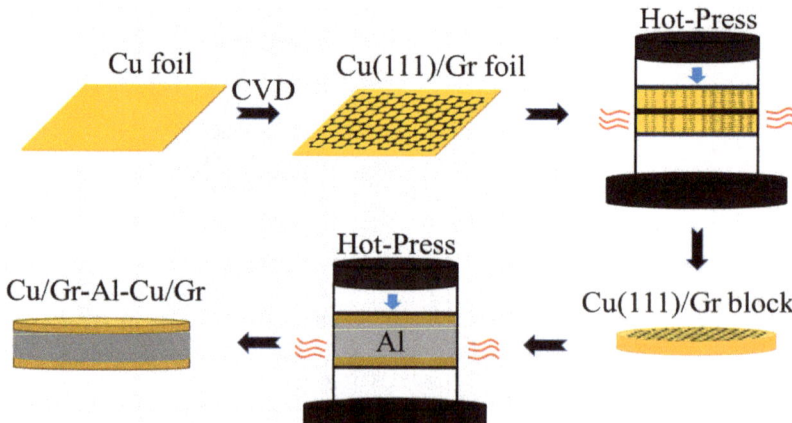

Figure 1. Schematic diagram of the Cu/Gr-Al-Cu/Gr preparation method.

2.2.2. Characterization

The sample was thrown longitudinally and polished by grinding. The X-Ray Diffractometer (XRD, Miniflex 600, Tokyo, Japan) with JADE 5 software was used to analyze the Cu foil Miller indices and diffusion layer phase species. The integrity of the graphene grown on the copper foil was characterized by Raman (Alpha300R, Ulm, Germany). The state of the interface and the diffusion layer were investigated, using an optical microscope (OM, BX53M, Tokyo, Japan) and a scanning electron microscope (SEM, Regulus 8100, Tokyo, Japan) equipped with energy dispersive spectroscopy (EDS).

The experimental specimens are shown in Figure 2a. The hot pressing samples were cleaned and smoothed by 1200 grit sandpaper and tested for conductivity by the van der Pauw method (vdP, Keithley 2182A, Cleveland, OH, USA). The principle is illustrated in Figure 2b. This measurement method was used on a small, flat-shaped sample with four terminals and a uniform thickness. A current is applied to the sample through two terminals and the voltage drop is measured through the opposite two terminals.

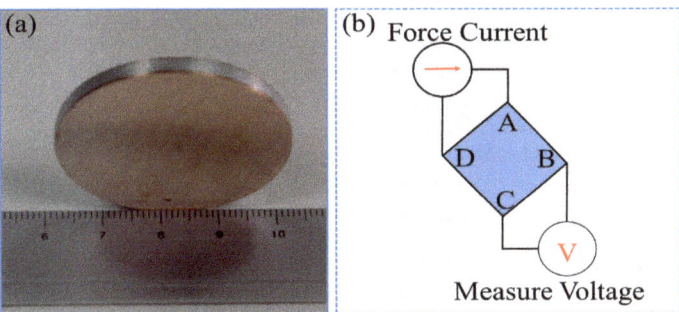

Figure 2. Schematic diagram of the sample for testing electrical conductivity. (**a**) Macroscopic photograph of the sample; (**b**) Schematic diagram of the van der Pauw method for measuring resistance.

3. Results and Discussion
3.1. Cu Miller Indices and Graphene State

Figure 3 shows the XRD result of the Miller indices of copper foil after annealing and copper block after hot pressing. The main peak appears at 2θ approximately 42°. It can be clearly seen that the Cu block remained in the single crystal state after being twice hot pressed. Cu(111) has the lowest surface energy of 1.387 J/m^2 calculated by the first natural principle [21], which results in the lowest Cu(111) formation energy and the easiest nucleation to Cu(111) when the crystal reaches the recrystallization temperature. A sample is obtained of Cu(111) foil and Al after two hot pressings. Due to the lowest Cu(111) Miller indices formation energy and the low temperature of the sample after two hot pressings, there was not enough energy to nucleate Cu(111) to other surfaces, resulting in the Cu sample remaining in a single crystal state. The presence of grain boundaries in polycrystals can lead to electron scattering, which affects the electrical conductivity of Cu, whereas single crystal Cu has only one grain without electron scattering at the grain boundaries, which helps the electrical properties of the sample. It can be seen from the figure that the Miller indices of the Cu foil do not change before or after hot pressing, ensuring a single crystal on the Cu(111) surface.

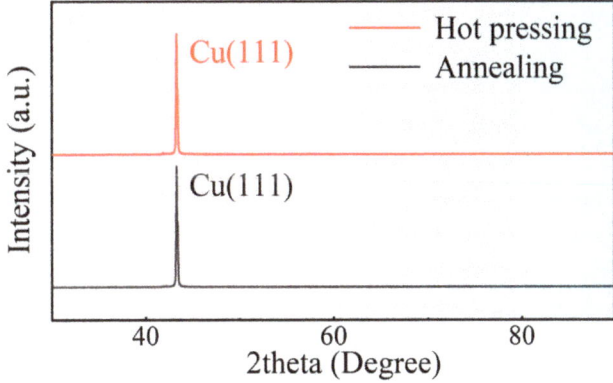

Figure 3. XRD tested the Miller indices of copper foil after annealing and copper block after hot pressing.

Figure 4 shows the Raman characterization of graphene mapping with a G peak at approximately 1580 cm^{-1} and a 2D peak at approximately 2700 cm^{-1} [28]. The Raman spectrum of graphene consists of several peaks, mainly G peak and 2D peak. The G peak is caused by the in-plane vibration of sp^2 carbon atoms and reflects the number of layers of graphene. The 2D peak is a double phonon resonance second-order Raman peak; it is used to characterize the interlayer stacking of carbon atoms in the graphene sample, and the intensity of the peak is also related to the laser power. This Raman graph reflects the growth of an intact layer of graphene on the surface of the Cu foil, which will bring out the high carrier mobility property of graphene.

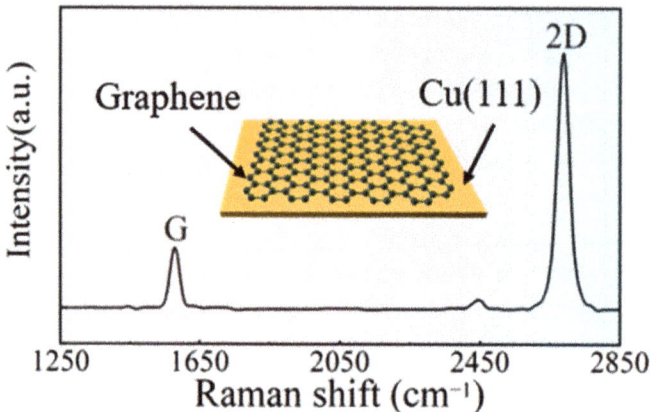

Figure 4. Raman characterization of graphene mapping and (inset) constructional view of Cu(111)/Gr foil grown by CVD.

3.2. Microstructure of the Composite Interface

The bonding interface and diffusion layer condition of the graphene reinforced Cu/Al/Cu (1 mm thickness Cu(111)/Gr) composites were observed, as shown in Figure 5. Figure 5a shows that the Cu/Al bond is free of pores and cracks, and Cu/Al forms a good metallurgical bond. Figure 5b shows the morphology of the Cu/Gr-Al-Cu/Gr diffusion layer. It can be seen that four different colors appeared between the Cu/Al diffusion layers. Figure 5c,d depict the microscopic views of the interface of Cu/Al top and bottom bonding. Figure 5 clearly confirmed that the interface bonding was tight and could form a metallurgical bond.

Figure 6 shows the XRD analysis results of the Cu/Gr-Al-Cu/Gr laminated composites sample cross section. It can be found that four different intermetallic phases, i.e., CuAl$_2$ ($2\theta \approx 25°, 36°$, etc.), CuAl ($2\theta \approx 15°, 29°$, etc.), Cu$_3$Al$_2$ ($2\theta \approx 37°, 63°$), and Cu$_9$Al$_4$ ($2\theta \approx 23°, 66°$) are obtained. Combined with the XRD analysis it can be seen that the four different color diffusion layers in Figure 5b are four different phases.

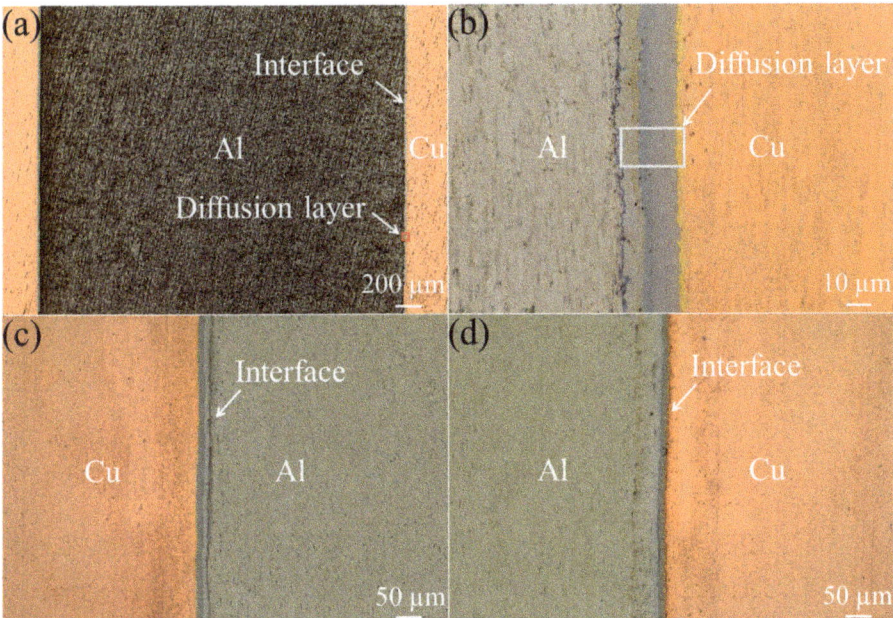

Figure 5. The bonding interface and diffusion layer condition of the composites. (**a**) Microscopic drawing of the sample. (**b**) Morphology of Cu/Al diffusion layer. (**c**,**d**) Microscopic views of Cu/Al upper and lower interfaces.

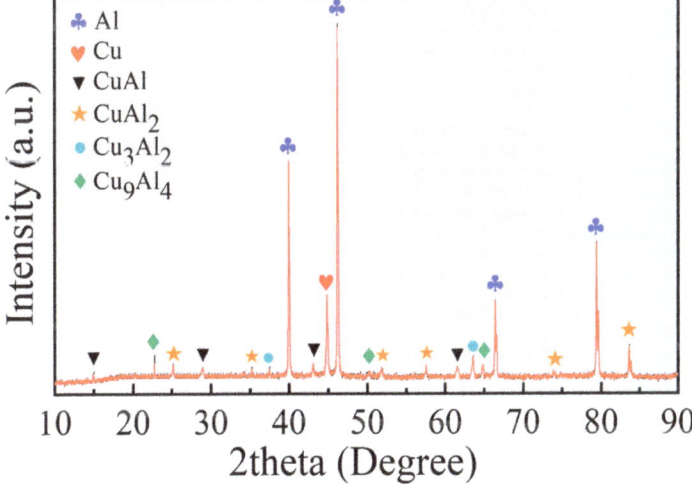

Figure 6. XRD test of Cu/Gr-Al-Cu/Gr laminated composites cross section.

Figure 7a shows the secondary electron images of the diffusion layer under the SEM of the Cu/Gr-Al-Cu/Gr laminated composites, and the obvious delamination at the diffusion layer is also found by SEM. Figure 7b shows the surface scanning area of the sample. The distribution mapping of Cu and Al elements can be found in Figure 7c,d, and the area with a large overlap of Cu and Al elements is the diffusion layer. This area was quantified by line scan for a total Cu/Al diffusion layer thickness of 27.19 µm. Figure 7e shows the results for the line scan area in Figure 7b. The line scan results also easily reveal the diffusion reaction between Cu and Al. The diffusion coefficient of Cu and Al can be expressed by Arrhenius, which can be calculated from Equation (1) [29].

$$D = D_0 e^{(-Q/RT)} \quad (1)$$

where D is the diffusion coefficient, D_0 is the diffusion constants, Q is the diffusion activation energy, R is the gas constant, and T is the thermodynamic temperature. The diffusion coefficient of Cu in Al is 4.9×10^{-16} m$^2 \cdot$s^{-1}, while that of Al in Cu is 3.76×10^{-19} m$^2 \cdot$s^{-1}, at 530 °C. The diffusion coefficient of Cu atoms in Al is much larger than that of Al atoms in Cu. Cu is regarded as the first limiting element. Therefore, the formation of the Cu/Al interface reaction layer is mainly through the diffusion of Cu atoms to the Al side. Moreover, the formation energies of $CuAl_2$ and Cu_9Al_4 are 0.78 eV and 0.83 eV. It is inferred that $CuAl_2$ forms first and then forms Cu_9Al_4. CuAl and Cu_3Al_2 are exhibited to form after the formation of the previous two phases [30]. The Cu/Al intermediate phase has the characteristics of high strength, low ductility and high resistance, but the generation of the Cu/Al intermetallics can bind the samples tightly and form a good interface to improve the force and deformation of the composite. Figure 7f shows a schematic diagram of the Cu/Al diffusion reaction.

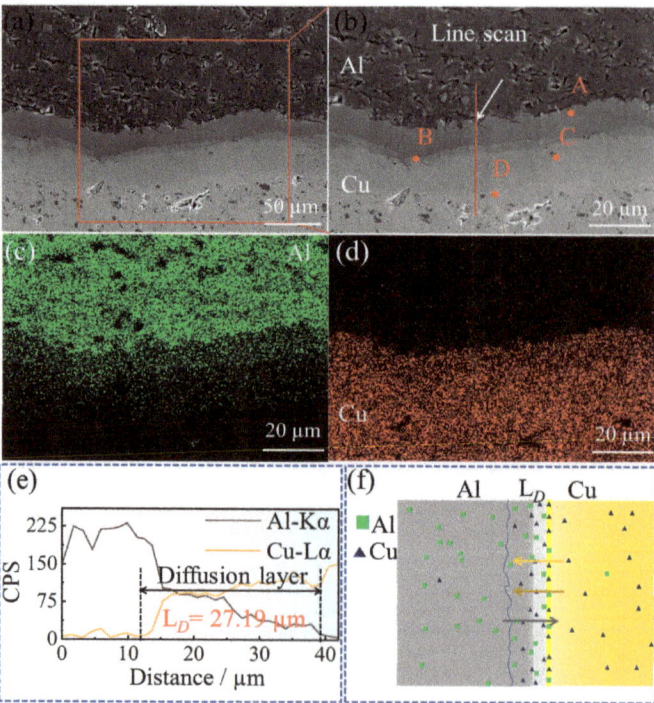

Figure 7. (a) SEM images of diffusion layer morphology; (b) SEM image of the sample surface scanning area and A, B, C, D are point scanning areas; (c) Distribution of Al elements; (d) Distribution of Cu elements; (e) Sample line scan area and energy spectrum; (f) Cu and Al diffusion diagram.

Figure 8 shows the EDS of points A–D in Figure 7b. Point A is close to the aluminum side, which can be identified as the $CuAl_2$ phase by combining the XRD and EDS results of point A. Point B is closer to the copper side than point A, and the CuAl phase can be determined from the XRD and EDS results. The EDS results and XRD analysis at point C confirm that this is the Cu_3Al_2 phase. Point D is close to the copper side, and the EDS results and XRD analysis determine that this is the Cu_9Al_4 phase. Therefore, the intermediate phases from the Al side to the Cu side are $CuAl_2$, $CuAl$, Cu_3Al_2, and Cu_9Al_4 in that order.

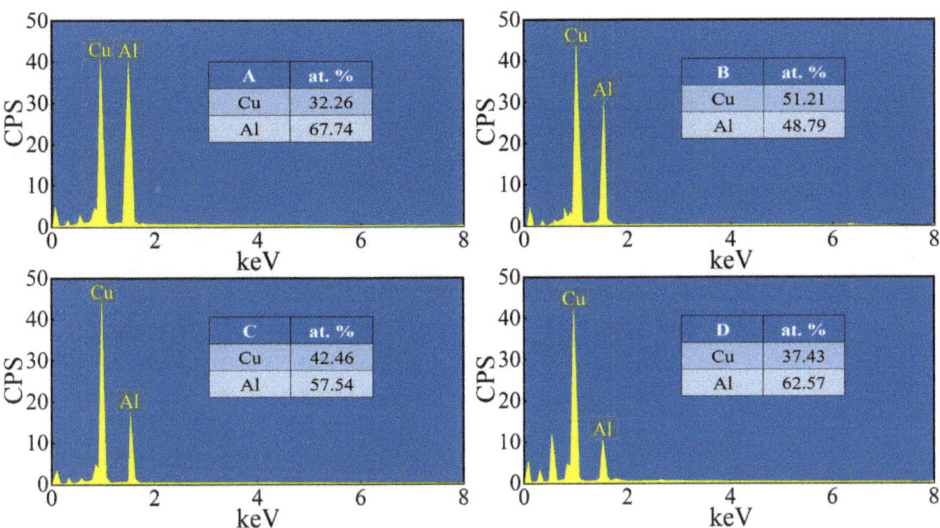

Figure 8. EDS result of point A-D in Figure 7b.

3.3. Electrical Conductivity

The vdP method of resistance measurement is repeated eight times around the edge of the sample, and the eight sets of voltages (U) and test currents (I) obtained are used to calculate the resistivity ρ, which can be calculated from Equations (2)–(4).

$$\rho = \frac{\pi \cdot d}{\ln 2} \times \left(\frac{U_{CD}}{I_{AB}} + \frac{U_{DA}}{I_{BC}} \right) \times \frac{1}{2} \times f \quad (2)$$

$$f \approx 1 - \left(\frac{R_{AB,CD} - R_{BC,DA}}{R_{AB,CD} + R_{BC,DA}} \right)^2 \frac{\ln 2}{2} - \left(\frac{R_{AB,CD} - R_{BC,DA}}{R_{AB,CD} + R_{BC,DA}} \right)^4 \quad (3)$$

$$R_{AB,CD} = \frac{U_{CD}}{I_{AB}}; R_{BC,DA} = \frac{U_{DA}}{I_{BC}} \quad (4)$$

where d is the sample thickness, f is the vdP factor [31], U_{CD} and U_{DA} are the measured voltages, I_{AB} and I_{BC} are the measured currents, and $R_{AB,CD}$ and $R_{BC,DA}$ are the resistance of the sample.

The electrical conductivity of the sample with different Cu(111) and Cu(111)/Gr thicknesses are shown in Figure 9. This work considered three different Cu(111) and Cu(111)/Gr thicknesses, i.e., 0 mm (Al), 1 mm (Cu/Al/Cu and Cu/Gr-Al-Cu/Gr), and 1.2 mm (Cu/Al/Cu and Cu/Gr-Al-Cu/Gr). It should be noted that as the thickness of Cu changes, the number of graphene layers changes. It can be seen that for the sample with a layer thickness of 1 mm, the electrical conductivity reached 68.8% IACS, which is 2.5% higher than the 67.2% IACS of the same sized sample without graphene. Meanwhile, for the sample with a layer thickness of 1.2 mm, the electrical conductivity is 71.9% IACS, which is 3.3% higher than that of Cu/Al/Cu without graphene (69.6% IACS). In addition, it is known that the number of graphene layers introduced by increasing the thickness of the Cu(111)/Gr layer also increased, and it was found that the sample's electrical conductivity improved considerably after graphene was added. For example, by comparing the sample of Cu/Gr-Al-Cu/Gr with the thickness 1mm and 1.2mm, the electrical conductivity is increased by approximately 4.5% from 68.8% IACS to 71.9% IACS, whereas for the Cu/Al/Cu samples without Gr, the electrical conductivity is increased by approximately 3.6% with thicknesses increased from 1 mm to 1.2 mm. Furthermore, it can also be seen from Figure 9 that the graphene-embedded Cu/Al composite exhibited high carrier mobility from controlled experiments, and the resulting samples showed a maximum enhancement of 20.5% over the raw material Al.

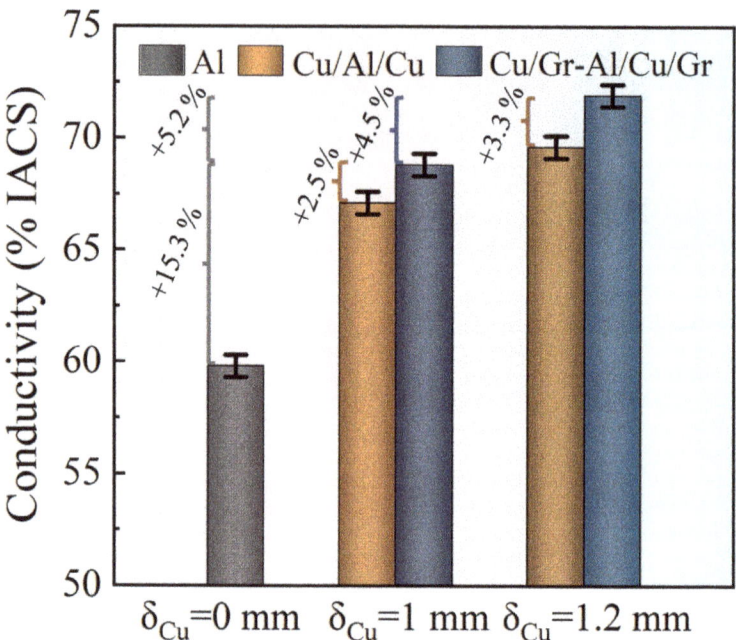

Figure 9. Comparison of the electrical conductivity of three different Cu(111) and Cu(111)/Gr thickness (0 mm, 1 mm and 1.2 mm) in the composites.

The high quality graphene is grown on Cu foil by CVD [32]. Moreover, the weak adhesion energy of graphene to Cu can be improved by the hot pressing process so that graphene and Cu can form a strong mechanical bond and Cu(111)/Gr blocks without inclusions and voids can be obtained [33,34]. Moreover, graphene and Cu(111) have the same triple symmetry and very similar lattice constants, which allows graphene to be grown more completely on Cu(111), thus exploiting the high carrier migration properties of graphene [21]. In many studies on the electrical properties of graphene, there is a trade-off between electron mobility and electron density when suspended graphene or graphene interacts with a well-designed substrate; the result is a less than high conductivity of graphene-reinforced metals. In contrast, graphene materials embedded with Cu metal achieve high electron density and maintain high electron mobility. In addition, the thickness of the diffusion layer of Cu and Al is opposite to the conductivity; the electrical conductivity decreases with the increase of the thickness of the diffusion layer. The thickness of the diffusion layer of the Cu/Al composite prepared by the hot pressing method is generally smaller than that of the casting method, which also improves the electrical properties of the Cu/Al composite [18]. Therefore, by hot pressing composites Cu(111)/Gr with Al, the resulting samples will have the combined properties of lightweight and high electrical conductivity.

4. Conclusions

In this study, an innovative Cu/Gr-Al-Cu/Gr laminated composite was prepared by the hot pressing method, and the graphene states, interface bonding, diffusion layer thickness, electrical conductivity, Miller indices and phase analysis of the composites were studied. The main conclusions are as follows:

1. The Cu(111)/Gr sample obtained by CVD was vacuum pressed twice at 900 °C and 530 °C. The Miller indices of the sample remained at the (111) crystal face, and the Cu block was still in the single crystal state;
2. The Cu/Gr-Al-Cu/Gr laminated composites were successfully prepared by hot pressing for 1 h at a temperature of 530 °C and a pressure of 10 MPa at a heating rate of 10 °C/min. It was found that the laminated composites were well bonded without pores or cracks, there was an obvious diffusion layer at the interface bond, and the transition layer generated by the diffusion reaction of Cu and Al connected the composites. The total thickness of the diffusion layer was found to be 27.19 μm by EDS spotting, line scan, surface scan, and the intermediate phases from the Al side to the Cu side are $CuAl_2$, $CuAl$, Cu_3Al_2, and Cu_9Al_4, in that order;
3. The Cu/Gr-Al-Cu/Gr laminated composites prepared by the hot pressing method were able to exploit the high carrier mobility of graphene to improve the electrical conductivity of the composites and the thickness of the Cu(111)/Gr layer from 1 mm to 1.2 mm; the electrical conductivity of the Cu/Gr-Al-Cu/Gr increased by 4.5%, while the increase in the thickness of the Cu(111) layer was from 1 mm to 1.2 mm, and the Cu/Al/Cu conductivity increased by only 3.6%.

Author Contributions: H.Z., investigation, data curation, and writing—original draft; R.Z., data curation and editing; Q.X., revision of original draft and funding acquisition; X.K., review and supervision; W.S., writing—review and editing; Y.F., investigation and funding acquisition; M.W. and K.L., methodology. All authors have read and agreed to the published version of the manuscript.

Funding: The project was supported by the Guangdong Major Project of Basic and Applied Basic Research (2021B0301030002), the Innovative Model Factory Project of Songshan Lake Materials Laboratory (Y1D1051C511/Y1Q1011C511), and the Educational Department of Henan Province (21A430010).

Institutional Review Board Statement: Not applicable.

Informed Consent Statement: Not applicable.

Data Availability Statement: All relevant data generated by the authors or analyzed during the study are included within the paper.

Conflicts of Interest: The authors declare no conflict of interest.

References

1. Kunčická, L.; Kocich, R.; Dvořák, K.; Macháčková, A. Rotary swaged laminated Cu-Al composites: Effect of structure on residual stress and mechanical and electric properties. *Mater. Sci. Eng. A* **2019**, *742*, 743–750. [CrossRef]
2. Wu, K.; Chang, H.; Maawad, E.; Gan, W.; Brokmeier, H.; Zheng, M. Microstructure and mechanical properties of the Mg/Al laminated composite fabricated by accumulative roll bonding (ARB). *Mater. Sci. Eng. A* **2010**, *527*, 3073–3078. [CrossRef]
3. Tayyebi, M.; Eghbali, B. Study on the microstructure and mechanical properties of multilayer Cu/Ni composite processed by accumulative roll bonding. *Mater. Sci. Eng. A* **2013**, *559*, 759–764. [CrossRef]
4. Abbas, M.; Rasheed, M. *Solid State Reaction Synthesis and Characterization of Cu Doped TiO_2 Nanomaterials*; Journal of Physics: Conference Series; IOP Publishing: Bristol, UK, 2021; pp. 12059–12063.
5. Mozaffari, A.; Manesh, H.D.; Janghorban, K. Evaluation of mechanical properties and structure of multilayered Al/Ni composites produced by accumulative roll bonding (ARB) process. *J. Alloy. Compd.* **2010**, *489*, 103–109. [CrossRef]
6. Ghalandari, L.; Mahdavian, M.; Reihanian, M.; Mahmoudiniya, M. Production of Al/Sn multilayer composite by accumulative roll bonding (ARB): A study of microstructure and mechanical properties. *Mater. Sci. Eng. A* **2016**, *661*, 179–186. [CrossRef]
7. Sun, Y.; Tsuji, N.; Fujii, H.; Li, F. Cu/Zr nanoscaled multi-stacks fabricated by accumulative roll bonding. *J. Alloy. Compd.* **2010**, *504*, S443–S447. [CrossRef]

8. Motevalli, P.D.; Eghbali, B. Microstructure and mechanical properties of Tri-metal Al/Ti/Mg laminated composite processed by accumulative roll bonding. *Mater. Sci. Eng. A* **2015**, *628*, 135–142. [CrossRef]
9. Rogachev, S.; Andreev, V.; Yusupov, V.; Bondareva, S.; Khatkevich, V.; Nikolaev, E. Effect of Rotary Forging on Microstructure Evolution and Mechanical Properties of Aluminum Alloy/Copper Bimetallic Material. *Met. Mater. Int.* **2022**, *28*, 1038–1046. [CrossRef]
10. Shayanpoor, A.; Ashtiani, H.R. Microstructural and mechanical investigations of powder reinforced interface layer of hot extruded Al/Cu bimetallic composite rods. *J. Manuf. Process.* **2022**, *77*, 313–328. [CrossRef]
11. Xing, B.-H.; Huang, T.; Song, K.-X.; Xu, L.-J.; Xiang, N.; Chen, X.-W.; Chen, F.-X. Effect of electric current on formability and microstructure evolution of Cu/Al laminated composite. *J. Mater. Res. Technol.* **2022**, *21*, 1128–1140. [CrossRef]
12. Chen, D.; Zhang, H.; Zhao, D.; Liu, Y.; Jiang, Z. Effects of annealing temperature on interface microstructure and element diffusion of ultra-thin Cu/Al composite sheets. *Mater. Lett.* **2022**, *322*, 132491. [CrossRef]
13. Kocich, R.; Kunčická, L.; Davis, C.F.; Lowe, T.C.; Szurman, I.; Macháčková, A. Deformation behavior of multilayered Al-Cu clad composite during cold-swaging. *Mater. Des.* **2016**, *90*, 379–388. [CrossRef]
14. Kunčická, L.; Kocich, R. Optimizing electric conductivity of innovative Al-Cu laminated composites via thermomechanical treatment. *Mater. Des.* **2022**, *215*, 110441. [CrossRef]
15. Lapovok, R.; Berner, A.; Qi, Y.; Xu, C.; Rabkin, E.; Beygelzimer, Y. The effect of a small copper addition on the electrical conductivity of aluminum. *Adv. Eng. Mater.* **2020**, *22*, 2000058. [CrossRef]
16. Kocich, R.; Kunčická, L.; Král, P.; Strunz, P. Characterization of innovative rotary swaged Cu-Al clad composite wire conductors. *Mater. Des.* **2018**, *160*, 828–835. [CrossRef]
17. Han, Y.-q.; Ben, L.-h.; Yao, J.-j.; Feng, S.-w.; Wu, C.-j. Investigation on the interface of Cu/Al couples during isothermal heating. *Int. J. Miner. Metall. Mater.* **2015**, *22*, 309–318. [CrossRef]
18. Yuan, H.; Chen, Y.-q.; Li, L.; Hu, H.-d.; Zhu, Z.-a. Microstructure and properties of Al/Cu bimetal in liquid–solid compound casting process. *Trans. Nonferrous Met. Soc. China* **2016**, *26*, 1555–1563.
19. Shamaila, Y.; Ahmad, B.A.; Adnan, A. Study of Carbon Nanbotubes and Boron Nanotubes Using Degree Based Topological Indices. *Polycycl. Aromat. Compd.* **2022**, *42*, 7724–7737.
20. Rasheed, M.; Shihab, S.; Sabah, O.W. *An Investigation of the Structural, Electrical and Optical Properties of Graphene-Oxide Thin Films Using Different Solvents*; Journal of Physics: Conference Series; IOP Publishing: Bristol, UK, 2021; p. 012052.
21. Xu, X.; Zhang, Z.; Dong, J.; Yi, D.; Niu, J.; Wu, M.; Lin, L.; Yin, R.; Li, M.; Zhou, J.; et al. Ultrafast epitaxial growth of metre-sized single-crystal graphene on industrial Cu foil. *Sci. Bull.* **2017**, *62*, 1074–1080. [CrossRef]
22. Zhao, Y.; Niu, M.; Yang, F.; Jia, Y.; Cheng, Y. Ultrafast electro-thermal responsive heating film fabricated from graphene modified conductive materials. *Eng. Sci.* **2019**, *8*, 33–38. [CrossRef]
23. Stankovich, S.; Dikin, D.A.; Dommett, G.H.; Kohlhaas, K.M.; Zimney, E.J.; Stach, E.A.; Piner, R.D.; Nguyen, S.T.; Ruoff, R.S. Graphene-based composite materials. *Nature* **2006**, *442*, 282–286. [CrossRef]
24. Yu, J.; Wang, L.; Liu, Z.; Xu, J.; Zong, Y. Electrodeposition-based fabrication of graphene/copper composites with excellent overall properties. *J. Alloy. Compd.* **2022**, *924*, 166610. [CrossRef]
25. Dong, L.; Chen, W.; Zheng, C.; Deng, N. Microstructure and properties characterization of tungsten–copper composite materials doped with graphene. *J. Alloy. Compd.* **2017**, *695*, 1637–1646. [CrossRef]
26. Cao, M.; Xiong, D.-B.; Tan, Z.; Ji, G.; Amin-Ahmadi, B.; Guo, Q.; Fan, G.; Guo, C.; Li, Z.; Zhang, D. Aligning graphene in bulk copper: Nacre-inspired nanolaminated architecture coupled with in-situ processing for enhanced mechanical properties and high electrical conductivity. *Carbon* **2017**, *117*, 65–74. [CrossRef]
27. Chen, F.; Ying, J.; Wang, Y.; Du, S.; Liu, Z.; Huang, Q. Effects of graphene content on the microstructure and properties of copper matrix composites. *Carbon* **2016**, *96*, 836–842. [CrossRef]
28. Cong, C.; Yu, T.; Ni, Z.; Liu, L.; Shen, Z.; Huang, W. Fabrication of graphene nanodisk arrays using nanosphere lithography. *J. Phys. Chem. C* **2009**, *113*, 6529–6532. [CrossRef]
29. Zhang, J.; Wang, B.-h.; Chen, G.-h.; Wang, R.-m.; Miao, C.-h.; Zheng, Z.-x.; Tang, W.-m. Formation and growth of Cu–Al IMCs and their effect on electrical property of electroplated Cu/Al laminar composites. *Trans. Nonferrous Met. Soc. China* **2016**, *26*, 3283–3291. [CrossRef]
30. Chen, C.Y.; Hwang, W.-S. Effect of annealing on the interfacial structure of aluminum-copper joints. *Mater. Trans.* **2007**, *48*, 0706110009. [CrossRef]
31. Yihui, G. *Mathematical Analysis of van der Pauw's Method for Measuring Resistivity*; Journal of Physics: Conference Series; IOP Publishing: Bristol, UK, 2022; Volume 2321.
32. Li, X.; Cai, W.; An, J.; Kim, S.; Nah, J.; Yang, D.; Piner, R.; Velamakanni, A.; Jung, I.; Tutuc, E.; et al. Large-Area Synthesis of High-Quality and Uniform Graphene Films on Copper Foils. *Science* **2009**, *324*, 1312–1314. [CrossRef] [PubMed]

33. Hwang, J.; Yoon, T.; Jin, S.H.; Lee, J.; Kim, T.S.; Hong, S.H.; Jeon, S. Enhanced mechanical properties of graphene/copper nanocomposites using a molecular-level mixing process. *Adv. Mater.* **2013**, *25*, 6724–6729. [CrossRef]
34. Yoon, T.; Shin, W.C.; Kim, T.Y.; Mun, J.H.; Kim, T.-S.; Cho, B.J. Direct measurement of adhesion energy of monolayer graphene as-grown on copper and its application to renewable transfer process. *Nano Lett.* **2012**, *12*, 1448–1452. [CrossRef] [PubMed]

Disclaimer/Publisher's Note: The statements, opinions and data contained in all publications are solely those of the individual author(s) and contributor(s) and not of MDPI and/or the editor(s). MDPI and/or the editor(s) disclaim responsibility for any injury to people or property resulting from any ideas, methods, instructions or products referred to in the content.

Article

Orientation Control for Nickel-Based Single Crystal Superalloys: Grain Selection Method Assisted by Directional Columnar Grains

Songsong Hu [1,†], Yunsong Zhao [2,†], Weimin Bai [1], Yilong Dai [1,*], Zhenyu Yang [2], Fucheng Yin [1] and Xinming Wang [1,*]

1 School of Materials Science and Engineering, Xiangtan University, Xiangtan 411105, China; songhu@xtu.edu.cn (S.H.); baiweimin@xtu.edu.cn (W.B.); fuchengyin@xtu.edu.cn (F.Y.)
2 Key Laboratory of Advanced High Temperature Structural Materials, AECC Beijing Institute of Aeronautical Materials, Beijing 100095, China; yunsongzhao@163.com (Y.Z.); xuanshangyiyi@163.com (Z.Y.)
* Correspondence: daiyilong@xtu.edu.cn (Y.D.); wangxm@xtu.edu.cn (X.W.)
† These authors contributed equally to this work.

Abstract: The service performance of single crystal blades depends on the crystal orientation. A grain selection method assisted by directional columnar grains is studied to control the crystal orientation of Ni-based single crystal superalloys. The samples were produced by the Bridgman technique at withdrawal rates of 100 μm/s. During directional solidification, the directional columnar grains are partially melted, and a number of stray grains are formed in the transition zone just above the melt-back interface. The grain selected by this method was one that grew epitaxially along the un-melted directional columnar grains. Finally, the mechanism of selection grain and application prospect of this grain selection method assisted by directional columnar grains is discussed.

Keywords: Ni-based single crystal superalloy; directional solidification; selection grain; columnar grains; crystal orientation

Citation: Hu, S.; Zhao, Y.; Bai, W.; Dai, Y.; Yang, Z.; Yin, F.; Wang, X. Orientation Control for Nickel-Based Single Crystal Superalloys: Grain Selection Method Assisted by Directional Columnar Grains. Materials 2022, 15, 4463. https:// doi.org/10.3390/ma15134463

Academic Editor: Andrey Belyakov

Received: 25 May 2022
Accepted: 23 June 2022
Published: 24 June 2022

Publisher's Note: MDPI stays neutral with regard to jurisdictional claims in published maps and institutional affiliations.

Copyright: © 2022 by the authors. Licensee MDPI, Basel, Switzerland. This article is an open access article distributed under the terms and conditions of the Creative Commons Attribution (CC BY) license (https:// creativecommons.org/licenses/by/ 4.0/).

1. Introduction

Nickel-based single crystal superalloys have excellent high-temperature mechanical properties and are the preferred materials for aero-engine and gas turbine blades [1]. Due to the anisotropy of mechanical properties, the single crystal blade has the best service performance when the <001> direction of the crystal is consistent with the maximum stress direction [2]. Directional solidification investment casting, combined with grain selection or seeding technology, is the main method for preparing single crystal blades [3,4]. The grain selection method is the earliest used to prepare single crystal blades, and it is also the most common method for the preparation of single crystal blades in industrial production. In this method, a spiral selector is usually set at the bottom of the casting. A large number of grains is nucleated firstly at the bottom of the spiral selector, and then the competitive grain growth occurs during the directional solidification process, and finally, only one grain survives at the outlet of the spiral selector [5].

The spiral selector can generally be divided into two parts: the starter block and the spiral part [6–8]. The functions of these two parts have received extensive attention and a consensus has been reached [5,7,9]. The main function of the starter block is obtaining well-oriented grains with the <001> texture [10]. Furthermore, the main function of the spiral part is to accelerate the competition grain's grain and ensure only one grain is close to the inner side of the spiral channel growing into the casting cavity [11]; therefore, the orientation of the final surviving grain is mainly controlled by the competitive grains growth process in the starter block [12]. Only the primary orientation of the surviving grain can be limited within a certain range, which is the degree of <001> direction deviating from the directional solidification, whereas the second orientation is randomly distributed due to the selection mechanism.

In order to improve the control level of the crystal orientation of the surviving grain obtained by the grain selection method, researchers have carried out related research from multiple perspectives, such as the competitive grain growth [13,14], the optimization of the starter block structure and size [10,12], and the effect of process parameters on grain selection [3,15]. Zhang and Zhou et al. found that a high withdrawal rate is beneficial for obtaining a sharper <001> texture and smaller grain size attributing to the competitive growth of a large amount of randomly oriented grains [16]. Wang et al. reported that the misorientation of the final grain can be decreased with the increase in the ratio of length to diameter of the starter block [10]. Rezaei et al. reported that the withdrawal rate, height of the starter block, and temperature gradient during directional solidification were inter-dependent and careful optimization of these process parameters was indeed needed to achieve a minimum misorientation of the final grain [17]. Some studies have also shown that by adjusting the structure and size of the spiral part carefully, the orientation of the final grain can also be optimized within a small scale [5,9,12]. At present, the misorientation of the final grain obtained by the grain selection method can be controlled within 15° for turbine blade applications in industrial production.

With the complexity and thinning tendency of turbine blades, the service performance of the single crystal blade is increasingly sensitive to the crystal orientation [18,19]; therefore, the misorientation of the final grain obtained by the grain selection method is required to be narrower. To address this, an adjusted grain selection method assisted by directional columnar grains for single crystal blades is proposed [20]. A directional columnar grain is set in the starter block during directional solidification, which can significantly reduce the misorientation of the final grain. Using Optical microscope and EBSD technology, the preparation process of the adjusted grain selection method assisted by directional columnar grains for a single crystal sample was investigated. It provides a new insight for optimizing the orientation control level of the grain selection method.

2. Experiment

The materials used in this paper were DZ125 and DD6, and their nominal composition are listed in Table 1. The cylinder-shaped directional columnar grains used in this work had a diameter of 10 mm and a length of 40 mm. The spiral part used in this study was a spiral part for the industrial production of single crystal blades. The directional columnar grains were plugged into a mold and the clearance between the columnar grains and the mold was less than 0.05 mm. Then, the mold with the directional columnar grains (as shown in Figure 1) was mounted on a water-cooled copper chill plate in a Bridgman furnace. The furnace chamber was evacuated to a partial pressure above 10^{-2} Pa, and then the mold was heated to 1520 °C for 5 min by a graphite heating element. Melton alloy was poured into the preheated mold cavity at 1500 °C and stabilized for 5 min. Finally, the mold was withdrawn to a cold element at a rate of 100 μm/s. During directional solidification, the main heat flow direction is consistent with the withdrawal direction, whereas it is the opposite for the directional solidification; they are all parallel to the columnar axis.

Table 1. The nominal compositions of the DZ125 and DD6 (wt.%).

Element	Cr	Co	W	Mo	Re	Al	Ti	Ta	Nb	C	Hf	B	Ni
DZ125	8.9	10.0	7.0	2.0	-	5.2	0.9	3.8	-	0.1	1.5	0.015	Bal.
DD6	4.3	9.0	8.0	2.0	2.0	5.6	-	7.5	0.5	-	0.1	-	Bal.

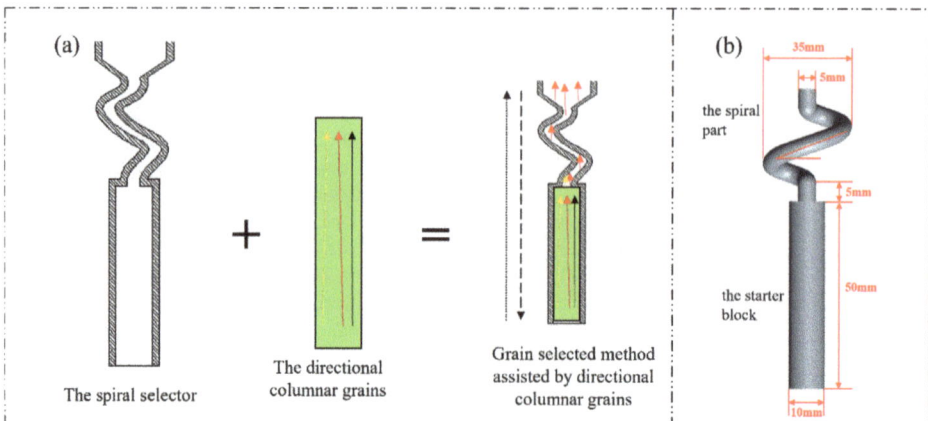

Figure 1. Schematic diagram of the grain selected method assisted by directional columnar grains: (**a**) the novel method; (**b**) the spiral selector used in this paper.

The microstructure of the directional columnar grains used in this study is shown in Figure 2. From Figure 2a, it can be seen that the typical dendrite morphology at the cross-section of the directional columnar grains and the arrangement of the dendrites were not completely consistent due to the difference in the orientation of these dendrites. According to the statistical results, the size of the grains in the directional columnar grains in the cross-section was between 0.8 mm and 5.1 mm, and the average primary dendrite spacing was about 304 μm. The maximum deviation angle of the <001> direction of the grains in the directional columnar grains from the columnar axis was not more than 8°. Figure 2b showed the typical microstructure in the longitudinal section; it can be seen that the <001> direction of the grain deviated from the columnar axial within a small angle.

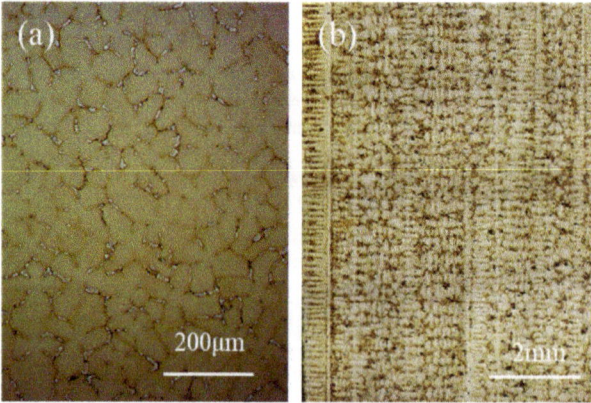

Figure 2. The cross-sectional (**a**) and longitudinal sectional (**b**) microstructure of the directional columnar grains.

Following directional solidification, the casting was cleared and macro-etched by a solution of 50% HCl and 50% H_2O_2 for the macroscopically inspecting grains evolution process. Casting was subsequently machined using wire electro-discharge machining (EDM), and then the samples were polished and etched with a mixture of 14% HNO_3, 28% HF and 58% H_3O_8 to display the microstructure. An optical microscope (Leica DM-4000, Wetzlar, Germany) and SEM (ZEIS SUPRA 55, Germany) equipped with EBSD were employed to obtain the microstructures and crystal orientation information of the samples. In order to accurately display the information of crystal orientation during directional solidification, an orthogonal coordinate system was established. The 0Z direction is parallel to the directional solidification direction, and the X0Y plane is perpendicular to the directional solidification direction. In particular, the 0X direction coincides with the projection of the secondary dendrite of the grain selected by this method on the X0Y plane.

3. Results

The macrostructure of the adjusted grain selection process assisted by directional columnar grains is shown in Figure 3a. It can be seen that well-oriented columnar grains were formed in the starter block, as the growth direction of the grains was basically parallel to the directional solidification. The columnar grains in the starter block were divided into two parts by the melt-back interface, namely the un-melted zone, which was located in the lower part, and the re-solidified zone in the upper part. Figure 3b shows a partial magnified view near the melt-back interface. A number of stray grains can be found to form exclusively at the surface of the directional columnar grains. The nucleation of stray grains was a transient phenomenon; it occurred only near the melt-back interface. Since the orientation of stray grains was random, which is attributed to nucleation behavior, only a few stray grains with an orientation advantage can grow up in the competitive growth process with the columnar grains. This phenomenon was consistent with the formation and growth of stray grains at the melt-back region of seeding for preparing single crystal casting. Subsequently, the number of grains in the starter block did not change significantly, whereas the number of grains decreased rapidly in the spiral part. A single crystal can be rapidly selected as the solidification front climbing to the spiral part; its cross-sectional and longitudinal microstructures are shown in Figure 3c,d, respectively. The <001> direction of the grain prepared by this method was closed to the direction of solidification, as shown in Figure 3d. It indicated that the grain selection method assisted by directional columnar grains could effectively prepare a single crystal with a small misorientation.

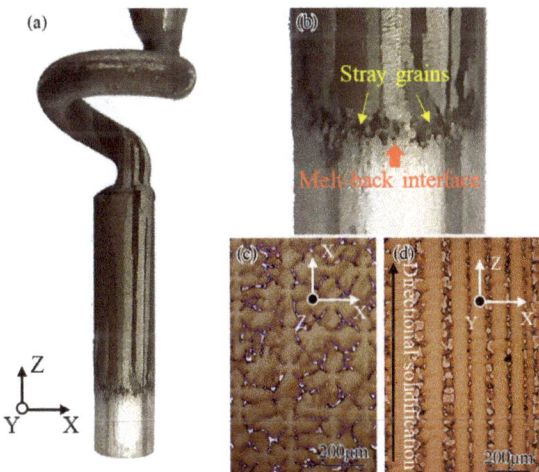

Figure 3. The macrostructure of the adjusted grain selection process assisted by directional columnar grains: (**a**) a magnification view near melt-back region; (**b**) cross-sectional; (**c**) longitudinal (**d**) microstructure of casting.

In order to further clarify the orientation control level of the single crystal castings prepared by the grain selection assisted by directional columnar grains, there were 16 single crystal castings fabricated using this method, and their orientation information is displayed in Figure 4. The <001> direction of the single crystal castings deviated from the directional solidification was all less than 8 degrees. Especially, the proportion of the single crystal casting whose <001> direction deviated from the directional solidification within 5 degrees was more than 50%. It showed that the grain selection assisted by directional columnar grains could more effectively control the orientation of the single crystal component than the traditional grain selection method.

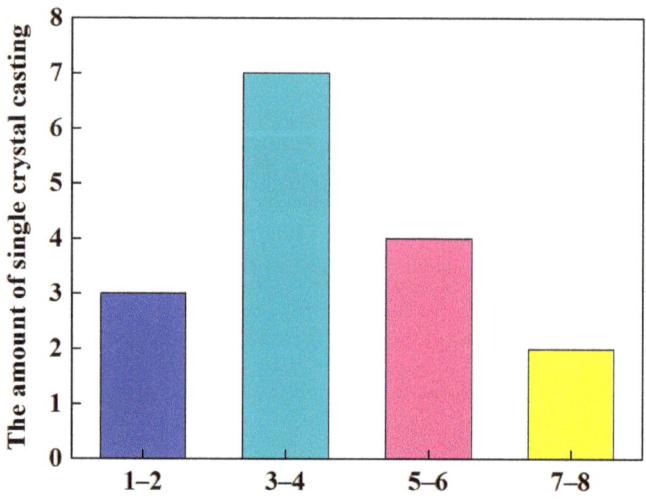

Figure 4. The statistical result of <001> direction of single crystal casting deviated from the directional solidification.

In order to further clarify the mechanism for obtaining a single crystal casting with a small misorientation using this adjusting grain selection method, the microstructure in the starter block was characterized, and the results are shown in Figure 5. From Figure 5a, it could be found that the starter block could be divided into four parts after directional solidification, namely the heat-affected zone, the mushy zone, the transition zone, and the directional growth zone. The heat-affected zone marked as I in Figure 5a was located at the lowest region of the starter block. From Figure 5b, it could be found that there was no obvious re-melting phenomenon in this zone, whereas the $(\gamma + \gamma')$ eutectic was almost completely eliminated, indicating that the microsegregation was significantly reduced. Mushy zone II was immediately adjacent to the heat-affected zone and there was no clear boundary between the two zones. Due to microsegregation, the $(\gamma + \gamma')$ eutectic reformed in the interdendritic region after directional solidification, as shown in Figure 5c. Transition zone III was clearly separated from the mushy zone by the melt-back interface, as shown in Figure 5a. Figure 5d shows atypical dendrites formed and some stray grains nucleated at the initial directional solidification stage. Finally, the directional growth zone IV could be found at the top part of the starter block, and a typical columnar microstructure was reformed, as shown in Figure 5a,e. It should be noted that Figure 5a displayed that the grain with a larger misorientation was overgrown by one with a small misorientation.

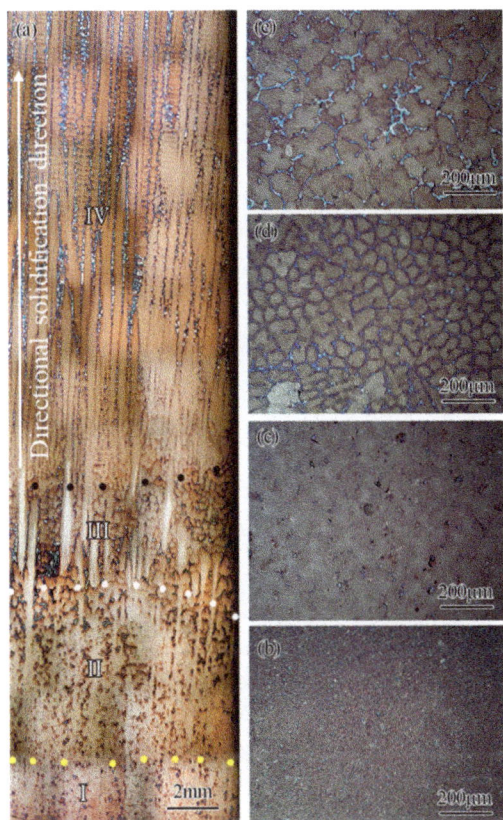

Figure 5. The microstructure of the directional columnar grains after preparing single crystal casting: (**a**) the longitudinal microstructure, and the cross-section microstructures in (**b**) heat-affected zone I; (**c**) mushy zone II; (**d**) transition zone III; (**e**) directional growth zone IV.

The crystal orientations at different regions of the longitudinal section were measured by EBSD—the results are shown in Figure 6. From Figure 6(b1,b2), it can be seen that the formation of stray grains began to occur at the transition zone just above the melt-back interface. At the same time, no stray grain was found below the melt-back interface. A number of equiaxed grains formed in the middle region of the transition zone, as shown in Figure 6(c1,c2); however, the grains aligned with the matrix (located un-melted region) orientation still dominated the transition zone. After the solidification interface was advanced to the directional growth zone, the nucleation of stray grains was not found by EBSD detection. Comparing Figure 6(b1,b2) and (d1,d2), it could be found that the crystal orientation of gains in this zone was basically the same as that of the matrix, which was consistent with the metallographic observation in Fig. 5. It indicated that the grains in the directional growth zone were mainly exitaxially grown along the un-melted columnar grains. This could explain why the maximum deviation angle of the single crystal obtained by the final selection method was consistent with the maximum orientation deviation of the columnar grains.

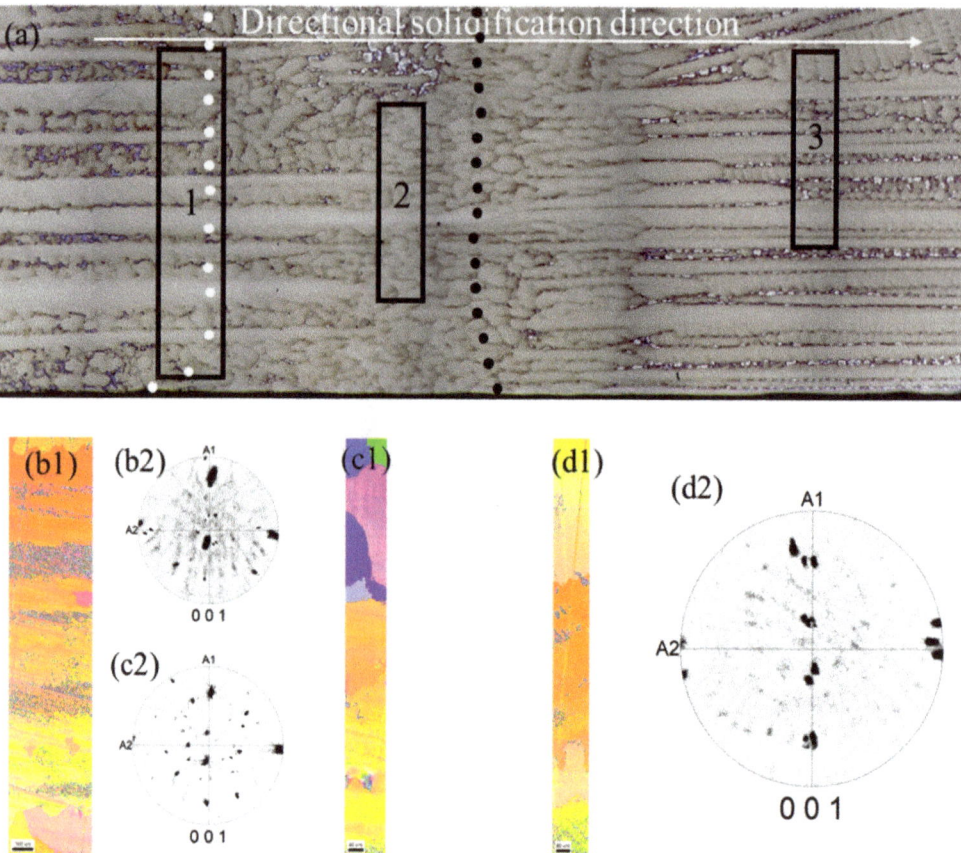

Figure 6. The metallographic microstructure and EBSD results detected by local area: (**a**) the longitudinal microstructure and EBSD maps and corresponding polar diagram tested in (**b1,b2**) zone 1, (**c1,c2**) zone 2, and (**d1,d2**) zone 3 in (**a**).

4. Discussion

As mentioned in the introduction, the grain selection and seeding methods are the main technique for preparing single crystal components [4,5]. The grain selection technique mainly relies on the competitive grain growth in the grain selector to obtain single crystal casting. In order to reduce the misorientation range of single crystal casting, a lot of effort is required to determine the grain selection mechanism [3,7,12], optimization of the selector structure [8,10], and directional solidification perimeter [16,17]; however, the random nature of the grain selection process has not changed, leading to the orientation range of the selected grain in industrial production still being quite large for advanced turbine blades. The nature of the seeding technique is epitaxial growth along an un-melted seed; therefore, high single crystal integrity is required for seeding to avoid solidification defects growing up from the un-melted seeding. At the same time, the process parameters, such as the gap between the seed and mold, the inner surface roughness of the mold, etc., should be strictly limited to avoid the formation of stray grains at the initial withdrawal stage during directional solidification [21,22]. Some studies have also shown that the seed oxidation occurred prior to melt pouring, which gives us a chance to form solidification defects at the withdrawal stage [23]; therefore, the seeding method is often combined with a selector partner between the seed and component to improve the success rate of the single crystal preparation [24].

The grain selection method for Ni-based single crystal blades relies on the competitive grain growth to obtain only one grain at the top of the spiral selector. Walton and Chalmers first proposed a competitive growth model based on the relationship between dendrite growth rate and undercooling [25]. After revealing this model with a schematic diagram by Rappaz [13], it is widely accepted and can explain a large number of experimental phenomena. Since then, great effort has been devoted to the optimization of the model to better predict the competitive growth process [14,26–28]. The grain selection process has been successfully predicted as the grain competitive growth model is applied to the temperature field evolution obtained by numerical simulation [6,7,12]. It has been found that in order to stably obtain single crystals with a small misorientation by the grain selection method, the length of the starter block should be very large, which is basically beyond the acceptable range of industrial production [10]. This is due to the fact that the overgrowth rate among the grains becomes very slow as the primary orientation between the grains is close [13]. Recently, we have proposed a grain selection method assisted by short seeding to obtain only one grain with the desired orientation through the competitive growth between the grain epitaxially grown along the seed crystals with other grains nucleated in the starter block [29]. The seeding will inevitably undergo a certain degree of oxidation, and the oxide layer is not easily broken by pouring melt during directional solidification, which causes the melt to be unable to epixially grow along the un-melted seeding; therefore, this method runs the risk of crystal orientation control failure.

As the roles of the starter block and the spiral part in the grain selection process have been clarified [5,7,11], we provide a grain selection method assisted by directional columnar grains to control the crystal orientation within a small misorientation. Using this novel grain selection method, the misorientation of single crystal casting is limited to a small angle. Similar to the seeding method, the upper part of the directional columnar grains was re-melted, whereas the lower part remained un-melted due to the temperature field characteristics during the heating and holding stage [30]. During the directional solidification process, the melt epitaxially grows on the un-melted directional columnar grains so that the <001> direction of the grains at the top of the starter block deviates from the directional solidification within a small range. Similar to the process of the seeding method, a certain number of randomly oriented stray grains are formed just above the melt-back interface, as shown in Figures 5 and 6. These stray grains are overgrown by the grains grown epitaxially along the directional columnar grains in the subsequent directional solidification process. The <001> direction of the grain selected by this method deviates from the directional solidification direction within 8 degrees, such as that shown in Figure 4, which is not greater than the maximum <001> direction deviation angle of the directional columnar grains used in this method. It showed that the final grain selected by this novel method was one that grew epitaxially along the un-melted directional columnar grains.

The grain selection assisted by the directional columnar grains method not only uses the competitive growth characteristics of the grain selection method to obtain single crystal casting but also takes advantage of the epitaxial growth of the seeding method; therefore, this method has unique characteristics compared to the grain selection and seeding methods, which are as follows: (1) Compared to the grain selection method, as shown in Figure 3, this method can control the orientation deviation within a little range of the single crystal casting by selecting an appropriate directional columnar grain used in this method because of the characteristic of the epitaxial growth. (2) Compared to the seeding method, the microstructure of the directional columnar grains used in this method is easier to obtain than that of the seeding. The nucleation and growth of stray grains can be tolerated due to the competitive growth characteristic, as shown in Figures 5 and 6. The grain selection assisted by the directional columnar grains method can be used as the main method in the transition stage of the grain selection method to the seeding method in actual industrial production for advanced single crystal turbine blades, taking into account product requirements, technical difficulty, and cost issues.

5. Conclusions

A grain selection assisted by directional columnar grains for single crystal casting was studied using the Bridgman technique at a withdrawal rate of 100 μm/s. The directional columnar grains partially melt-back during directional solidification, and it was divided into four parts after directional solidification, namely the heat-affected zone, the mushy zone, the transition zone, and the directional growth zone. A lot of stray grain appeared in the transition zone just above the melt-back interface, and almost all of them did not grow up during the directional solidification process due to competitive growth with grains growing epitaxially along the un-melted directional columnar grains. The grain selected by this novel method was one that grew epitaxially along the un-melted directional columnar grains; therefore, the <001> direction of the single crystal casting prepared by this method deviated from directional solidification within a small misorientation. This method can be used for advanced single crystal turbine blades in actual industrial production in the transition stage of the grain selection method to the seeding method.

Author Contributions: Conceptualization, Z.Y.; Investigation, S.H. and Y.Z.; Supervision, Y.D., Z.Y., F.Y. and X.W.; Validation, W.B.; Writing—original draft, S.H. and Y.Z.; Writing—review & editing, F.Y., Y.D. and X.W. All authors have read and agreed to the published version of the manuscript.

Funding: This work was supported by the science and technology innovation Program of Hunan Province (No: 2021RC2088), Hunan Provincial Natural Science Foundation of China (2022JJ40439), the Guangdong Basic and Applied Research Foundation (No: 2021A1515110791), National Natural Science Foundation of China (5210011310), Natural Science Foundation of Shaanxi Province (2021JQ-604).

Institutional Review Board Statement: Not applicable.

Informed Consent Statement: Not applicable.

Conflicts of Interest: The authors declare no conflict of interest.

References

1. Harris, K.; Erickson, G.L.; Schwer, R.E. *Metals Handbook*, 10th ed.; ASM Internationa: Material Park, OH, USA, 1990.
2. Zhang, Y.; Qiu, W.; Shi, H.-J.; Li, C.; Kadau, K.; Luesebrink, O. Effects of secondary orientations on long fatigue crack growth in a single crystal superalloy. *Eng. Fract. Mech.* **2015**, *136*, 172–184. [CrossRef]
3. Dong, H.B. Analysis of Grain Selection during Directional Solidification of Gas Turbine Blades. In Proceedings of the World Congress on Engineering 2007, London, UK, 2–4 July 2007; pp. 1257–1262.
4. Stanford, N. Seeding of single crystal superalloys—Role of seed melt-back on casting defects. *Scr. Mater.* **2004**, *50*, 159–163. [CrossRef]
5. Raza, M.H.; Wasim, A.; Hussain, S.; Sajid, M.; Jahanzaib, M. Grain Selection and Crystal Orientation in Single-Crystal Casting: State of the Art. *Cryst. Res. Technol.* **2019**, *54*, 1800177. [CrossRef]
6. Zhang, H.; Zhu, X.; Wang, F.; Ma, D. 2-D Selector Simulation Studies on Grain Selection for Single Crystal Superalloy of CM247LC. *Materials* **2019**, *12*, 3829. [CrossRef]
7. Gao, S.F.; Liu, L.; Wang, N.; Zhao, X.B.; Zhang, J.; Fu, H.Z. Grain Selection During Casting Ni-Base, Single-Crystal Superalloys with Spiral Grain Selector. *Metall. Mater. Trans. A-Phys. Metall. Mater. Sci.* **2012**, *43*, 3767–3775. [CrossRef]
8. Zhu, X.; Yang, Q.; Wang, F.; Ma, D. Grain Orientation Optimization of Two-Dimensional Grain Selector during Directional Solidification of Ni-Based Superalloys. *Materials* **2020**, *13*, 1121. [CrossRef]
9. Sadeghi, F.; Kermanpur, A.; Norouzi, E. Optimizing Grain Selection Design in the Single-Crystal Solidification of Ni-Based Superalloys. *Cryst. Res. Technol.* **2018**, *53*, 1800108. [CrossRef]
10. Wang, N.; Liu, L.; Gao, S.; Zhao, X.; Huang, T.; Zhang, J.; Fu, H. Simulation of grain selection during single crystal casting of a Ni-base superalloy. *J. Alloys Compd.* **2014**, *586*, 220–229. [CrossRef]
11. Seo, S.-M.; Kim, I.-S.; Lee, J.-H.; Jo, C.-Y.; Miyahara, H.; Ogi, K. Grain structure and texture evolutions during single crystal casting of the Ni-base superalloy CMSX-4. *Met. Mater. Int.* **2009**, *15*, 391–398. [CrossRef]
12. Meng, X.; Li, J.; Jin, T.; Sun, X.; Sun, C.; Hu, Z. Evolution of Grain Selection in Spiral Selector during Directional Solidification of Nickel-base Superalloys. *J. Mater. Sci. Technol.* **2011**, *27*, 118–126. [CrossRef]

13. Rappaz, M.; Gandin, C.A.; Desbiolles, J.L.; Thévoz, P. Prediction of grain structures in various solidification processes. *Metall. Mater. Trans. A* **1996**, *27*, 695–705. [CrossRef]
14. Li, J.; Wang, Z.; Wang, Y.; Wang, J. Phase-field study of competitive dendritic growth of converging grains during directional solidification. *Acta Mater.* **2012**, *60*, 1478–1493. [CrossRef]
15. Liu, J.-l.; Jin, T.; Luo, X.-h.; Feng, S.-b.; Zhao, J.-z. Effects of Solidification Conditions on the Crystal Selection Behavior of an Al Base Alloy During Directional Solidification. *Microgravity Sci. Technol.* **2016**, *28*, 109–113. [CrossRef]
16. Zhang, X.; Zhou, Y.; Jin, T.; Sun, X.; Liu, L. Effect of Solidification Rate on Grain Structure Evolution During Directional Solidification of a Ni-based Superalloy. *J. Mater. Sci. Technol.* **2013**, *29*, 879–883. [CrossRef]
17. Rezaei, M.; Kermanpur, A.; Sadeghi, F. Effects of withdrawal rate and starter block size on crystal orientation of a single crystal Ni-based superalloy. *J. Cryst. Growth* **2018**, *485*, 19–27. [CrossRef]
18. Arakere, N.K.; Swanson, G. Effect of Crystal Orientation on Fatigue Failure of Single Crystal Nickel Base Turbine Blade Superalloys. *J. Eng. Gas Turbines Power* **2002**, *124*, 161–176. [CrossRef]
19. Qiu, W.; He, Z.; Fan, Y.-N.; Shi, H.-J.; Gu, J. Effects of secondary orientation on crack closure behavior of nickel-based single crystal superalloys. *Int. J. Fatigue* **2016**, *83*, 335–343. [CrossRef]
20. Yang, Z.; Chen, H.; Cao, J.; Zheng, S.; Luo, K.; Luo, Y.; Dai, S. A Novel Grain Selection Method Assisted by Columnar Grains for Ni-based Single Crystal Superalloy. *Fail. Anal. Prev.* **2021**, *16*, 173–176, 196.
21. Hu, S.; Yang, W.; Li, Z.; Xu, H.; Huang, T.; Zhang, J.; Su, H.; Liu, L. Formation mechanisms and control method for stray grains at melt-back region of Ni-based single crystal seed. *Prog. Nat. Sci.-Mater. Int.* **2021**, *31*, 624–632. [CrossRef]
22. Hu, S.; Liu, L.; Yang, W.; Sun, D.; Huo, M.; Li, Y.; Huang, T.; Zhang, J.; Su, H.; Fu, H. Inhibition of stray grains at melt-back region for re-using seed to prepare Ni-based single crystal superalloys. *Prog. Nat. Sci.-Mater. Int.* **2019**, *29*, 582–586. [CrossRef]
23. Aveson, J.W.; Tennant, P.A.; Foss, B.J.; Shollock, B.A.; Stone, H.J.; D'Souza, N. On the origin of sliver defects in single crystal investment castings. *Acta Mater.* **2013**, *61*, 5162–5171. [CrossRef]
24. Li, Y.; Liu, L.; Huang, T.; Zhang, J.; Fu, H. The process analysis of seeding-grain selection and its effect on stray grain and orientation control. *J. Alloys Compd.* **2016**, *657*, 341–347. [CrossRef]
25. Walton, D.; Chalmer, B. The origin of the preferred orientation in the columnar zone of ingots. *Trans. Metall. Soc. AIME* **1959**, *215*, 447–457.
26. Yang, C.; Liu, L.; Zhao, X.; Wang, N.; Zhang, J.; Fu, H. Competitive grain growth mechanism in three dimensions during directional solidification of a nickel-based superalloy. *J. Alloys Compd.* **2013**, *578*, 577–584. [CrossRef]
27. Clarke, A.J.; Tourret, D.; Song, Y.; Imhoff, S.D.; Gibbs, P.J.; Gibbs, J.W.; Fezzaa, K.; Karma, A. Microstructure selection in thin-sample directional solidification of an Al-Cu alloy: In situ X-ray imaging and phase-field simulations. *Acta Mater.* **2017**, *129*, 203–216. [CrossRef]
28. Wang, H.; Zhang, X.; Meng, J.; Yang, J.; Yang, Y.; Zhou, Y.; Sun, X. A new model of competitive grain growth dominated by the solute field of the Nickel-based superalloys during directional solidification. *J. Alloys Compd.* **2021**, *873*, 159794. [CrossRef]
29. Yang, W.; Hu, S.; Huo, M.; Sun, D.; Zhang, J.; Liu, L. Orientation controlling of Ni-based single-crystal superalloy by a novel method: Grain selection assisted by un-melted reused seed. *J. Mater. Res. Technol.* **2019**, *8*, 1347–1352. [CrossRef]
30. Hu, S.; Zhao, Y.; Bai, W.; Wang, X.; Yin, F.; Yang, W.; Liu, L. Temperature Field Evolution of Seeding during Directional Solidification of Single-Crystal Ni-Based Superalloy Castings. *Metals* **2022**, *12*, 817. [CrossRef]

Article

Predictions and Experiments on the Distortion of the 20Cr2Ni4A C-ring during Carburizing and Quenching Process

Yongming Yan [1,*], Yanjun Xue [1], Wenchao Yu [1], Ke Liu [2], Maoqiu Wang [1], Xinming Wang [3] and Liuqing Ni [3]

1. Central Iron & Steel Research Institute Company Limited, Beijing 100081, China; 15110562925@163.com (Y.X.); yuwemchao@nercast.com (W.Y.); maoqiuwang@hotmail.com (M.W.)
2. Jianglu Machinery and Electronics Group Co., Ltd., Xiangtan 411100, China; liuke820x@126.com
3. School of Material Science and Engineering, Xiangtan University, Xiangtan 411105, China; wangxm@xtu.edu.cn (X.W.); 202121551539@smail.xtu.edu.cn (L.N.)
* Correspondence: yanyongming@nercast.com; Tel.: +86-010-6218-2728

Abstract: This paper focuses on the effect of gear steel on distortion due to phase transformation in carburizing and quenching. The carburizing and quenching process of C-rings under suspension was studied by using the finite element method based on the thermo-mechanical theory, considering phase transformation. The phase transformation kinetics parameters, depending on different carbon contents, were measured by Gleeble-3500. The distortion behavior of the carburized C-ring during the cooling stage was analyzed, as well as the carbon concentration distribution and martensite volume fractions. The accuracy of the simulation was also verified by comparing the experimental data with the simulated result of the distortion and microstructure. A reliable basis is provided for predicting the distortion mechanism of gear steels in carburizing and quenching.

Keywords: martensite transformation; carburizing and quenching process; simulation; distortion

Citation: Yan, Y.; Xue, Y.; Yu, W.; Liu, K.; Wang, M.; Wang, X.; Ni, L. Predictions and Experiments on the Distortion of the 20Cr2Ni4A C-ring during Carburizing and Quenching Process. *Materials* 2022, 15, 4245. https://doi.org/10.3390/ma15124245

Academic Editor: Edward Bormashenko

Received: 16 May 2022
Accepted: 11 June 2022
Published: 20 June 2022

Publisher's Note: MDPI stays neutral with regard to jurisdictional claims in published maps and institutional affiliations.

Copyright: © 2022 by the authors. Licensee MDPI, Basel, Switzerland. This article is an open access article distributed under the terms and conditions of the Creative Commons Attribution (CC BY) license (https://creativecommons.org/licenses/by/4.0/).

1. Introduction

Carburizing and quenching are often used to improve the mechanical properties and the service life of a gear [1,2]. However, the deformation caused by carburizing and quenching leads to dimensional deviation and structural instability of the gear, increasing the machining workload and reducing the production efficiency. Controlling deformation has always been a difficult problem in the manufacturing of gear parts, especially for thin-walled parts. There are many factors that affect distortion during carburizing and quenching, such as the carburizing and quenching temperature, quenching medium, material composition, size of the gear, etc. [3,4]. Ingredient differences between the surface and core after carburizing make the stress–strain derivation law during the quenching process become very complicated. With the development of computer simulation technology, theoretical models and finite element analysis methods have gradually improved, providing new ways to control heat treatment distortion. Some scholars have used numerical simulation to conduct in-depth analysis of carburizing and quenching [5–10]. Silva et al. simulated the quenching process of an AISI 4140 steel C-ring in oil, covering the analysis of the distortion caused by both thermal contraction and phase transformation [6]. Farivar et al. [9] investigated the effects of a modified hardening temperature and different soaking times on the core microstructure, the final dimensional stability, and the mechanical properties of 20MoCr4 steel.

Only with correct and reliable material parameters can reliable numerical simulation results be obtained. However, there are many problems in the simulation of material parameters at present, such as inconsistent data grades, imperfect material data, or even no data. Among them, the K-M formula is often used for the calculation model of martensite transformation [11,12], and the parameter α in the formula is generally selected as 0.011, which is obviously not rigorous enough. The error of martensite transition temperature calculated by the empirical formula is large. Different materials should be calculated by corresponding experimental measurements. The method was developed to obtain the phase transformation expansion and transformation plasticity of steel during phase transformation and heat treatment, and the coefficients of phase transformation expansion and transformation plasticity can be calculated based on the kinetics theory of phase transformation [13–15]. The kinetics of the martensitic transformation in three carbon steels (C60, C70 and C80) have been studied using the acoustic emission (AE) technique during continuous cooling in the Gleeble 1500 by van Bohemen [16]. Therefore, in order to improve the accuracy of the simulation of the carburizing and quenching process, this study firstly conducted thermal expansion experiments on smelted materials with different carbon contents to obtain the martensitic transformation parameters. Then, the finite element method was used to predict the properties and distortion of the C-ring, suggesting optimization of the carburizing and quenching process.

2. Numerical Simulation Theory of Carburizing and Quenching

The carburizing and quenching process involves a complex continuous medium thermodynamic theory and requires consideration of the coupling between the carbon concentration diffusion field, temperature field, phase transformation kinetics, and tissue distribution, as well as the inelastic stress–strain field [10]. A concrete structure of the heat treatment simulation system is shown in Figure 1. Firstly, the diffusion of carbon atoms causes changes in material composition, material properties, and phase transformation kinetics. Secondly, temperature changes affect the nucleation and growth of phase and the temperature field, due to the generation of latent heat from the phase transformation. The stress–strain field changes due to the phase transition, which, in turn, induces expansion or contraction of the material. Conversely, the stress–strain can also inhibit or induce the formation of the phase transformation. Thirdly, due to the inevitable changes in the temperature field, expansion or contraction of the material, i.e., thermal strain, would occur. When large distortions occur within the material, as a result of processing and heat treatment, heat generation also occurs, leading to a change in the temperature field. It can be seen that the phenomenon of multi-field coupling should be considered in the carburizing and quenching simulation.

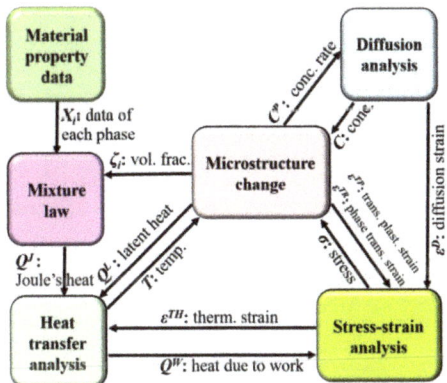

Figure 1. Structure of the heat treatment simulation system and flow of data.

2.1. Theoretical Model of Non-Diffusion-Type Transformation (Martensite)

The martensitic transformation rate is related to the carbon concentration and the cooling rate, and the degree of transformation is higher when the degree of undercooling is greater. The K-M equation used for martensitic transformation can be described by Equation (1).

$$\xi_M = 1 - \exp[-\alpha(M_s - T)] \tag{1}$$

In the formula, ξ_M is the volume fraction of the martensite, M_s is the initiation temperature of the martensite transformation, and α is a constant, usually taken as 0.011 in many calculations for most steels.

2.2. Theoretical Model of the Stress–Strain Field

The Thermal Prophet module based on the finite element method was utilized in the present simulations. The part distortion was predicted, taking into account the deformation based on the following strain components:

$$d\varepsilon_{ij} = d\varepsilon_{ij}^e + d\varepsilon_{ij}^p + d\varepsilon_{ij}^{th} + d\varepsilon_{ij}^{tr} + d\varepsilon_{ij}^{tp} \quad (2)$$

where e, p, th, tr, and tp represent the contributions from elastic, plastic, thermal, phase transformation, and transformation plasticity, respectively. The strain due to transformation plasticity $d\varepsilon_{ij}^{tr}$ was considered in this calculation and can be described as Equation (3) [17,18].

$$d\varepsilon_{ij}^{tp} = 3K\sigma_{ij}(1 - V) \cdot dV \quad (3)$$

where K is the transformation plasticity coefficient related to the microstructure, carbon content, and temperature of the steel; V is the volume fraction of the new phase structure; and dV is the volume formation rate of the new phase.

In order to obtain adequate simulation results, the material data should be accurate. The following information on materials properties have to be known for distortion prediction caused by quenching through numerical simulations [19]:

- Phase transformation kinetics, i.e., TTT and CCT diagrams.
- Temperature-dependent thermophysical properties for each phase formed, such as density, Young's modulus, thermal expansion coefficient, and thermal conductivity.
- Temperature-dependent mechanical properties of each phase formed, including tensile strength, yield strength, and hardness.

The kinetics parameters of martensite transformation were identified by a thermal expansion experiment in the present work. The partial mechanical and thermophysical properties of the various materials and constituents (austenite, martensite, bainite, ferrite and pearlite) are taken into consideration and provided by JMatPro (Sente Sofeware Ltd., Surrey, UK) [20].

3. Methodology

3.1. Experimental Procedure of Heat Treatment

The material employed for the C-ring was a 20Cr2Ni4A steel, with a nominal composition of 0.17 wt.% C, 0.32 wt.% Si, 0.39 wt.% Mn, 1.35 wt.% Cr, and 3.57 wt.% Ni. The geometry and dimensions of the C-rings are given in Figure 2.

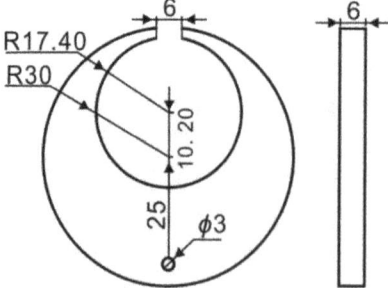

Figure 2. C-ring geometry (all dimensions in mm).

The C-rings were prepared via the process shown in Figure 3. The continuous gas carburizing temperature was 930 °C and the total time was 10 h in an Ipsen atmosphere furnace, followed by high-temperature tempering at 620 °C for 4 h. The samples were held at 800 °C for 1 h, then quenched through fast vertical movement of the ring (thickest part of the ring at the bottom and ring gap at the top) into the quenching oil (HQK) at 25 °C, where it was held for at least 300 s. After quenching, low-temperature tempering at 150 °C for 4 h was carried out. The carburizing and quenching process was simulated according to the actual process.

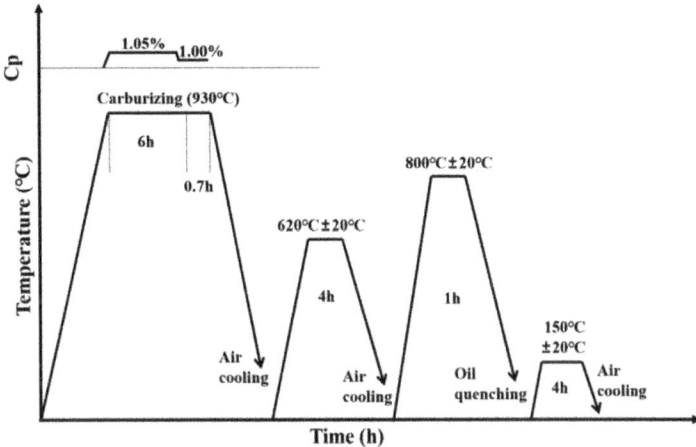

Figure 3. Preparation process of the C-Nary rings.

3.2. Phase Transformation Behavior Depending on Carbon Content

The carbon content of the 20Cr2Ni4A steel after carburizing changes significantly, which leads to changes in material parameters, such as the phase transformation kinetic parameters, elastic modulus, thermal expansion coefficient, and other parameters. The accuracy of these parameters plays a key role in the accuracy of the numerical simulation of the quenching heat treatment of the magic detective. Therefore, the thermal expansion experiment was carried out to obtain the transformation kinetics and thermal expansion coefficients of martensite and austenite under different carbon contents.

Five samples that differed only in carbon content were smelted by an electromagnetic induction furnace, and their chemical compositions, which were tested by an Inductively Coupled Plasma Emission Spectrometer iCAP 6300 Radial (Thermo Fisher Scientific, Walthamm, MA, USA), are shown in Table 1. The samples were processed for thermal expansion experiments with different cooling rates of 17.2, 8.6, 4.3, 1.72, 0.86, 0.28, 0.14, and 0.06 °C/s by using a thermal simulation testing machine Gleeble-3500 (Dynamic Systems Inc Corporation, New York, NY, USA).

Table 1. The composition of the XXCr2Ni4Asteel (wt.%).

	C	Si	Mn	Cr	Ni
20Cr	0.18	0.31	0.52	1.46	3.31
40Cr	0.38	0.33	0.51	1.38	3.44
60Cr	0.57	0.35	0.52	1.37	3.43
80Cr	0.77	0.36	0.51	1.35	3.42
100Cr	1.02	0.36	0.52	1.34	3.40

3.3. Experimental Results of Martensitic Transformation Kinetics

The thermal expansion curves and CCT curves were obtained as shown in Figures 4 and 5. According to the expansion curves, the phase transition point, phase expansion coefficient of the martensite and austenitic, and thermal expansion coefficient were obtained by the tangent method. The parameters of the phase transformation of the XXCr2Ni4A samples were obtained and are listed in Table 2. It can be seen that the austenitizing temperature Ac_3, M_s point, and thermal expansion coefficient of the martensite α_M decreased, while the phase expansion coefficient of the martensite β_M^0 increased with the increase in carbon content. The constant α in the K-M equation used for martensitic transformation is calculated based on Equation (1), and the results are greater than 0.011. The K-M equations for five kinds of carbon content are shown as Equations (4)–(8), which were used for the simulation.

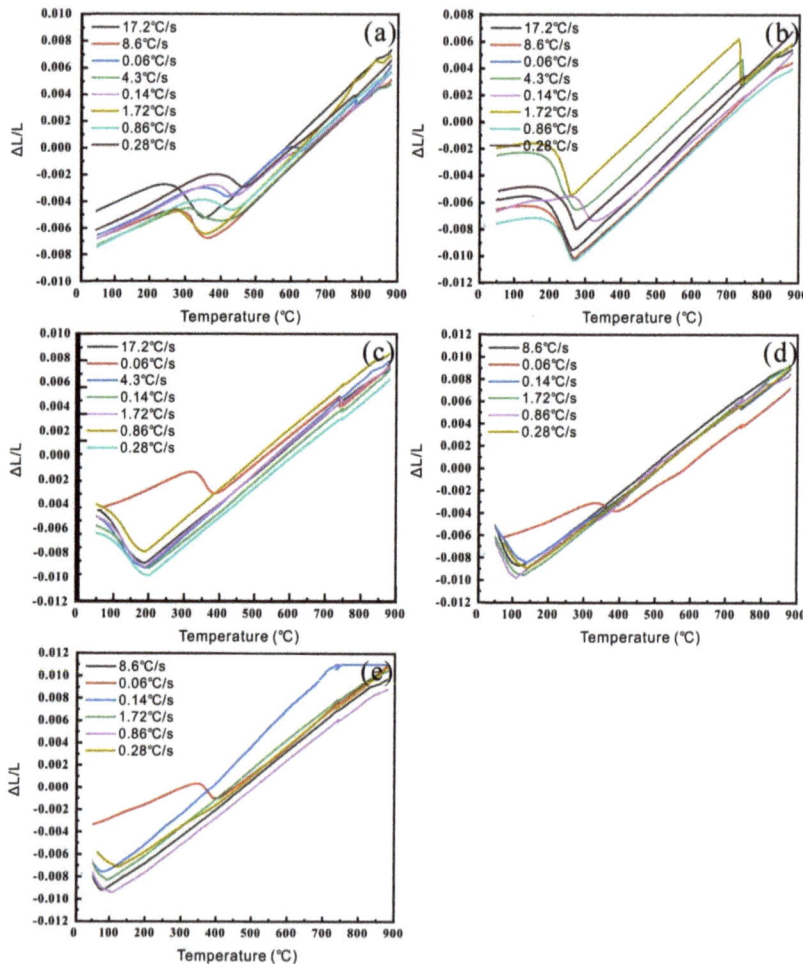

Figure 4. The thermal expansion curves with different samples at different cooling rates: (**a**) 20Cr; (**b**) 40Cr; (**c**) 60Cr; (**d**) 80Cr; (**e**) 100Cr.

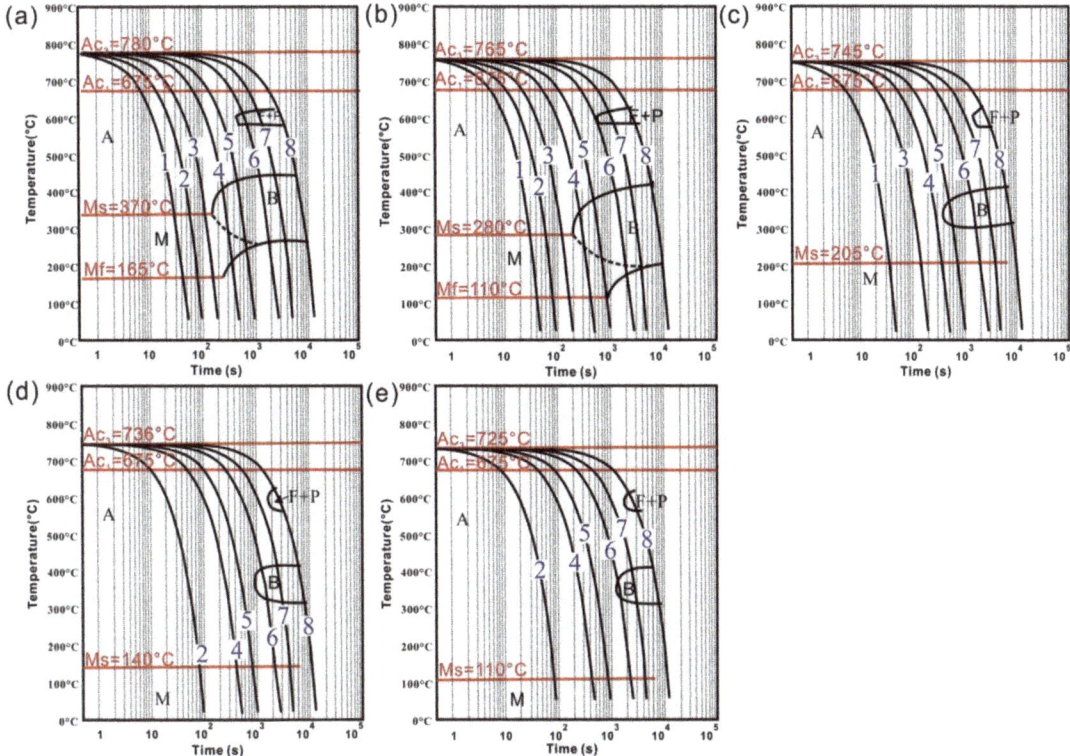

Figure 5. The CCT curves of samples with different carbon contents: (**a**) 20Cr; (**b**) 40Cr; (**c**) 60Cr; (**d**) 80Cr; (**e**) 100Cr (cooling rates: 1–17.2 °C/s, 2–8.6 °C/s, 3–4.3 °C/s, 4–1.72 °C/s, 5–0.86 °C/s, 6–0.28 °C/s, 7–0.14 °C/s and 8–0.06 °C/s; Abbreviation for Microstructure: A—Austenite, B—Bainite, F—Ferrite, P—Pearlite, M—Martensite).

Table 2. The parameters obtained from the expansion experiment.

	20Cr	40Cr	60Cr	80Cr	100Cr
$Ac_3/°C$	780	765	745	736	725
$M_s/°C$	370	280	205	140	110
α_A	2.402×10^5	2.456×10^5	2.320×10^5	2.494×10^5	2.319×10^5
α_M	1.242×10^5	1.143×10^5	1.044×10^5	0.945×10^5	0.846×10^5
β_M^0	0.008631	0.009824	0.01107	0.012210	0.013403
α	0.0288	0.0271	0.0295	0.0281	0.0249

In the formula, α is the linear expansion coefficient, L is the length after heat treatment, and L_0 is the size before heat treatment.

$$\xi_{M0.2\%} = 1 - \exp\left[-0.0288(M_s - T)\right] \qquad (4)$$

$$\xi_{M0.4\%} = 1 - \exp\left[-0.0271(M_s - T)\right] \qquad (5)$$

$$\xi_{M0.6\%} = 1 - \exp\left[-0.0295(M_s - T)\right] \qquad (6)$$

$$\xi_{M0.8\%} = 1 - \exp\left[-0.0281(M_s - T)\right] \qquad (7)$$

$$\xi_{M1.0\%} = 1 - \exp\left[-0.0249(M_s - T)\right] \qquad (8)$$

where $\xi_{M\ x\%}$ is the volume fraction of the martensite (x = 0.2, 0.4, 0.6, 0.8, 1.0), M_s is the initiation temperature of the martensite transformation, and T is the temperature.

3.4. Experimental Results of Heat Transfer Coefficient

The quenching medium is the key factor affecting the degree of distortion and mechanical properties of the components during the quenching process. HQG rapid quenching oil was used for quenching. Its operating temperature is 60 °C and the maximum cooling rate reaches up to 90 °C/s. In order to improve the authenticity and accuracy of the quenching simulation results, the actual heat transfer coefficient is obtained by an inverse heat transfer calculation [21]. Figure 6 shows the variation in the heat transfer coefficient with temperature between 20Cr2Ni4A steel and HQG oil.

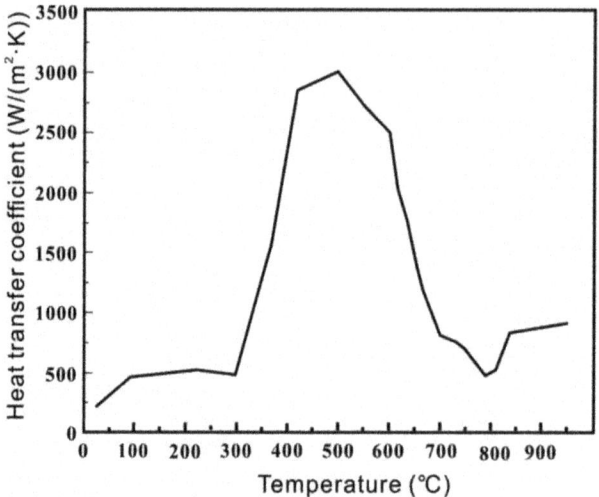

Figure 6. Heat transfer coefficient between 20Cr2Ni4A steel and HQG oil.

4. Results of Simulation

4.1. Time-Dependent Temperature and Phase Transformation

In Figure 7, we can see the variation in and distribution of temperature clouds from 1, 5, 10, 15, 20, and 126 s with time during cooling of the C-ring specimen. As shown in Figure 7, the temperature at the notch of the C-ring cools the fastest, and the center part of the C-ring cools the slowest, due to the different thinness of the specimen and the different distance to the surface. From Figure 7b, it can be seen that after cooling for 5 s, the temperature of the notch position has been cooled to near 252 °C, but the center temperature of the specimen still reached 496 °C. The temperature difference between the notch position and the center position of the specimen is large at this time, resulting in relatively large deformation. The temperature difference between the surface and core decreases with the increase in quenching time. When cooling to 126 s, the overall temperature of the C-ring is 60 °C, which is the same as the oil.

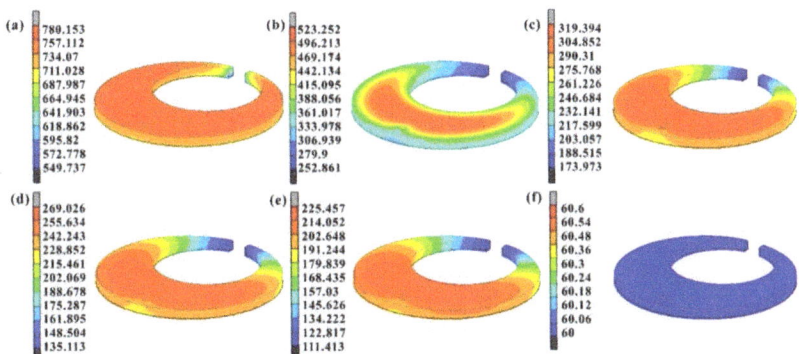

Figure 7. Temperature distribution of simulated predicted 20Cr2Ni4A steel C-type specimens with time: (**a**) 1 s, (**b**) 5 s, (**c**) 10 s, (**d**) 15 s, (**e**) 20 s and (**f**) 126 s.

Figure 8a shows the cross-sectional distribution of carbon concentration of the C-ring of 20Cr2Ni4A steel after gas carburizing at 930 °C and quenching at 800 °C. It can be seen that the carbon concentration on the surface of the specimen can reach up to 0.93%, and from the local enlargement of the cross-section, it can be seen that the carbon concentration was 0.42~0.34%, showing a slow gradient trend; finally, the carbon concentration in the core was 0.20%. The carbon concentration profiles of the actual samples were also measured using a glow discharge photoemission spectrometer-LECO GDS850A. As shown in Figure 8b, the actual measured carbon concentration profile has good agreement with the simulated values. Figure 8c shows the simulated evolution of the martensite volume fraction of the C-ring after heat treatment, from which it can be seen that less martensite was formed on the surface after quenching due to the increase in surface carbon content after carburizing, leading to a decrease in the surface martensite M_s point from 370 to 140 °C. The simulated residual austenite content of the surface was 16.32%. The XRD diffraction method was used to determine the retained austenite content, and its XRD diffraction pattern of the surface is shown in Figure 8d. Based on Equation (9), the retained austenite content was calculated to be 18.75%, which is a little higher than 14% in ref [22] with a slightly lower carbon content on the surface. The experimental results in the present work fit well with the simulation.

$$V_\gamma = \frac{1.4 I_\gamma}{I_\alpha + 1.4 I_\gamma} \qquad (9)$$

where $V\gamma$ is the volume fraction of retained austenite; $I\gamma$ is the mean integrated intensities of the austenite peaks, including $\gamma(111)$, $\gamma(200)$ and $\gamma(220)$; and $I\alpha$ is the mean integrated intensities of the martensite peaks, including $M/\alpha(110)$, $M/\alpha(200)$ and $M/\alpha(211)$.

As shown in Figure 9a, five different typical locations from the notched end to the core of the C-ring were selected in turn, marked as Node 1, Node 2, Node 3, Node 4 and Node 5, where Node 1 is the outer surface. The variation in martensite content with time at each point during quenching was simulated, as shown in Figure 9b. It can be seen that the martensitic transformation at point 1 occurred at the beginning of 12.53 s, which was more consistent with the temperature field shown in Figure 7c,d. At this time, the carbon concentration ranged from 0.788 to 0.935 wt.%, and the corresponding M_s temperature was 140–110 °C. This indicates that the simulation accuracy of the relationship between carbon concentration, temperature field, and martensite content made by 20Cr2Ni4A steel is better for C-rings.

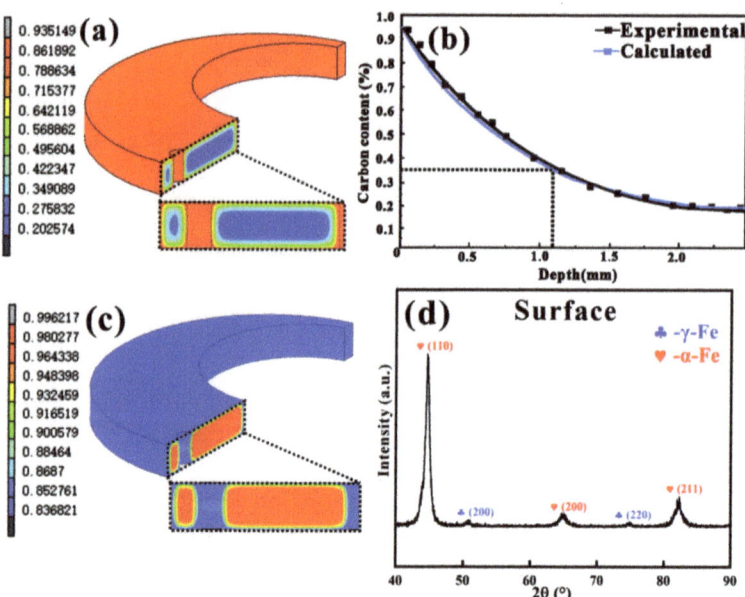

Figure 8. (**a**) Carbon concentration distribution; (**b**) the calculated and experimental carbon concertation distribution of the carburized layer; (**c**) martensite content distribution at different positions; (**d**) XRD diffraction pattern measured by the experiment on the surface layer of the specimen.

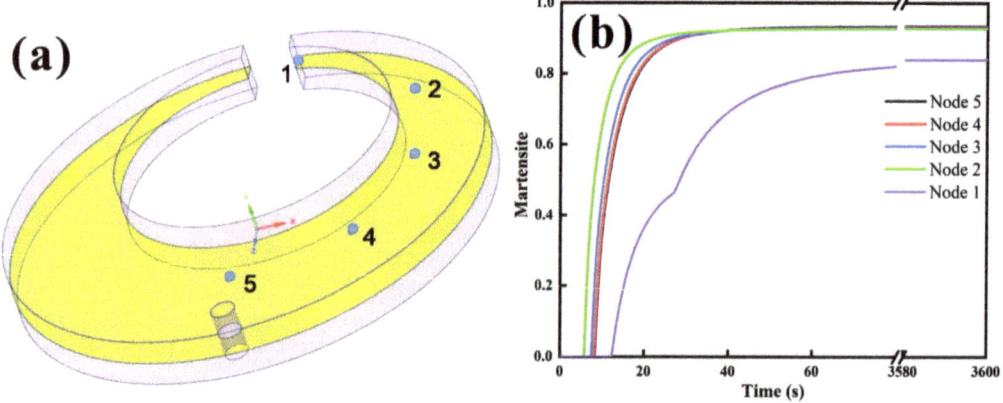

Figure 9. (**a**) Schematic diagram of simulated sites; (**b**) martensite volume versus time graph.

The microstructure of the C-ring at different points is shown in Figure 10. As can be seen from Figure 10a, the surface layer has typical carburized layer organization, consisting of black high-carbon needle-like martensite, white-bright residual austenite blocks, and carbides. The matrix is all low-carbon slate-like martensite, with slight differences in the size of the martensite slat bundles depending on the temperature field, but no changes in the tissue composition were found. This indicates that the actual tissue composition was consistent with the simulated martensite content.

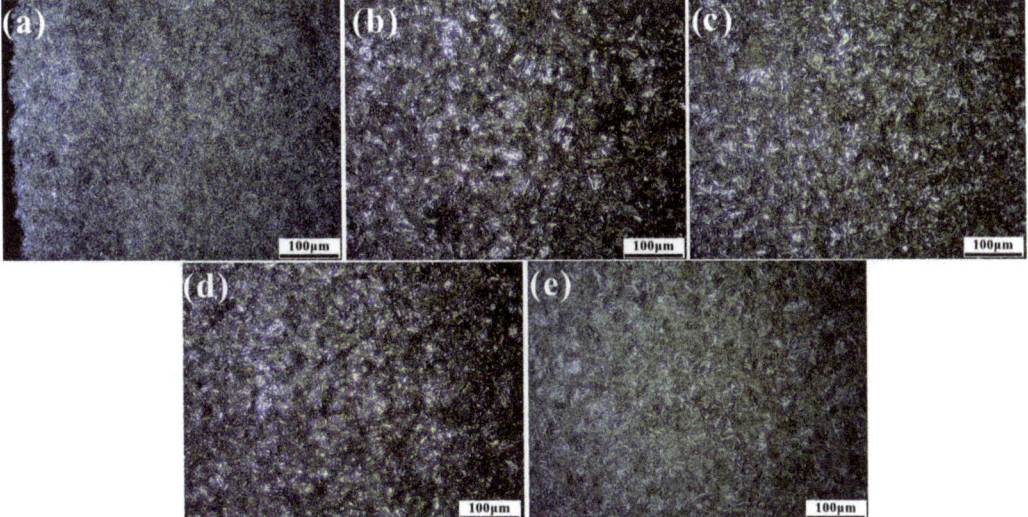

Figure 10. OM microstructure of 20Cr2Ni4A steel C-ring specimens at different locations: (**a**) Node 1, (**b**) Node 2, (**c**) Node 3, (**d**) Node 4 and (**e**) Node 5.

4.2. Distortion and Displacement Distribution

The distortion after carburizing and quenching is an important component of a quality check. If the amount of transformation is too large, it will lead to gear transmission unstable and produce a large noise level, which can affect the assembly of the transmission [12]. Figure 11a is a 20-fold magnification of the final state deformation of the C-ring after carburizing and quenching. Figure 11b displays the simulated quenching deformation with time at a typical position of the carburized C-ring for a period of 100 s after quenching. The measurements of the ring dimensions before and after quenching were performed in an ATOS Core 135 with an accuracy of 0.01 mm. Two dimensional changes were analyzed: gap opening (G) and outside diameter (OD). Based on Figure 12, the dimensional changes may be expressed as follows [6]:

$$G = n' - n \tag{10}$$

$$OD = m' - m \tag{11}$$

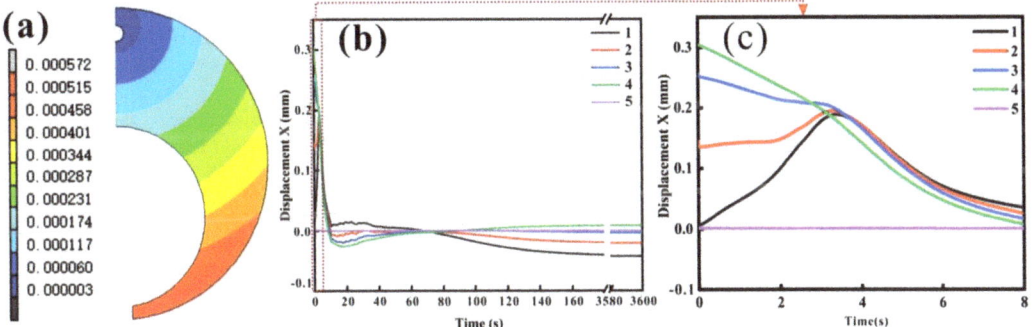

Figure 11. Simulated quenching deformation (**a**) in the x direction of the points in the longitudinal cross-section of the C-ring (displacement magnified 20×); (**b**) with time at typical position of C-Nary rings; (**c**) enlarged view of the position marked with dotted line in the (**b**).

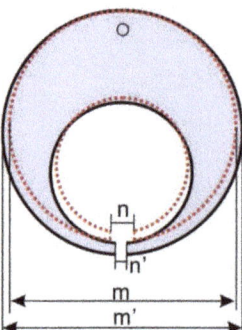

Figure 12. C-ring specimen before heat treatment (m and n dimensions) and after heat treatment (m' and n' dimensions).

The ring gap closed 0.129 mm and the outer diameter expanded about 0.032 mm. As shown in Table 3, the relative differences between gap opening and outside diameter increase were smaller than 8%. The calculated distortions of the gap and the outer diameter were in good agreement with the measured value. The amount of expansion was also different at the start of quenching for the five positions, while the deformation of each position showed different changing laws as the quenching time increased.

Table 3. Comparison of the experimental and simulated results for the geometric distortions of the C-ring.

Position	Experimental Result	Simulation Result	Prediction Difference
Gap opening (mm)	−0.14	−0.129	7.8%
Outside diameter (mm)	0.03	0.032	6.25%

A large number of studies have shown that deformation from carburizing and quenching is mainly caused by the unsynchronized and uneven martensitic transformation stress between the carburized layer and the core matrix. The martensitic transformation stress is closely related to the transformation process. Figure 13 shows the schematic diagram of the phase transformation process of the carburized layer (surface) and the center (Center) of the carburized sample, from which we can understand the deformation law of carburizing and quenching. Due to the big difference in carbon content between the core (only about 0.20%) and the carburized layer (the maximum value), the martensitic transformation temperature M_s changed from 370 to 140 °C, and the end temperature of the martensitic transformation M_f decreased from 165 to −20 °C, as shown in Figure 13. As we can see, the M_f of the 0.8 wt.% carbon content sample was lower than room temperature, leading to an increase in the retained austenite content. In addition, there was a temperature difference between the core and the surface. The core underwent martensitic transformation and expanded at the initial stage of quenching, while the carburized layer was still in the austenite state. In addition, the volume expansion due to the phase transformation of the surface with high carbon content was larger than that of the core, which eventually led to the formation of stress. The surface of the C-ring was under compressive stress, and exhibited shrinkage deformation along the diameter direction after quenching.

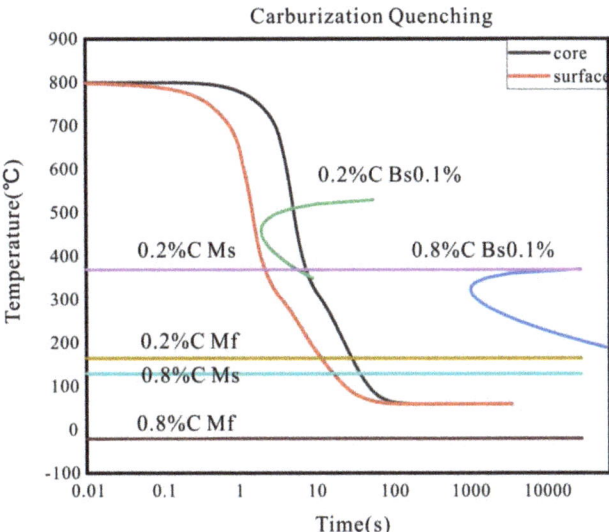

Figure 13. Schematic diagram of the cooling phase transformation process of the surface and core of the carburized sample.

In summary, the model simulated the gear steel after carburizing and quenching, considering the multi-field coupling effect of temperature field, stress–strain field, and phase transformation field. The distributions of carbon concentration, microstructure, and deformation were predicted, which were in good agreement with the experimental results. This indicates that the thermophysical parameters used were accurate and the model could be used as a guide for practical application. These provide a prerequisite for the subsequent realization of the microstructure, as well as performance optimization and micro deformation control of gear steel.

5. Conclusions

The carburizing and quenching process of the 20Cr2Ni4A steel was simulated by using a finite element simulation. Based on the simulation and experimental results, we obtained the following conclusions:

(1) The phase transformation parameters, depending on the carbon content and heat transfer coefficient between the steel and HQG oil, were obtained to improve the accuracy of the carburizing and quenching simulation.
(2) The distortion and microstructure of the C-ring after carburizing and quenching were predicted by considering the effect of phase transformation strain. The measured results concerning distortion and microstructure were in good agreement with the simulated values.
(3) The methodology used to predict the distortion during carburizing and quenching may be applied to parts of various shapes and materials. Therefore, the heat treatment process may be included in the process design to obtain the final product dimension.

Author Contributions: Conceptualization, Y.Y.; data curation, Y.X.; funding acquisition, Y.Y. and W.Y.; investigation, Y.Y., X.W., M.W. and W.Y.; methodology, W.Y. and K.L.; project administration, M.W. and Y.Y.; validation, X.W.; writing—original draft, Y.X. and L.N.; writing—review and editing, Y.Y. and L.N. All authors have read and agreed to the published version of the manuscript.

Funding: This work was supported financially by the National Key Research and Development Program under the subject no. 20T60860B.

Institutional Review Board Statement: Not applicable.

Informed Consent Statement: Not applicable.

Data Availability Statement: The raw/processed data required to reproduce these findings cannot be shared at this time as the data also forms part of an ongoing study.

Conflicts of Interest: The authors declare no conflict of interest.

References

1. Freborg, Z.; Hansen, M.A.; Srivatsan, D.B.; Srivatsan, S.T. Modeling the Effect of Carburization and Quenching on the Development of Residual Stresses and Bending Fatigue Resistance of Steel Gears. *J. Mater. Eng. Perform.* **2013**, *22*, 664–672.
2. Wei, L.; Liu, B. Experimental investigation on the effect of shot peening on contact fatigue strength for carburized and quenched gears. *Int. J. Fatigue* **2017**, *106*, 103–113.
3. Lopez-Garcia, R.D.; Garcia-Pastor, F.A.; Castro-Roman, M.J.; Alfaro-Lopez, E.; Acosta-Gonzalez, F.A. Effect of Immersion Routes on the Quenching Distortion of a Long Steel Component Using a Finite Element Model. *Trans. Indian Inst. Met.* **2016**, *69*, 1–12. [CrossRef]
4. Li, Z.; Grandhi, R.V.; Srinivasan, R. Distortion minimization during gas quenching process. *J. Mater. Processing Technol.* **2006**, *172*, 249–257. [CrossRef]
5. Song, G.-S.; Liu, X.-H.; Wang, G.-D.; Xu, X.-Q. Materials, Numerical Simulation on Carburizing and Quenching of Gear Ring. *J. Iron Steel Res. Int.* **2007**, *14*, 6. [CrossRef]
6. Silva, A.; Pedrosa, T.A.; Gonzalez-Mendez, J.L.; Jiang, X.; Cetlin, P.R.; Altan, T. Distortion in quenching an AISI 4140 C-ring—Predictions and experiments. *Mater. Des.* **2012**, *42*, 55–61. [CrossRef]
7. Miao, S.; Ju, D.Y.; Chen, Y.; Liu, Y.Q. Optimization Based on Orthogonal Experiment Design and Numerical Simulation for Carburizing Quenching Process of Helical Gear. *Mater. Perform. Charact.* **2018**, *7*, 20180019. [CrossRef]
8. Arimoto, K. A Brief Review on Validation for Heat Treatment Simulation. *HT2021* **2021**, 71–80. [CrossRef]
9. Farivar, H.; Prahl, U.; Hans, M.; Bleck, W. Microstructural adjustment of carburized steel components towards reducing the quenching-induced distortion. *J. Mater. Processing Technol.* **2018**, *264*, S0924013618303820. [CrossRef]
10. Rohde, J.; Jeppsson, A. Literature review of heat treatment simulations with respect to phase transformation, residual stresses and distortion. *Scand. J. Metall.* **2000**, *29*, 47–62. [CrossRef]
11. Inoue, T.; Nagaki, S.; Kishino, T.; Monkawa, M. Description of transformation kinetics, heat conduction and elastic-plastic stress in the course of quenching and tempering of some steels. *Arch. Appl. Mech.* **1981**, *50*, 315–327. [CrossRef]
12. Wang, J.; Yang, S.; Li, J.; Ju, D.; Li, X.; He, F.; Li, H.; Chen, Y. Mathematical Simulation and Experimental Verification of Carburizing Quenching Process Based on Multi-Field Coupling. *Coatings* **2021**, *11*, 1132. [CrossRef]
13. Inoue, T.; Wakamatsu, H. Unified theory of transformation plasticity and the effect on quenching simulation. *Strojarstvo Časopis za Teoriju i Praksu u Strojarstvu* **2011**, *53*, 11–18.
14. Inoue, T. Mechanism of transformation plasticity and the unified constitutive equation for transformation-thermomechanical plasticity with some applications. *Int. J. Microstruct. Mater. Prop.* **2010**, *5*, 319–327.
15. Inoue, T. Mechanics and characteristics of transformation plasticity and metallo-thermo-mechanical process simulation. *Procedia Eng.* **2011**, *10*, 3793–3798. [CrossRef]
16. Zhao, H.Z.; Lee, Y.K.; Liu, X.H.; Wang, D.G. On the Martensite Transformation Kinetics of AISI 4340 Steel. *J. Northeast. Univ.* **2006**, *27*, 650–653.
17. Li, H. Effect of Transformation Plasticity on Gear Distortion and Residual Stresses in Carburizing Quenching Simulation. *Coatings* **2021**, *11*, 1224. [CrossRef]
18. Ju, D.Y.; Zhang, W.M.; Matsumoto, Y.; Mukai, R. Modeling and Experimental Verification of Plastic Behavior of the Martensitic Transformation Plastic Behavior in A Carbon Steel. In *Solid State Phenomena*; Trans Tech Publications Ltd.: Freinbach, Switzerland, 2006.
19. Esfahani, A.K. Numerical Simulation of Heat Treatment Process by Incorporating Stress State on Martensitic Transformation to Investigate Microstructure and Stress State of 1045 Steel Gear Parts. *Metall. Mater. Trans. B* **2021**, *52*, 4109–4129. [CrossRef]
20. Saunders, N.; Guo, U.; Li, Z.; Miodownik, A.P.; Schillé, J. Using JMatPro to model materials properties and behavior. *JOM J. Miner. Met. Mater. Soc.* **2003**, *55*, 60–65. [CrossRef]
21. Kanamori, H.; Ju, D.Y. Identification of Heat Transfer Coefficients and Simulation of Quenching Distortions on Disk Probe. *Mater. Trans.* **2020**, *61*, 884–892. [CrossRef]
22. Wei, S.; Wang, G.; Zhao, X.; Zhang, X.; Rong, Y. Experimental study on vacuum carburizing process for low-carbon alloy steel. *J. Mater. Eng. Perform.* **2014**, *23*, 545–550. [CrossRef]

Article

Effect of Y on Recrystallization Behavior in Non-Oriented 4.5 wt% Si Steel Sheets

Jing Qin [1,*], Haibin Zhao [2], Dongsheng Wang [1,*], Songlin Wang [1] and Youwen Yang [3]

1. College of Mechanical Engineering, Tongling University, Tongling 244000, China; wsl-hf@126.com
2. Faculty of Materials Metallurgy and Chemistry, Jiangxi University of Science and Technology, Ganzhou 341000, China; zhaohaibin1030@163.com
3. School of Mechanical and Electrical Engineering, Jiangxi University of Science and Technology, Ganzhou 341000, China; yangyouwen@jxust.edu.cn
* Correspondence: qinjing301@hotmail.com (J.Q.); wangdongsheng@tlu.edu.cn (D.W.)

Abstract: 4.5 wt% Si steel sheets with four different yttrium (Y) contents (0, 0.006, 0.012 and 0.016 wt%) were fabricated by hot rolling, normalizing, warm rolling and a final annealing process. Y addition greatly weakened the γ-fiber ($\langle 111 \rangle$//ND) texture and enhanced the $\{001\} \langle 130 \rangle$ and $\{114\} \langle 481 \rangle$ texture components, and the magnetic properties were improved related to the effects of Y on the recrystallized grain nucleation. Y segregation at the grain boundaries inhibited the nucleation of $\{111\}$ oriented grains at grain boundaries, which was beneficial to the nucleation and growth of other oriented grains elsewhere. At the same rolling reduction, Y_2O_2S inclusion caused more stress concentration than Al_2O_3 inclusion. Y_2O_2S in deformed grains with low energy storage provided more preferential nucleation sites for $\{001\} \langle 130 \rangle$ and $\{114\} \langle 481 \rangle$ grains. Strong $\{001\} \langle 130 \rangle$ and $\{114\} \langle 481 \rangle$ recrystallization textures due to the high mobility were obtained in samples containing 0.012 wt% Y.

Keywords: 4.5 wt% Si steel; yttrium; recrystallization textures; segregation; inclusion

Citation: Qin, J.; Zhao, H.; Wang, D.; Wang, S.; Yang, Y. Effect of Y on Recrystallization Behavior in Non-Oriented 4.5 wt% Si Steel Sheets. *Materials* **2022**, *15*, 4227. https://doi.org/10.3390/ma15124227

Academic Editor: Csaba Balázsi

Received: 24 May 2022
Accepted: 12 June 2022
Published: 15 June 2022

Publisher's Note: MDPI stays neutral with regard to jurisdictional claims in published maps and institutional affiliations.

Copyright: © 2022 by the authors. Licensee MDPI, Basel, Switzerland. This article is an open access article distributed under the terms and conditions of the Creative Commons Attribution (CC BY) license (https://creativecommons.org/licenses/by/4.0/).

1. Introduction

Steel sheets with 4.5 wt% Si have excellent magnetic properties and good ductility compared with 6.5 wt% silicon steel, and they are suitable for manufacturing iron cores of motors at high frequency [1,2]. The motor operation urges the magnetic properties of electrical steels to be more stringent, and thus the steel purification, grain-size control and texture optimization for magnetic property improvement become more important. As is well known, the amount and size of inclusions directly affect the grain size and magnetic properties of non-oriented electrical steels [3].

It is demonstrated in a number of studies that rare earth (RE) can control the amount and distribution of inclusions by deoxygenation and desulfurization and can influence the grain size as well as optimize the recrystallization textures and finally improve the magnetic properties of silicon steel [4,5]. It has been reported that the inclusions density in the size range of 0–1 μm of the Fe-1.15%Si electrical steels with 0.011 wt% Ce was much lower than that of samples without Ce, and the average grain size of samples containing 0.011 wt% Ce was the largest. Therefore, Ce plays the role of reducing the number of fine inclusions and coarsening the size of inclusions [4].

The magnetic induction of silicon steel increased after adding 0.0066 wt% La, which was attributed to the enhancement of $\{001\} \langle 110 \rangle$ and $\{110\} \langle 110 \rangle$ textures [6]. La addition changed the micro-orientation relationship between the recrystallized grain and the matrix, and the nucleation sites of $\{001\}\langle 100 \rangle$ (Cube) grains existed at the $\{112\}\langle 132 \rangle$ and $\{110\}\langle 110 \rangle$ matrix [7].

The intensity of the η-fiber ($\langle 100 \rangle$//RD) texture increased, and the magnetic properties improved after adding 0.012 wt% Y in Fe-6.5%Si electrical steels as Y provided more nucleation sites for η-fiber texture by promoting the occurrence of shear bands [8]. Moreover, the tendency of RE segregation at the grain boundaries also affects the recrystallized grain nucleation. The moderate additions of La and Ce enhanced the favorable {113} ⟨361⟩ and λ-fiber (⟨100⟩//ND) textures and reduced the γ-fiber texture due to their segregation at grain boundaries inhibiting the nucleation of γ grains [9].

Much work thus far has focused on the effects of RE on the recrystallization behavior of silicon steel. However, the mechanism of texture optimization with RE addition remained unclear, and the effect of RE modification on recrystallization texture was rarely studied. In the present work, the recrystallization behavior of 4.5 wt% Si steels are studied by adding different Y content, and the influence mechanism of Y segregation and inclusion modification on recrystallization texture of 4.5 wt% Si steel is analyzed.

2. Experimental

To improve the magnetic properties of 4.5 wt% Si electrical steels, an appropriate amount of solid-state 99.99 wt% pure Y was added after refining to ensure the yield of rare earth by protective casting. Four kinds of 4.5 wt% Si steel ingots with 0, 0.006, 0.012 and 0.016 wt% Y were produced by a vacuum induction melting furnace, and the chemical composition was detected by inductively coupled plasma-atomic emission spectrometry (ICP-AES) as shown in Table 1.

Table 1. The chemical composition of the ingots (wt%).

Sample	Y	Si	C	S	O	N	Fe
0 Y	0	4.5	0.0025	0.0023	0.020	0.0014	Bal.
0.006 Y	0.006	4.5	0.0029	0.0024	0.0017	0.0017	Bal.
0.012 Y	0.012	4.5	0.0028	0.0018	0.0017	0.0017	Bal.
0.016 Y	0.016	4.5	0.0025	0.0019	0.0016	0.0016	Bal.

The ingots were forged into 21 mm thick and 65 mm wide slabs at 1050–900 °C, hot-rolled at 1100–800 °C to 2 mm-thick bands (80 mm width and 280 mm length), and then normalized at 950 °C and pickled. Afterwards, the normalized bands were cut and warm-rolled at 200 °C to 0.3 mm-thick sheets (80 mm width and 620 mm length). Finally, the sheets were annealed at 900 °C for 2 min under a mixed H_2 and N_2 atmosphere. The final sheets were tailored to rectangular samples (30 mm width and 300 mm length) for magnetic testing. The short-time annealing was to obtain fine grains to improve the strength of silicon steel sheets, which meets the requirements of high-frequency drive motor on mechanical properties. In order to study the stress concentration around different inclusions, the normalized sheets without Y and containing 0.012 wt% Y were cold-rolled to the same thickness of 1.6 mm.

The textures and local stress concentration around inclusions were measured by an Oxford Instruments HKL-Channel 5 electron backscattered diffraction (EBSD). The inclusions were analyzed by Zeiss ΣIGMA scanning electron microscope (SEM) and energy-dispersive spectroscopy (EDS). The transgranular fractures obtained at room temperature and intergranular fractures obtained at liquid nitrogen temperature were analyzed by X-ray photoelectron spectroscopy (XPS). The magnetic induction B_{50} (at 5000 A/m), and the core loss $P_{10/400}$ (at a magnetic flux density of 1.0 T and 400 Hz) of samples were measured by an electrical steel tester (MPG200D).

3. Results and Discussion

3.1. Effect of Y Content on the Magnetic Properties and Textures of Final Sheets

Figure 1 shows the magnetic properties of the final sheets with different Y contents. It can be seen that, with the increase of Y content, the magnetic induction increased, and the core loss decreased to a certain extent. When the Y content was 0.012 wt%, the magnetic induction B_{50} reached the highest value (1.6785 T), and the core loss $P_{10/400}$ reached the lowest value (16.905 W/kg). The magnetic properties of the 0.012 Y sample had the best magnetic properties among these samples.

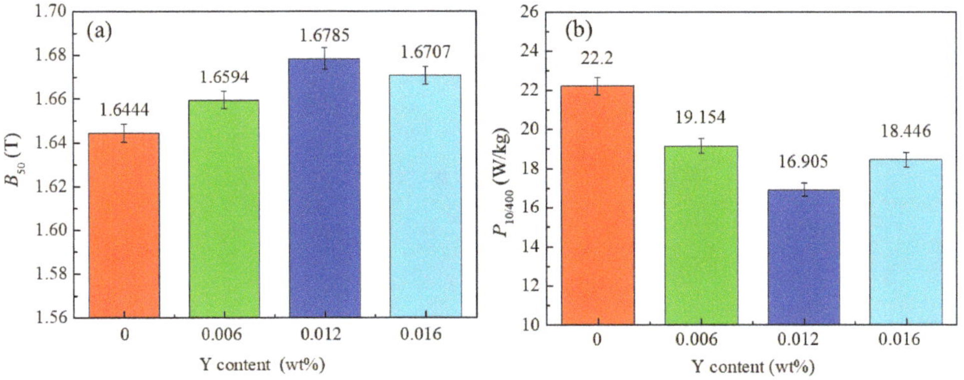

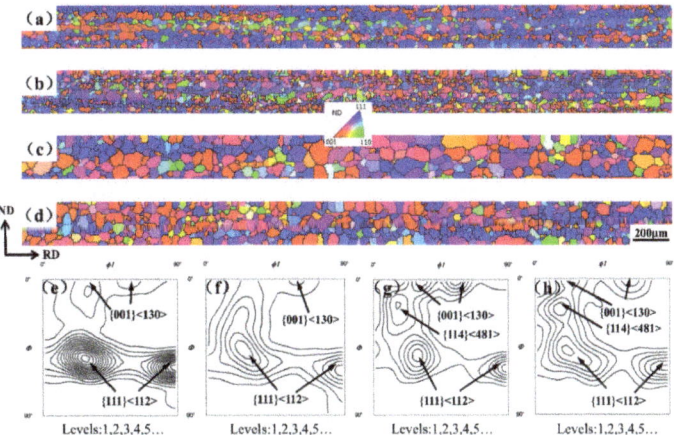

Figure 1. Magnetic properties of final sheets with different Y contents: (a) B_{50} and (b) $P_{10/400}$.

The IPF maps and ODF maps of the final sheets are shown in Figure 2. It can be seen that all the samples have been completely recrystallized. The average grain sizes of four samples with different Y contents were 34, 39, 57 and 50 μm, respectively. Adding an appropriate amount of Y can promote the grain growth in 4.5 wt% Si steel sheets; however, the grain growth was inhibited once the Y content was 0.016 wt%. The recrystallization texture of the 0Y sample was mainly composed of a strong {111} ⟨112⟩ texture and a weak {001} ⟨130⟩ texture.

Figure 2. IPF maps, ODF at $\varphi_2 = 45°$ section of final sheets with different Y: (a,e) 0 Y; (b,f) 0.006 Y; (c,g) 0.012 Y; and (d,h) 0.016 Y.

With adding 0.006 wt% Y, the {111} ⟨112⟩ texture was greatly weakened and the α* ({h,1,1} ⟨1/h,1,2⟩) texture closing to λ texture was enhanced to a degree. With adding 0.012 wt% Y, the γ texture intensity continued to weaken, and the {001} ⟨130⟩ texture and {114} ⟨481⟩ texture components increased significantly. When the content of Y reached 0.016 wt%, the γ texture components increased to a certain extent, and the {001} ⟨130⟩ texture was weakened; however, the {114}⟨481⟩ texture was further enhanced.

3.2. Effect of Y Content on Texture Evolution

In order to investigate the role of Y on recrystallization texture development, the origin of the nucleation of {111} grains and other competitive oriented grains in a partially recrystallized specimen was analyzed, and the textures were compared between 0 Y and 0.012 Y samples in Figure 3. As shown in Figure 3a,b, a large number of {111} ⟨112⟩ recrystallized grains nucleated at the grain boundaries and interior of the {111} ⟨110⟩ deformed grains, while the amount of {001} ⟨130⟩ and {114} ⟨481⟩ recrystallized grains was small, and the nucleation sites were scattered.

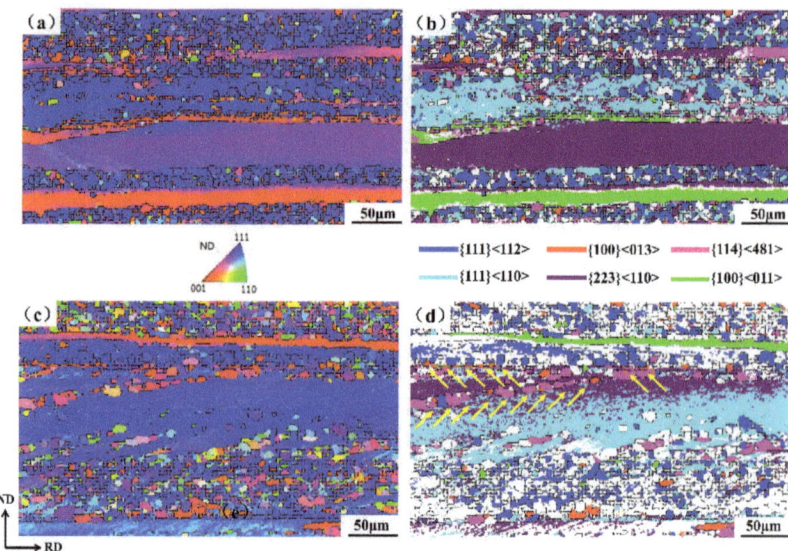

Figure 3. The IPF maps and texture component figures: (**a,b**) 0 Y and (**c,d**) 0.012 wt% Y.

However, in the 0.012 Y sample as shown in Figure 3c,d, {111} ⟨112⟩ grains were no longer the main recrystallized grains. Clearly, the number of the {100} ⟨130⟩ and {114} ⟨481⟩ recrystallized grains increased, and their nucleation sites were more diverse, nucleating at grain boundaries and within shear bands of {223} ⟨110⟩ deformed grains (indicated by yellow arrows in Figure 3d). Generally, γ grains nucleate at {111} deformed grains and at the grain boundaries between {111} and {hkl} ⟨110⟩ deformed grains [10,11].

The {111} recrystallized grains with nucleation advantage at the early recrystallization stage can consume α (⟨110⟩//RD) and λ deformed grains with low energy storage after consuming {111} deformed grains. As shown in Figure 3b, a large number of {111} recrystallized grains nucleated at the grain boundaries between {111} and {100} ⟨110⟩ deformed grains, and even some {111} recrystallized grains penetrated into the {100} matrix. In the 0.012 Y sample, there were only a few other oriented grains nucleating at the boundary between {111} and {100} ⟨110⟩ deformed grains, which was related to the segregation of Y at grain boundaries.

Due to the fact that the atomic diameter of Y (1.81 Å) is much larger than that of Fe (1.24 Å), and the solid solubility of Y in steel is low, it is more likely to segregate at grain boundaries with defects [8,12]. Figure 4a,b presents the intergranular fracture under impact at liquid nitrogen temperature and the transgranular fracture at room temperature of the final sheets containing 0.012 wt% Y. The XPS result of Y is shown in Figure 4c. According to the peak area, the Y content in the intergranular fracture was higher than that in the transgranular fracture. By comparison, it can be speculated that the Y atoms were enriched at the grain boundaries. Therefore, we considered that Y segregation at the grain boundaries restrained the preferential nucleation of γ grains and then reduced the γ texture.

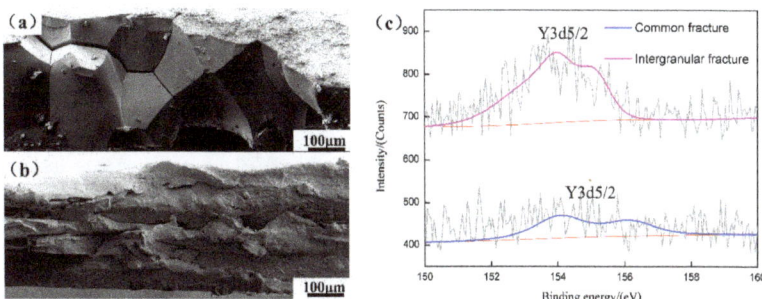

Figure 4. XPS in fracture of the final sheets containing 0.012 wt% Y: (**a**) intergranular fracture; (**b**) transgranular fracture; and (**c**) XPS map.

3.3. Effect of Y Inclusion on Recrystallization Behavior

An appropriate amount of Y can reduce the number of fine inclusions and increase the number of large-size inclusions, which significantly changes the grain size of the final sheets; however, effects on texture development were rarely found [8]. The static recrystallization process will be affected by the size, morphology and dispersion degree of the second phase particles in the matrix. We considered that large non-deformable particles promote recrystallization by particle-stimulated nucleation (PSN) and then affect the texture evolution [13].

Therefore, it is necessary to further study the effects of Y inclusion and modification on the texture development. Figure 5a,b shows the micrograph and EDS results around the inclusion in 2 μm in the 0.012 Y sample. Figure 5c–e shows the IPF map, the {200} pole figure and the recrystallization map, respectively. We found that the inclusion was Y_2O_2S, and the recrystallized grains mainly nucleated at the grain boundary and around the inclusion. The new grains at the grain boundary and inside the {100} deformed matrix were {001} ⟨130⟩ grains (indicated by a black frame in Figure 5c).

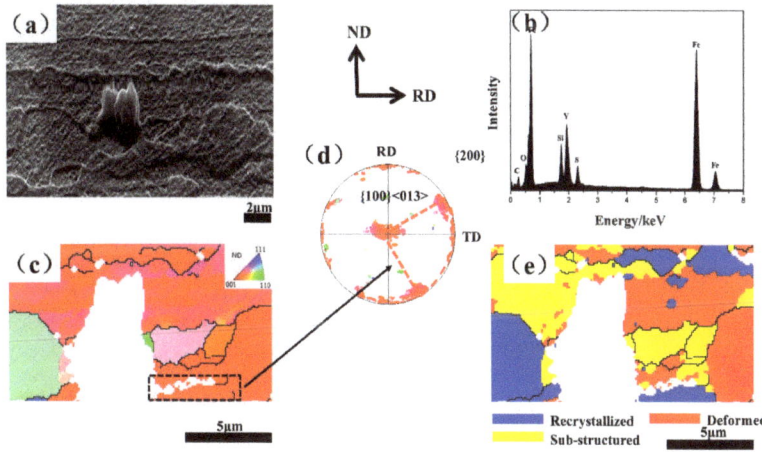

Figure 5. Recrystallization around a Y_2O_2S inclusion. (**a**) SEM micrograph; (**b**) EDS result; (**c**) IPF map; (**d**) {200} pole figure; and (**e**) recrystallization map.

Similarly, as shown in Figure 6, {001} ⟨130⟩ grain nucleation was found in the interior of {113} ⟨110⟩ deformed matrix near Y_2O_2S inclusion at the early recrystallization stage in the 0.012 Y sample. It is generally believed that the energy storage of different deformed grains decreases with the sequence of $E_{\{110\}}$, $E_{\{111\}}$, $E_{\{112\}}$ and $E_{\{100\}}$; thus, the deformed grains with low energy storage cannot be the main nucleation sites, and they are always consumed by γ grains at the later recrystallization stage [14].

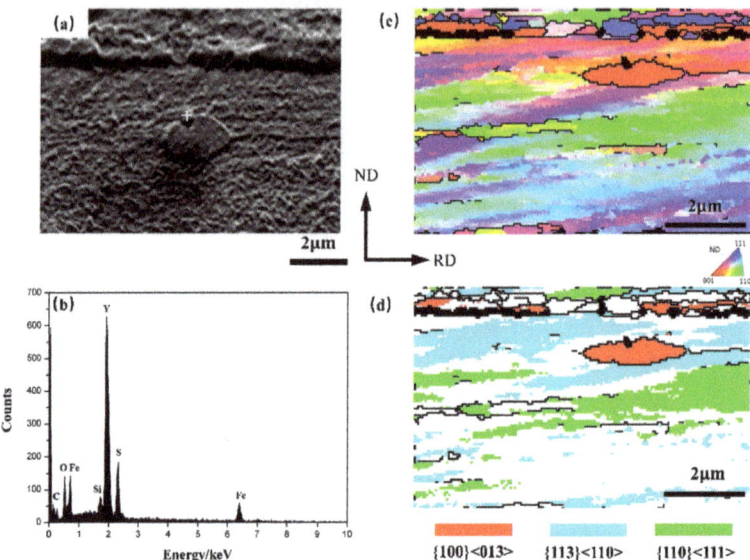

Figure 6. Recrystallization around a Y_2O_2S inclusion. (**a**) SEM micrograph; (**b**) EDS result; (**c**) IPF map; and (**d**) main texture components.

However, the nucleation of {001} ⟨130⟩ recrystallized grains in the {100} and {113} deformed matrix interior was observed in the 0.012Y sample. The sub-structured grains around the inclusions at the region of the high dislocation density can grow without the energy from the deformed matrix. Once sub-structured grains are formed around the large inclusions, they can grow by grain boundary migration and form a crystal nucleus [15]. It has been claimed that the preferred nucleation of grain has a particular crystallographic orientation relationship with the deformed matrix.

At present, a large number of studies show that the {001} grain originates from λ oriented grains [16,17]. The large-size RE inclusions in {001} deformed grains with low energy storage are conducive to the recrystallization of favorable oriented grains.

The driving force of recrystallization is mainly related to the disappearance of dislocations in the sub-boundary, and the inhomogeneous deformation zone is more likely to appear around the large-size particles, which can effectively promote recrystallization [18]. In order to quantitatively further explain the distribution of dislocation density around different inclusions during plastic deformation, the RD-ND (roll direction–normal direction) planes of 0 Y and 0.012 Y samples after 20% reduction were scanned by SEM and EBSD.

Figure 7a–d shows the micrographs and EDS results of different inclusions with the same size. IPF maps and the kernel average misorientation (KAM) are shown in Figure 7e–h. The stress concentration around Y_2O_2S was clearly higher than that of Al_2O_3 in the same {100} matrix. KAM can qualitatively reflect the degree of homogenization of plastic deformation. The relationship between geometrically necessary dislocation density (GND) and KAM is as follows [19]:

$$\rho^{GND} = \frac{2KAM_{ave}}{\mu b} \quad (1)$$

where μ is the EBSD step size (0.1 μm), b is the Burgers vector (0.248 nm) and KAM_{ave} is the average local misorientation. All KAM values used for dislocation density calculation exclude KAM values greater than 3°, which are caused by grain boundaries rather than dislocation accumulation. By calculating the GND of the identical areas around the inclusion, we found that the dislocation density of 0Y sample was 3.215×10^4 cm^{-2}, which was less than the dislocation density of 8.596×10^4 cm^{-2} of the 0.012Y sample. The dislocation density around the Y inclusion was much higher than for Al_2O_3.

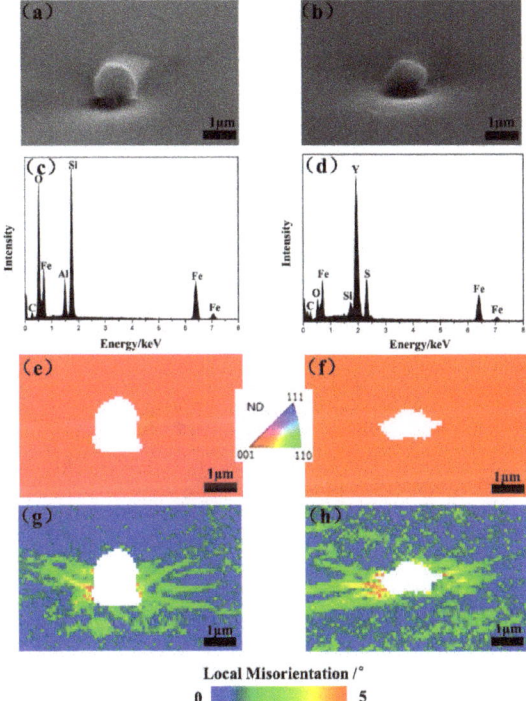

Figure 7. Micrographs around the Al_2O_3 (a,c,e,g) and Y_2O_2S (b,d,f,h) inclusion. (a,b) SEM micrograph; (c,d) EDS results; (e,f) IPF map; and (g,h) local misorientation map.

The inclusion size distribution of the final sheets is shown in Figure 8. The addition of Y reduced the number of fine inclusions, which promoted the grain growth. With the increase of the Y content, the number of inclusions larger than 1 μm continued to increase, which generated orientation gradients that provided sites for PSN. The addition of rare earth Y significantly coarsened and increased the number of large inclusions in the final sheets. RE can adsorb other harmful elements to form composite inclusions, and the hardness of RE oxysulfides is higher than that of Al_2O_3 and MnS [20].

Therefore, it is more likely to see rapid sub-boundary migration around coarse RE inclusions and the accumulation of more misorientation during the rolling process as well as high angle grain boundary (HAGB) formation during recrystallization [21]. Similarly, {114} ⟨481⟩ had widespread nucleation sites and was often found at the grain boundaries of α-fiber deformed grains [22], and the coarse Y inclusions provided additional nucleation sites for {114} ⟨481⟩. The increasing number of {114} ⟨481⟩ nucleation at the early stage of recrystallization impelled {114} ⟨481⟩, thereby, becoming one of the main textures in the 0.012 Y sample.

3.4. Nucleation and Growth of {001} ⟨130⟩ and {114} ⟨481⟩ Grains

Figure 9 shows the partial recrystallization maps of the 0.012 Y sample and the misorientation relationship between recrystallized grains and adjacent deformed grains. A large number of {114} ⟨481⟩ grains and a small amount of {001} ⟨130⟩ grains nucleated in the interior of {223} ⟨110⟩ deformed grains, and almost no {111} ⟨112⟩ grains appeared in {223} ⟨110⟩ deformed grains. Considering the strong {001} ⟨130⟩ and {114} ⟨481⟩ recrystallization textures obtained in the 0.012 Y sample, it is necessary to investigate the misorientation relationship between the newly nucleated grains and the deformed matrix.

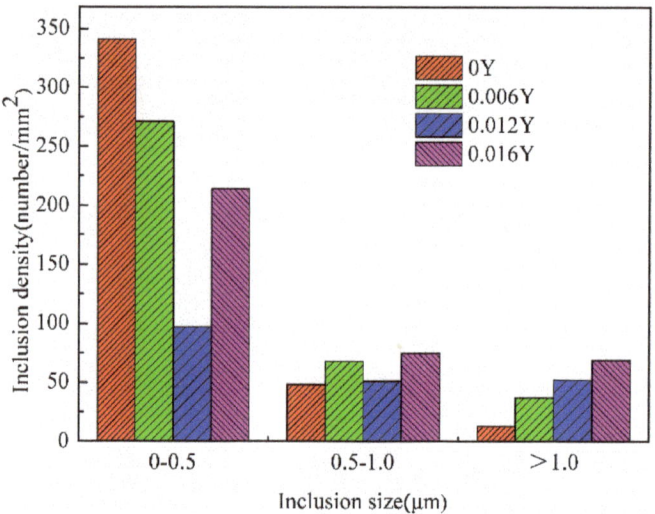

Figure 8. The inclusion size distribution of final sheets with different Y contents.

G1, G2 and G3 represented {111} ⟨112⟩, {114} ⟨481⟩ and {001} ⟨130⟩ recrystallized grains, respectively. D1 and D2 represented {223} ⟨110⟩ and {100} ⟨011⟩ deformed grains, respectively. We found that the misorientation relationships of G1, G2, G3 and D1 were 35.1° ⟨413⟩, 25.0° ⟨616⟩ and 47.4° ⟨322⟩, respectively. The misorientation relationship of {114} ⟨481⟩ recrystallized grains was close to the high mobility of Σ 19 grain boundary [23]. The {114} ⟨481⟩ grains had the advantage of nucleation site and quantity in the interior of the {223} ⟨110⟩ deformed grains, which were easier to grow.

The misorientation relationships of G1, G2, G3 and D2 were 57.0° ⟨110⟩, 28.2° ⟨323⟩ and 26.0° ⟨001⟩, respectively. Since the 20–45° grain boundaries in BCC metal have high mobilities [24,25], {001} ⟨130⟩ and {114}⟨481⟩ recrystallized grains can grow rapidly in the {100} ⟨011⟩ deformed matrix. The oriented growth became the main mechanism for the formation of {001} ⟨130⟩ and {114} ⟨481⟩ recrystallization textures in the 0.012 Y sample.

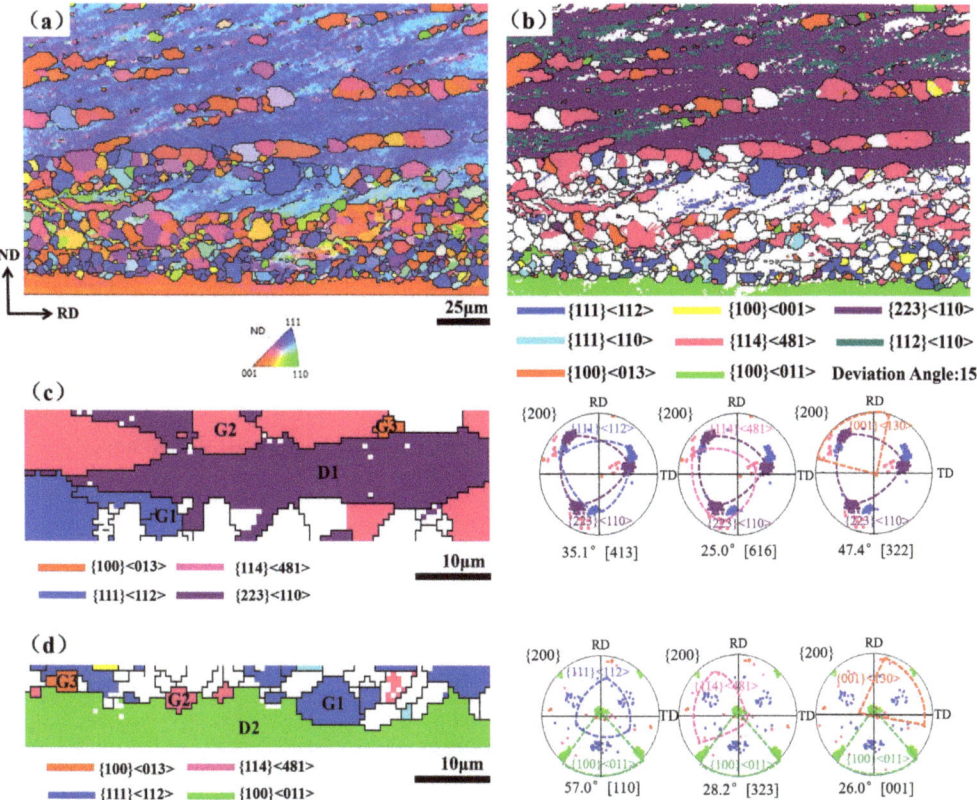

Figure 9. EBSD orientation maps in the partially recrystallized 0.012 Y sample. (**a**) IPF map; (**b**) texture component figure; (**c**) map of relationship between G1, G2 and G3 recrystallized grains and D1 deformed grains; and (**d**) map of relationship between G1, G2 and G3 recrystallized grains and D2 deformed grains.

4. Conclusions

The addition of Y decreased the $\{111\}\langle112\rangle$ and enhanced the $\{001\}\langle130\rangle$ and $\{114\}\langle481\rangle$ texture components in annealed 4.5 wt% Si steel sheets. The segregation of Y restrained the advantage of $\{111\}$ nucleation at grain boundaries, which was beneficial to the nucleation of other oriented grains elsewhere. Compared with Al_2O_3 inclusion, Y_2O_2S inclusion with the same size produced more distortion in the matrix under the same strain and accumulated larger misorientation, making the deformed grains with low energy storage nucleate faster at early recrystallization stage, which provided more preferential nucleation sites of $\{223\}\langle110\rangle$ and $\{001\}$ deformed grains for $\{001\}\langle130\rangle$ and $\{114\}\langle481\rangle$ grains. Strong $\{001\}\langle130\rangle$ and $\{114\}\langle481\rangle$ recrystallization textures formed in the samples containing 0.012 wt% Y due to more nucleation sites and the high mobility of grain boundaries.

Author Contributions: Conceptualization, J.Q. and H.Z.; methodology, J.Q.; software, H.Z.; validation, J.Q. and H.Z.; formal analysis, J.Q.; investigation, H.Z.; resources, D.W.; data curation, J.Q.; writing—original draft preparation, H.Z.; writing—review and editing, J.Q.; visualization, S.W.; supervision, Y.Y.; project administration, J.Q.; funding acquisition, J.Q., S.W. and D.W. All authors have read and agreed to the published version of the manuscript.

Funding: This research was funded by the National Natural Science Foundation of China (No. 51704131), Anhui Provincial Natural Science Foundation (2008085ME149), Anhui Provincial Top Academic Aid Program for Discipline (Major) Talents of Higher Education Institutions (gxbjZD2020087), Key Research and Development Project of Tongling City (20200201010), and Academic Leader and Backup Candidate Research Project of Tongling University (2020tlxyxs03).

Institutional Review Board Statement: Not applicable.

Informed Consent Statement: Not applicable.

Data Availability Statement: Not applicable.

Conflicts of Interest: The authors declare no conflict of interest.

References

1. Zhang, B.; Liang, Y.F.; Wen, S.B.; Wang, S.; Shi, X.J.; Ye, F.; Lin, J.P. High-strength low-iron-loss silicon steels fabricated by cold rolling. *J. Magn. Magn. Mater.* **2019**, *474*, 51–55. [CrossRef]
2. Zu, G.Q.; Zhang, X.M.; Zhao, J.W.; Wang, Y.Q.; Yan, Y.; Li, C.G.; Cao, G.M.; Jiang, Z.Y. Fabrication and properties of strip casting 4.5 wt% Si steel thin sheet. *J. Magn. Magn. Mater.* **2017**, *424*, 64–68. [CrossRef]
3. Li, F.; Li, H.; Wu, Y.; Zhao, D.; Peng, B.; Huang, H.; Zheng, S.; You, J. Effect of precipitates on grain growth in non-oriented silicon steel. *J. Mater. Res.* **2017**, *32*, 2307–2314. [CrossRef]
4. Hou, C.K.; Liao, C.C. Effect of Cerium Content on the Magnetic Properties of Non-oriented Electrical Steels. *ISIJ Int.* **2008**, *48*, 531–539. [CrossRef]
5. Wan, Y.; Chen, W.Q.; Wu, S.J. Effects of Lanthanum and Boron on the Microstructure and Magnetic Properties of Non-oriented Electrical Steels. *High Temp. Mater. Proc.* **2014**, *33*, 115–121. [CrossRef]
6. Wan, Y.; Chen, W.Q.; Wu, S.J. Effect of lanthanum content on microstructure and magnetic properties of non-oriented electrical steels. *J. Rare Earth.* **2013**, *31*, 89–95. [CrossRef]
7. Shi, C.J.; Jin, Z.L.; Ren, H.P.; You, J.L. Effect of lanthanum on recrystallization behavior of non-oriented silicon steel. *J. Rare Earth.* **2017**, *35*, 309–314. [CrossRef]
8. Qin, J.; Zhou, Q.Y.; Zhao, H.B.; Zhao, H.J. Effects of yttrium on the microstructure, texture, and magnetic properties of non-oriented 6.5 wt% Si steel sheets by a rolling process. *Mater. Res. Express* **2021**, *8*, 066103. [CrossRef]
9. He, Z.H.; Sha, Y.H.; Gao, Y.K.; Chang, S.T.; Zhang, F.; Zuo, L. Recrystallization texture development in rare-earth (RE)-doped non-oriented silicon steel. *J. Iron Steel Res. Int.* **2020**, *27*, 1339–1346. [CrossRef]
10. Park, J.T.; Szpunar, J.A. Evolution of recrystallization texture in nonoriented electrical steels. *Acta Mater.* **2003**, *51*, 3037–3051. [CrossRef]
11. Barnett, M.R.; Kestens, L. Formation of {111} <110> and {111} <112> Textures in Cold Rolled and Annealed IF Sheet Steel. *ISIJ Int.* **1999**, *39*, 923–929. [CrossRef]
12. Wang, L.M.; Qin, L.; Ji, J.W.; Lan, D.N. New study concerning development of application of rare earth metals in steels. *J. Alloy. Compd.* **2006**, *408–412*, 384–386. [CrossRef]
13. Li, L.F.; Yang, W.Y.; Sun, Z.Q. Dynamic Recrystallization of Ferrite with Particle-Stimulated Nucleation in a Low-Carbon Steel. *Met. Mater. Trans. A* **2013**, *44*, 2060–2069. [CrossRef]
14. Zu, G.Q.; Zhang, X.M.; Zhao, J.W.; Wang, Y.Q.; Cui, Y.; Yan, Y.; Jiang, Z.Y. Analysis of {411}<148>recrystallisation texture in twin-roll strip casting of 4.5 wt% Si non-oriented electrical steel. *Mater. Lett.* **2016**, *180*, 63–67. [CrossRef]
15. Wang, X.Y.; Jiang, J.T.; Li, G.A.; Wang, X.M.; Shao, W.Z.; Zhen, L. Particle-stimulated nucleation and recrystallization texture initiated by coarsened Al2CuLi phase in Al–Cu–Li alloy. *J. Mater. Res. Technol.* **2021**, *10*, 643–650. [CrossRef]
16. Xu, Y.B.; Zhang, Y.X.; Wang, Y.; Li, C.G.; Cao, G.M.; Liu, Z.Y.; Wang, G.D. Evolution of cube texture in strip-cast non-oriented silicon steels. *Scr. Mater.* **2014**, *87*, 17–20. [CrossRef]
17. Cheng, L.; Zhang, N.; Yang, P.; Mao, W.M. Retaining {100} texture from initial columnar grains in electrical steels. *Scr. Mater.* **2012**, *67*, 899–902. [CrossRef]
18. Rios, P.R.; Jr, F.S.; Sandim, H.R.Z.; Plaut, R.L.; Padilha, A.F. Nucleation and growth during recrystallization. *Mat. Res.* **2005**, *8*, 225–238. [CrossRef]
19. Liu, D.F.; Qin, J.; Zhang, Y.H.; Wang, Z.G.; Nie, J.C. Effect of yttrium addition on the hot deformation behavior of Fe–6.5 wt%Si alloy. *Mat. Sci. Eng. A* **2020**, *797*, 140238. [CrossRef]
20. Yang, C.Y.; Luan, Y.K.; Li, D.Z.; Li, Y.Y. Effects of rare earth elements on inclusions and impact toughness of high-carbon chromium bearing steel. *J. Mater. Sci. Technol.* **2019**, *35*, 1298–1308. [CrossRef]
21. Robson, J.D.; Henry, D.T.; Davis, B. Particle effects on recrystallization in magnesium–manganese alloys: Particle-stimulated nucleation. *Acta Mater.* **2009**, *57*, 2739–2747. [CrossRef]
22. Li, H.Z.; Liu, Z.Y.; Wang, X.L.; Ren, H.M.; Li, C.G.; Cao, G.M.; Wang, G.D. {114} <481> Annealing texture in twin-roll casting non-oriented 6.5 wt% Si electrical steel. *J. Mater. Sci.* **2017**, *52*, 247–259. [CrossRef]
23. Sanjari, M.; He, Y.L.; Hilinski, E.J.; Yue, S.; Kestens, L.A.I. Development of the {113}<uvw>texture during the annealing of a skew cold rolled non-oriented electrical steel. *Scr. Mater.* **2016**, *124*, 179–183. [CrossRef]

24. Rajmohan, N.; Szpunar, J.A. An analytical method for characterizing grain boundaries around growing goss grains during secondary recrystallization. *Scr. Mater.* **2001**, *44*, 2387–2392. [CrossRef]
25. Zhang, Y.H.; Yang, J.F.; Qin, J.; Zhao, H.B. The effect of grain size before cold rolling on the magnetic properties of thin-gauge non-oriented electrical steel. *Mater. Res. Express* **2021**, *8*, 016303. [CrossRef]

Article

Effect of Rolling Treatment on Microstructure, Mechanical Properties, and Corrosion Properties of WE43 Alloy

Bo Deng [1], Yilong Dai [1,*], Jianguo Lin [2] and Dechuang Zhang [2,*]

[1] School of Materials Science and Engineering, Xiangtan University, Xiangtan 411105, China; dengmxbo1123@163.com

[2] Key Laboratory of Materials Design and Preparation Technology of Hunan Province, Xiangtan University, Xiangtan 411105, China; lin_j_g@xtu.edu.cn

* Correspondence: daiyilong@xtu.edu.cn (Y.D.); dczhang@xtu.edu.cn (D.Z.)

Citation: Deng, B.; Dai, Y.; Lin, J.; Zhang, D. Effect of Rolling Treatment on Microstructure, Mechanical Properties, and Corrosion Properties of WE43 Alloy. *Materials* 2022, 15, 3985. https://doi.org/10.3390/ma15113985

Academic Editor: Daolun Chen

Received: 7 May 2022
Accepted: 2 June 2022
Published: 3 June 2022

Publisher's Note: MDPI stays neutral with regard to jurisdictional claims in published maps and institutional affiliations.

Copyright: © 2022 by the authors. Licensee MDPI, Basel, Switzerland. This article is an open access article distributed under the terms and conditions of the Creative Commons Attribution (CC BY) license (https://creativecommons.org/licenses/by/4.0/).

Abstract: Magnesium alloys show broad application prospects as biodegradable implanting materials due to their good biocompatibility, mechanical compatibility, and degradability. However, the influence mechanism of microstructure evolution during forming on the mechanical properties and corrosion resistance of the magnesium alloy process is not clear. Here, the effects of rolling deformation, such as cold rolling, warm rolling, and hot rolling, on the microstructure, mechanical properties, and corrosion resistance of the WE43 magnesium alloy were systematically studied. After rolling treatment, the grains of the alloy were significantly refined. Moreover, the crystal plane texture strength and basal plane density decreased first and then increased with the increase in rolling temperature. Compared with the as-cast alloy, the strength of the alloy after rolling was significantly improved. Among them, the warm-rolled alloy exhibited the best mechanical properties, with a tensile strength of 346.7 MPa and an elongation of 8.9%. The electrochemical experiments and immersion test showed that the hot working process can greatly improve the corrosion resistance of the WE43 alloy. The hot-rolled alloy had the best corrosion resistance, and its corrosion resistance rate was 0.1556 ± 0.18 mm/year.

Keywords: WE43 magnesium alloy; rolling; grain size; mechanical properties; corrosion properties

1. Introduction

Magnesium alloys show broad application prospects as biodegradable implanting materials due to their good biocompatibility, mechanical compatibility, and degradability [1–5]. It has been reported that the MgYREZr devices, such as the bioabsorbable interference screws (MAG-NEZIX®, and Magmaris by BIOTRONIK), have already been successfully commercialized [6]. However, too rapid a corrosion speed of magnesium alloys in the body fluid will cause the severe degradation of their strength early in their service, leading to the failure of the magnesium alloy before the bones are healed. Moreover, the corrosion of magnesium alloys in the human body is often accompanied by the production of H_2 to produce airbags around the implant, which may increase the chance of an inflammatory reaction [7,8]. Therefore, the improvement of the corrosion resistance of magnesium alloys in body fluid is the key to promote its clinical application [9–11]. In the past decade, extensive efforts have been made to improve the corrosion resistance of magnesium alloys, and many techniques have been employed to improve the corrosion resistance and mechanical properties of the magnesium alloys, including the development of a new alloy with the required properties by alloying [12], surface coating process, and surface physical or chemical modification [13].

Normally, magnesium alloys need a series of plastic deformation processing before their application, and the different processes can lead to differences in the microstructure of magnesium alloys. It has been well documented that the mechanical properties and corrosion resistance of magnesium alloys can be significantly improved through thermomechanical treatment to optimize the microstructure and refine the grain size. Argade et al. [14] investigated the effects of different grain sizes on the corrosion resistance of the Mg-Y-Re alloy, and the results showed that reducing the grain size could greatly promote the corrosion resistance of the Mg-Y-Re alloy. Li et al. [15] also indicated that hot rolling could improve the mechanical properties and the corrosion resistance of the MgCa alloy. Furthermore, Wang et al. [16] found that the rolling temperature had a significant impact on the microstructure of the Mg-2Zn-0.4Y alloy, and fully dynamic recrystallization occurred at the relatively high rolling temperature to produce a more uniform and finer microstructure in the Mg-2Zn-0.4Y, which was facilitated by the improvement of the mechanical properties and the corrosion resistance of the alloy. Subsequently, some new thermomechanical processes were applied to the processing of magnesium alloys. For example, Sabat et al. [17] applied one-way rolling and multi-step cross-rolling to the process of the as-cast WE43 magnesium alloy, and they found that, after one-way rolling or multi-step cross-rolling at 400 °C, the average grain size of the as-cast alloy (150 μm) was significantly reduced to 7 μm and 18 μm due to the continuous dynamic recrystallization, respectively. Zhu et al. [18] prepared the ZK60 magnesium alloy plate with ultra-fine grains by a high-strain-rate rolling technique, and the plate exhibited a high strength and ductility due to its ultra-fine grains and low density of dislocations and twins.

It is reported that some magnesium alloys have been studied in clinical applications, and the WE43 alloy is considered to be a promising biodegradable scaffold material. The WE43 alloy contains about 4 wt % of Y and 3 wt % of REEs, which includes Nd, Gd, Dy, and Zr. Such materials are usually prepared by casting and subsequent thermomechanical processing [19]. However, there is still little literature on the microstructure evolution, mechanical properties, and corrosion resistance of the WE43 alloy during rolling. In this paper, the plastic deformation of the WE43 alloy was carried out by cold rolling, warm rolling, and hot rolling, and the influence mechanism of rolling on the microstructure, mechanical properties, and corrosion resistance was studied to provide reference for further clinical medicine of the WE43 alloy.

2. Materials and Methods

2.1. Material Preparation

Nominal compositions of the WE43 magnesium alloy (Suzhou Rare Metal Company, Suzhou, China) in the present work are listed in Table 1. The alloy was solution-treated at 520 °C for 8 h, and then the samples were cut with the size of 60 × 20 × 10 mm^3 by the electric spark cutting method. A schematic diagram of the rolling routes is shown in Figure 1. The samples were rolled on a twin-roll mill at different temperatures. Cold-rolled (CR) samples were obtained at room temperature. Warm-rolled (WR) samples were obtained at 200 °C. Hot-rolled (HR) samples were obtained at 400 °C.

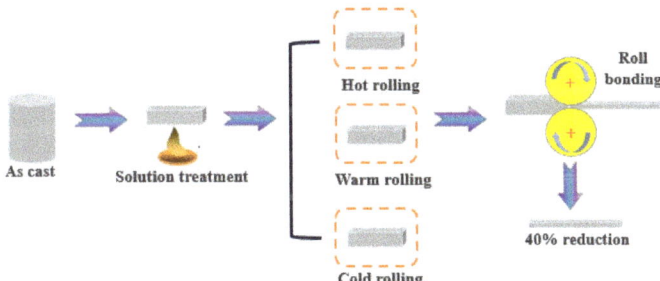

Figure 1. Schematic diagram of the three processing routes.

Table 1. The chemical composition of WE43.

Element	Y	Nd	Gd	Zr	Mg
Proportion (%)	4.1	3.2	1.2	0.53	Balance

2.2. Microstructure Characterization

The phase constituents of all the samples were identified by X-ray diffraction analysis (XRD, U1tima IV, Tokyo, Japan) with Cu-kα radiation (λ = 1.5406 nm) and a scanning rate of 4 min^{-1}. The microstructures and the compositions of the samples were examined by a scanning electron microscope (SEM, MIRA3, LMH, Oxford, UK) equipped with an energy-dispersive spectrometer (EDS, X-Max20, Oxford, UK) at 15 kV. The grain orientation of the WE43 alloy sample was analyzed by using electron backscatter diffraction (EBSD).

2.3. Mechanical Test

The Vickers hardness measurements were carried out on a ZHVST-30F microhardness tester with an applied load of 1 kg and holding time of 10 s. A total of 10 points on the sample surface were selected for indentation testing, and the results were averaged from these 10 points.

The sample for the tensile tests was prepared according to ASTM E8/E 8M–16 [20]. The samples in a dog-bone shape were cut from rolled sheets along the rolling direction by using an electric spark cutting machine, and the dimension of the samples was 8 mm in the gauge length of 2.5 mm in width. Tensile tests were carried out on a universal material testing machine (Instron 3369, Boston, MA, USA) with a strain rate of 8×10^{-4} s^{-1} at room temperature. Three samples were used for each test and the results were the average of the three samples. The fractographs of the samples after the tensile test were observed by SEM.

2.4. Electrochemical Test

The rectangular samples with the dimensions of 8 mm × 8 mm × 5 mm were cut from the as-cast and rolled sample, which were molded into epoxy resin with a surface area of 0.64 cm^2 for electrochemical tests. The tests were conducted in Hank's solution at 37 ± 1 °C by using a Gamry Reference 600+ electrochemical workstation. A three-electrode cell was used for potentiodynamic polarization tests, where a saturated calomel electrode (SCE) was used as the reference electrode, a Pt plate was used as the counter electrode, and the sample was the working electrode. All tests were carried out at a constant scan rate of 1 mV/s starting from −0.3 V to 0.8 V. The corrosion current density was measured by linear Tafel extrapolation and the corrosion rate was then estimated assuming that the number of valence electrons for Mg is 2. Electrochemical impedance tests were performed over a range of frequencies from 10^{-2} Hz to 10^5 Hz with an amplitude voltage of 10 mV, and the experimental results were fitted by using the ZSimpWin software.

2.5. Immersion Test

The immersion tests were carried out at 37 °C and three parallel samples were used in each test. The test period was 1 week. The samples were immersed in Hank's solution with the ratio of solution volume to the sample surface area of 40 mL/cm^2 [21]. The samples were taken out every 24 h, ultrasonically cleaned in a chromate-mixed solution (200 g/L CrO_3 + 10 g/L $AgNO_3$) to remove the corrosion products on their surfaces [22], and dried in air for 2 h. Then, an electronic balance was used to measure the weight of the sample before and after immersion, and calculate the corrosion rate of the sample according to the weight loss corrosion rate formula. The pH of the solution was measured and samples were taken every 24 h during the experiment. In order to analyze the surface of the sample, the sample was embedded in epoxy resin, the corrosion morphology after immersion was observed by a scanning electron microscope, and the chemical composition and weight percentage of the corrosion product were determined by an energy-dispersive spectrometer and XRD.

3. Results
3.1. Microstructure Characterization

Figure 2 shows the XRD patterns of the as-cast, cold-rolled, warm-rolled, and hot-rolled WE43. It can be observed from the XRD pattern of the as-cast alloy that the as-cast alloy was mainly composed of α-Mg and $Mg_{24}Y_5$ phases with a small amount of $Mg_{41}Nd_5$. After rolling treatment, the diffraction peaks of the $Mg_{24}Y_5$ phase disappeared, but the diffraction peaks of the $Mg_{41}Nd_5$ phase can be observed on the XRD pattern of the cold-rolled sample, implying that the eutectic $Mg_{24}Y_5$ phases were dissolved in the α-Mg matrix during the rolling process.

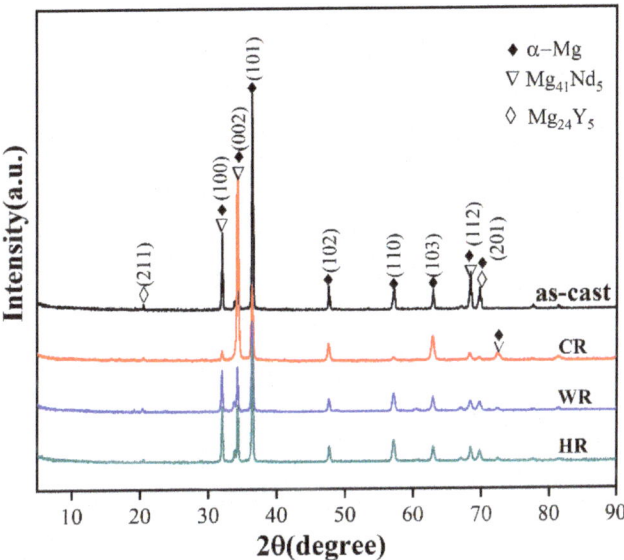

Figure 2. XRD patterns of WE43 in as-cast, cold-rolled, warm-rolled, and hot-rolled condition.

Figure 3 shows the SEM images of the microstructures of as-cast, cold-rolled, warm-rolled, and hot-rolled WE43. It can be seen that the as-cast alloy was mainly composed of equiaxed α-Mg phases with the eutectic $Mg_{24}Y_5$ phases at the grain boundaries, and the average grain size was about 70 μm. After cold rolling, the grains of the alloy were slightly elongated along the rolling direction and apparently refined. Close observation revealed that the eutectic phases at the grain boundaries disappeared, and a small amount of regular fine Gd-rich phases still remained at the grain boundaries. For the warm-rolled sample, large amounts of the fine $Mg_{41}Nd_5$ phase precipitated inside grains. It can be seen from the morphology of hot-rolled samples that the $Mg_{41}Nd_5$ phase was basically broken and dissolved in the matrix, and a small amount was precipitated on the grain boundary.

To clearly illustrate the effects of the rolling temperature on the microstructure of the alloy, EBSD analysis was conducted on the samples after rolling at different temperatures. Figure 4 shows the EBSD maps showing the grain structures of as-cast, cold-rolled, warm-rolled, and hot-rolled WE43. It can be seen that the grain size of the as-cast sample was greatly reduced after rolling treatment. The average grain size of the cold-rolled sample was 7.75 μm. With the rolling temperature increasing, the grain size of the alloy was slightly increased. The average grain sizes of the warm-rolled and hot-rolled samples were 8.88 μm and 9.64 μm, respectively.

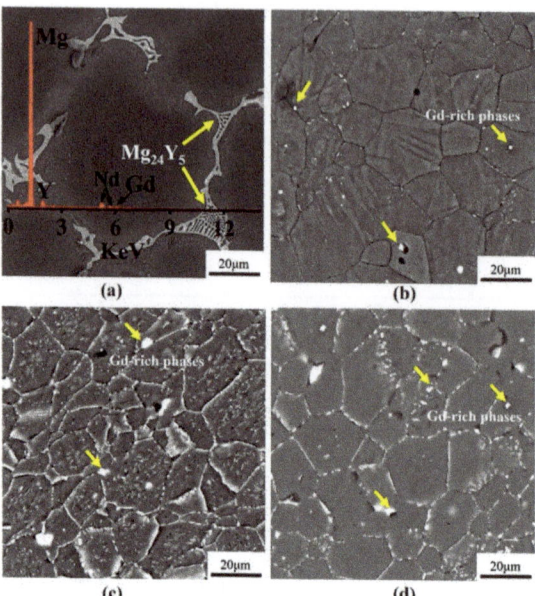

Figure 3. The SEM images of WE43 in (**a**) as-cast, (**b**) cold-rolled, (**c**) warm-rolled, and (**d**) hot-rolled condition.

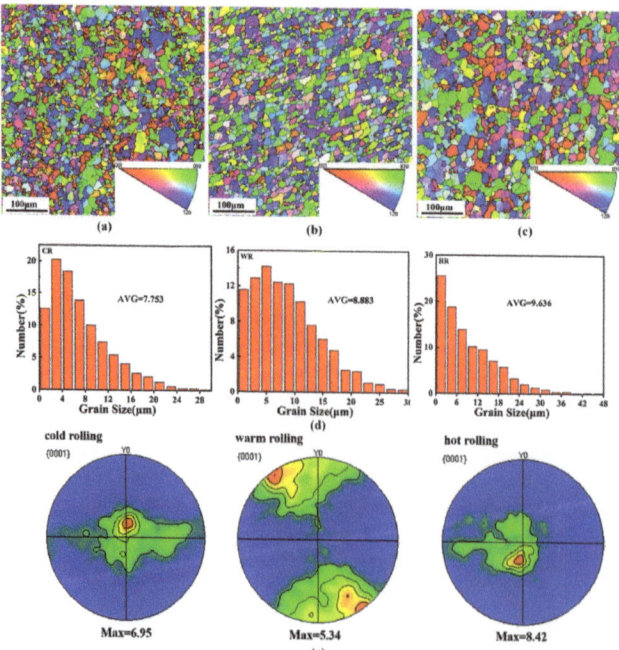

Figure 4. The IPF maps in the RD-SD plane (distribution of crystallographic direction parallel to CD) showing the microstructures of the WE43 alloy after rolling. (**a**) Cold-rolled, (**b**) warm-rolled, and (**c**) hot-rolled, and (**d**) distribution map of different rolled grain sizes. (**e**) Basal (0001) recalculated pole figures of the WE43 alloy.

Figure 4e shows the (0001) pole diagrams of samples after rolling at different temperatures. The rolling samples exhibited a (0001) plane texture with a moderate intensity, and the texture intensities of the texture of the cold-rolled, the warm-rolled, and the hot-rolled samples were 6.95, 5.34, and 8.42, respectively. Moreover, the pole density distribution of the basal plane also changed with rolling temperature. The (0001) base plane pole density of the cold-rolled and hot-rolled samples exhibited the strengthening points in both the rolling and the transverse directions, while the basal plane pole density of the warm-rolled sample expanded to the transverse direction with a symmetrical distribution. The effect of rolling temperature on the resolved shear stress for which the slip systems are responsible is the difference in the texture intensity and distribution of the samples rolled at different temperatures.

3.2. Mechanical Properties

Figure 5 and Table 2 show the tensile properties and hardness of as-cast, cold-rolled, warm-rolled, and hot-rolled WE43. The yield strength (σYS), ultimate strength (σUTS), elongation (η), and hardness (HV) are listed as histograms. It can be seen that the yield strength, ultimate strength, and elongation of the as-cast alloy were 99.8 MPa, 130.7 MPa, 9.1%, and 67 HV. respectively. After rolling at different temperatures, the strengths of the alloy were significantly promoted. For the cold-rolled sample, its yield strength was 279.8 MPa, more than twice as large as that of the as-cast alloy, but its elongation was reduced to 6.1%. In contrast, the warm-rolled sample exhibited a more increased strength and slightly decreased ductility, whose yield strength and elongation reached 327.5 MPa and 8.9%, respectively. Meanwhile, the hot-rolled sample exhibited a relatively low strength and high ductility in comparison with the cold-rolled and the warm-rolled samples. Its yield stress and elongation were 178.4 MPa and 11.3%, respectively, which were much higher than those of the as-cast alloy. Therefore, the warm-rolled sample exhibited excellent comprehensive mechanical properties.

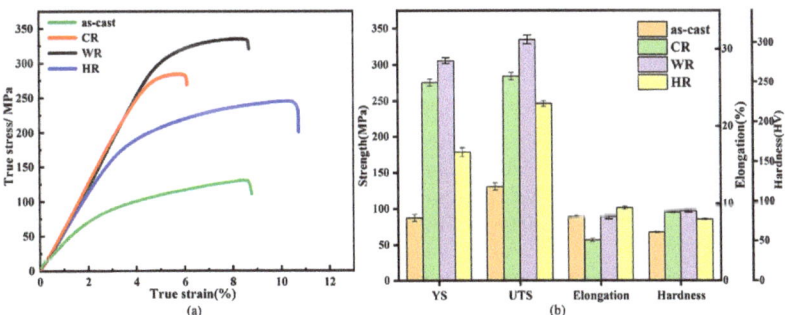

Figure 5. Mechanical properties of as-cast, CR, WR, and HR samples of WE43 alloy: (**a**) tensile stress–strain curves (**b**) yield strength, ultimate tensile strength, elongation, and hardness.

Table 2. Mechanical properties and hardness of WE43 alloy as-cast, CR, WR, and HR samples.

Processing Methods	σ_{YS} (MPa)	σ_{UTS} (MPa)	δ_k (%)	Hardness (HV)
as-cast	99.8 ± 1.1	130.7 ± 1.5	9.1	67 ± 1.9
CR	279.8 ± 1.9	284.1 ± 2.1	6.3	95.3 ± 2
WR	327.5 ± 2.3	346.7 ± 3.9	8.9	97 ± 1.7
HR	178.4 ± 3.1	251.2 ± 3.3	11.3	85.5 ± 3

Figure 6 shows the fracture morphology of as-cast, cold-rolled, warm-rolled, and hot-rolled WE43. The fracture of the as-cast alloy showed obvious shear lips, few pits, and a river pattern. The cleavage planes of cold-rolled samples were mostly thick, and those of warm-rolled and hot-rolled samples were obviously refined. This was consistent with the changing trend in the tensile results, indicating the enhanced ultimate tensile strength of all the rolled samples. Therefore, the fracture mode of the four states was obviously cleavage fracture.

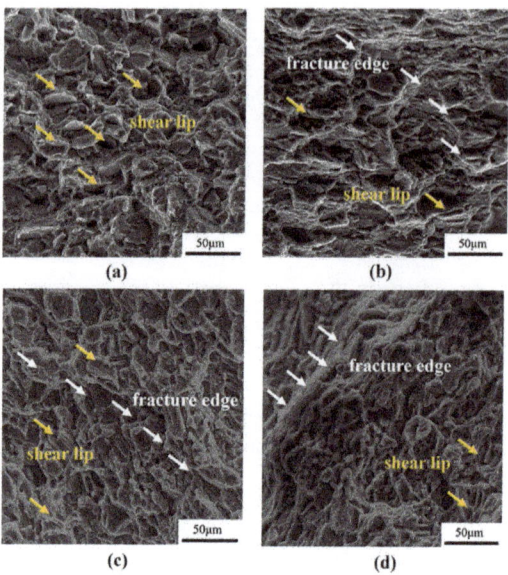

Figure 6. Fracture morphology of WE43 magnesium alloy under tension at room temperature in different states: (**a**) as-cast, (**b**) cold-rolled, (**c**) warm-rolled, and (**d**) hot-rolled.

3.3. Corrosion Behavior

3.3.1. Electrochemical Properties

The open-circuit potential was measured for the first 10 min of immersion in the Hank's solution, as shown in Figure 7a. The OCP curves increased rapidly at the initial immersion period and then increased slowly with the further extension in immersion time. The fast increase in the OCP in the first 10 min suggests the formation of the passivation film on the surface of all the samples. Figure 7b shows the potentiodynamic polarization curves of the as-cast alloy and the samples rolled at different temperatures. The corrosion potentials, corrosion current densities measured by linear Tafel extrapolation, and the estimated corrosion rates are shown in Table 3. The as-cast alloy exhibited a corrosion rate of 0.42 ± 0.29 mm/y. After cold rolling, the corrosion rate was greatly increased and reached 0.55 ± 0.33 mm/y. Meanwhile, the corrosion rates of the warm-rolled and hot-rolled alloy samples were estimated to be 0.28 ± 0.16 mm/y and 0.16 ±0.18 mm/y, respectively, which were much lower than those of the as-cast and cold-rolled samples. As a result, warm rolling and hot rolling can effectively promote the corrosion resistance of the WE43 alloy In Hank's solution.

The Nyquist, bode, and bode phase plots of all the samples are shown in Figure 7c,d. The Nyquist plots of all the samples exhibited two capacitor loops in both the high-frequency and low-frequency regions. One of them is related to the passive layer generated near the sample surface and indicates its corrosion resistance. The other one is related to charge transfer at the Mg-SBF electrolyte interface (double layer). As previously mentioned, the experimental EIS data were fitted by using the zSimWin 3.21 software. The simulated equivalent circuit diagram is shown in Figure 7c. Among them, R_s is the solution resistance, R_t is the charge transfer resistance at the electrolyte solution interface, R_f is the corrosion product resistance, and CPE_d and CPE_f, respectively, represent the corrosion product film capacity and galvanic double-layer capacity at the interface between the Mg matrix and the electrolyte solution, and are generally used to characterize the high-frequency capacitor loop [23]. According to the equivalent circuit, the Nyquist diagrams of the WE43 samples before and after rolling were obtained by using ZsimpWin 3.50 software to obtain the fitted values. The fitted values are shown in Table 4. R_t and R_f are two important parameters for evaluating corrosion resistance. Larger R_t and R_f usually correspond to smaller CPE_d and CPE_f to indicate better corrosion resistance. By comparing the electrochemical impedance diagram and the equivalent circuit fitting data, the hot-rolled corrosion performance was best.

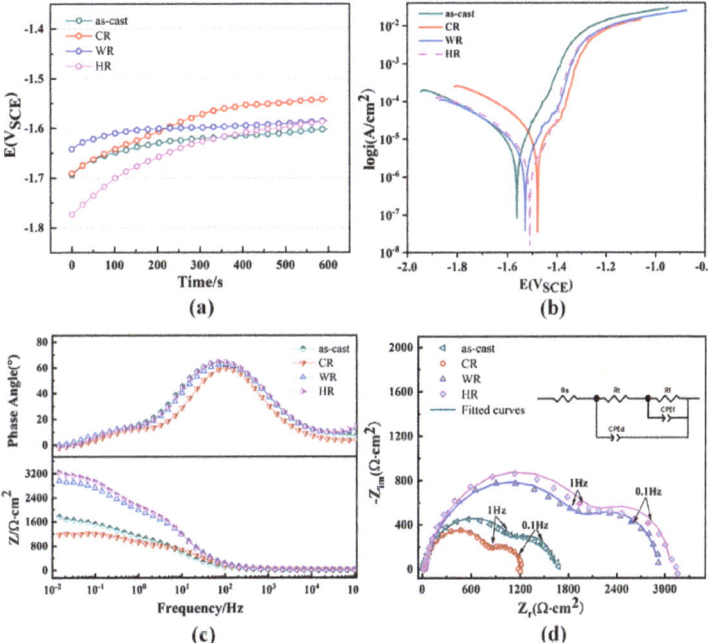

Figure 7. Electrochemical performance of WE43 magnesium alloy in Hanks' solution. (a) Open circuit potential,, (b) potentiodynamic polarization curves, (c) bode plots, and (d) Nyquist plots and equivalent circuits.

Table 3. Electrochemical performance parameters of the Mg alloys in Hank's solution.

Samples	Corrosion Potential (Ecorr), V vs. SCE	Corrosion Current Density (Icorr), µA/cm^2	Corrosion Rate (Vcorr), mm/y
as-cast	−1.54 ± 0.04	15.2 ± 8.1	0.42 ± 0.29
CR	−1.45 ± 0.03	23.1 ± 5.8	0.55 ± 0.33
WR	−1.51 ± 0.02	9.1 ± 5.2	0.28 ± 0.16
HR	−1.49 ± 0.03	6.1 ± 4.5	0.16 ± 0.18

Table 4. The fitted electrochemical parameters for WE43 alloys in Hank's solution.

	as-Cast	CR	WR	HR
R_s (Ωcm^2)	14.09	30.42	31.39	28.18
CPE_d ($\Omega^{-2} cm^{-2} s^{-2}$)	4.293×10^{-5}	2.171×10^{-5}	2.007×10^{-5}	2.032×10^{-5}
n_1	0.8272	0.8727	0.8262	0.8362
R_t (Ωcm^2)	1214	863.1	2074	2275
CPE_f ($\Omega^{-2} cm^{-2} s^{-2}$)	1.693×10^{-3}	1.066×10^{-3}	8.248×10^{-4}	8.029×10^{-4}
n_2	0.9836	0.975	0.9412	0.994
R_f (Ωcm^2)	406.9	310.8	820.4	771.4

3.3.2. Degradation Properties

It has been documented that the evaluation of the corrosion rates from polarization curves is unreliable for Mg alloys owing to the complexity of the corrosion reactions and usually disagree with the results from the hydrogen evolution rate and weight loss. Therefore, the immersion tests of all the samples were conducted in the present work, and the change in pH value of the solution and weight loss of the samples as a function of the immersion time was measured to assess the corrosion rates. The results are shown in Figure 8. It can be seen that the pH values of the immersion solution increased rapidly in the initial stage of the immersion due to the magnesium dissolution reactions to produce the hydrogen and OH^- for the samples, indicating a rapid increase in the corrosion rates of all the samples. With the further extension in the immersion time, the pH values of the immersion solutions increased slowly due to the passive films on the alloy surfaces thickening and the concentration of OH^- ions reaching a stable value. Moreover, the weight loss of the samples after immersing in Hank's solution for 1 day and 7 days was measured, from which the corrosion rate of all the samples was also estimated, and the results are also illustrated in the inset of Figure 8. It can be seen that the corrosion rates of all the samples were arranged in increasing order: HR < WR < as-cast < CR. The result was in good agreement with that obtained by electrochemical tests. Li et al. [24] and Yue et al. [25] also reported that the corrosion rate obtained from the electrochemical polarization method was approximately twice as high as that obtained from an immersion corrosion method due to the electrochemical polarization method being able to shorten the corrosion cycle, promote electron transferring, and accelerate the mass transferring process.

The phase structures and the compositions of the corrosion products on the surface of WE43 alloy samples after being immersed in Hank's solution for 7 days were determined by XRD and EDS analysis, and the results are shown in Figure 9. From the results obtained by EDS analysis (Table 5), the corrosion products on the surfaces of all the samples contained Mg, O, C, Cl, Nd, and Y elements. The XRD analysis revealed that the corrosion products were mainly composed of $Mg(OH)_2$. When the alloy was in a simulated body fluid, water penetrated through the magnesium oxide film to penetrate the surface of the magnesium alloy and corrode, forming porous $Mg(OH)_2$ and landing on the surface.

Table 5. Chemical composition of the surface corrosion products on WE43 Mg alloys.

	Chemical Composition (at.%)					
	Mg	O	C	Cl	Y	Nd
as-cast	39.83	46.7	7.59	5.52	0.64	0
CR	41.10	47.67	8.97	2.59	0.08	0.6
WR	39.2	49.77	6.47	1.55	1.34	1.67
HR	33.54	45.47	7.8	6.01	3.3	3.88

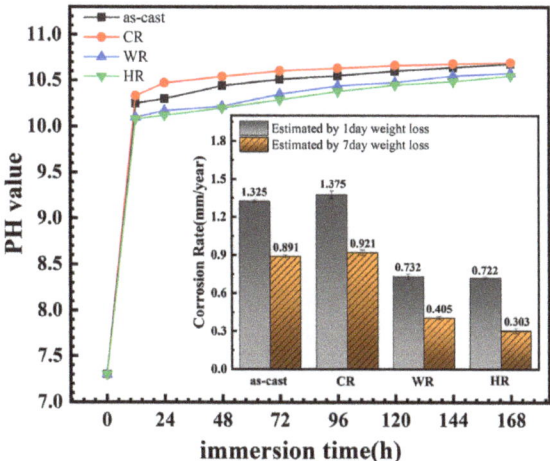

Figure 8. Variation in solution pH of the magnesium alloys with different states during 7 days of immersion in Hank's solution and the corresponding corrosion rate obtained by weight loss.

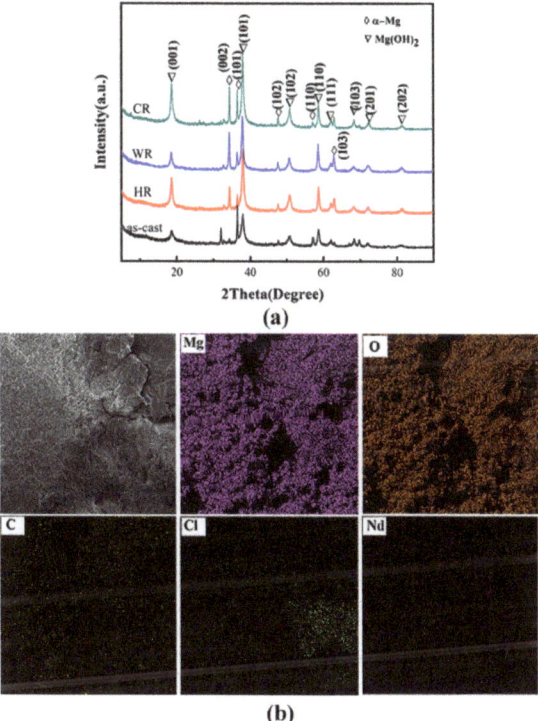

Figure 9. XRD pattern: (**a**) energy spectrum analysis (**b**) and energy spectrum map of the surface corrosion products on WE43 alloy after immersion for 7 days in Hanks' solution.

The corrosion product morphology of the cast and rolled WE43 after being immersed in the Hank's solution at 37 °C for 7 days is shown in Figure 10. The corrosion rate results obtained by weight loss were further verified from the corrosion topography map. The cold-rolled state had a higher corrosion rate in the simulated body fluid, and a large number of filiform corrosion and corrosion pits appeared on the surface. The hot-worked WE43 alloy had small surface corrosion pits and no large corrosion pits. The corrosion pits were relatively uniform, so its corrosion rate was lower than that of the cold-worked alloy.

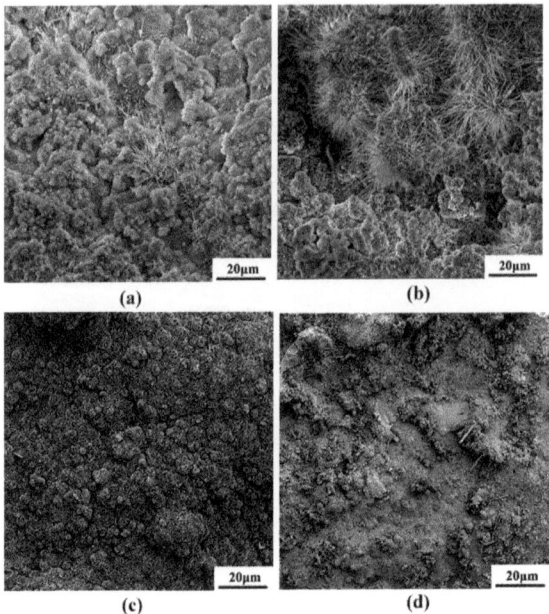

Figure 10. The SEM morphologies of the surface corrosion products on WE43 alloy after immersion for 7 days in Hanks' solution: (**a**) as-cast, (**b**) cold-rolled, (**c**) warm-rolled, and (**d**) hot-rolled.

4. Discussion

According to the experimental results, the grains are remarkably refined in different rolling temperatures. After cold rolling, the eutectic phase at the grain boundary disappears, resulting in a large number of dislocation defects and twins. Dynamic recovery during warm rolling can accelerate the precipitation of the second phase. A large amount of fine $Mg_{41}Nd_5$ second phase is precipitated and dispersed in the grain boundary, which also plays a role in the strengthening effect of the alloy. From being hot-rolled at 400 °C, dynamic recrystallization occurs, and most of the rare-earth second phase is dissolved in the matrix. The atomic thermal motion, dislocation motion, and grain boundary migration in the alloy become easier under high-temperature rolling deformation, and finally the dynamic recrystallization is accelerated [26].

The (0001) base density of cold-rolled and hot-rolled samples shows a strengthening effect in the rolling direction and transverse direction. The base density of the hot-rolled sample gradually expands horizontally and distributes symmetrically. It can be seen that the change in basal polar density is consistent with the change in basal texture intensity, and both decrease and then increase with the increase in temperature. This is mainly due to the effect of temperature on the critical shear stress of different slip systems [13].

Due to work hardening, the strength of cold-rolled samples is significantly improved. However, due to some dislocation defects and twins in the cold rolling process, the elongation decreases. For the warm-rolled sample, the alloy undergoes dynamic recovery, grain refinement, and subgrain formation. At the same time, the size of the second phase decreases and the distribution is uniform, and the strengthening effect on the alloy is more obvious. Dynamic recrystallization occurs during hot rolling, and defects such as dislocations in the alloy are eliminated. The microstructure is more uniform, and the plasticity of the alloy is greatly improved.

The current density of the anode branch of the WE43 magnesium alloy increases faster than that of the cathode branch, indicating that the magnesium alloy matrix is dissolved [27]. Both warm rolling and hot rolling alloys show large capacitance rings, indicating that warm rolling and hot rolling can improve the corrosion resistance of the WE43 magnesium alloy. The capacitance loop of cold rolling is smaller than that of the as-cast state, indicating that cold rolling reduces the corrosion resistance of the magnesium alloy. Two time constants in Figure 7b are caused by the double-layer capacitor and the corresponding charge transfer resistance of the oxide film and the corrosion product produced in the corrosion process [28]. Normally, a $Mg(OH)_2$ protective film is formed during the immersion in corrosive solution. The Rt and Rf of the cast and cold-rolled WE43 are relatively small, which is related to the poor protection ability of the $Mg(OH)_2$ layer.

The corrosion rate of the magnesium alloy after hot processing is low. After hot rolling, the segregation degree in the micro-region of the sample decreases, the grains are refined, and the second phase is dissolved in the matrix, resulting in a uniform microstructure and lower corrosion rate. For warm-rolled samples, there are many fine second phases on the magnesium alloy matrix, and the corrosion rate is higher than that of the hot-rolled magnesium alloy. For cold-rolled samples, the second phase of the magnesium alloy is coarser than those of other rolling samples, resulting in a large number of dislocation defects, and then forming a large number of galvanic double layers, prone to pitting corrosion, and corrosion performance is greatly reduced.

Overall, all those mentioned above can contribute to the improvement of ultimate tensile strength by rolling, and the hot working can greatly enhance the corrosion resistance. The warm-rolled WE43 alloy exhibits good mechanical properties and degradation properties for implant application.

5. Conclusions

In this paper, the influence of cold rolling, warm rolling, and hot rolling on the microstructure, mechanical properties, and corrosion properties of the WE43 magnesium alloy is comprehensively evaluated. The main conclusions are as follows:

(1) After rolling, the alloy grains are significantly refined, and the rare-earth phase is distributed in the grain along the deformation direction. The basal surface texture strength and basal pole density distribution in the alloy gradually weaken and then increase with the increase in the rolling temperature.

(2) The strength and hardness of the rolled WE43 alloy are significantly improved, and the yield strength of the warm rolling can reach up to 327.5 Mpa. The fracture mode of the as-cast and rolled alloy is the cleavage fracture mechanism.

(3) The corrosion resistance of the WE43 magnesium alloy can be greatly improved through hot working. The corrosion rate of the warm-rolled and hot-rolled samples is reduced by 3 times, compared with that of the as-cast state.

Author Contributions: Conceptualization, B.D. and D.Z.; methodology, B.D.; validation, B.D.; formal analysis, Y.D.; investigation, D.Z.; resources, J.L.; data curation, D.Z.; writing—original draft preparation, B.D.; writing—review and editing, Y.D.; visualization, J.L.; supervision, Y.D.; project administration, D.Z.; funding acquisition, J.L. and D.Z. All authors have read and agreed to the published version of the manuscript.

Funding: This research is financially supported by the National Natural Science Foundation of China (Grant No. 11872053 and No. 51971190), and the Scientific research project of Hunan Education Department (20C1796).

Conflicts of Interest: The authors declare no conflict of interest.

References

1. Virtanen, S. Biodegradable Mg and Mg alloys: Corrosion and biocompatibility. *Mater. Sci. Eng. B* **2011**, *176*, 1600–1608. [CrossRef]
2. Zheng, Y.F.; Gu, X.N.; Witte, F. Biodegradable metals. *Mater. Sci. Eng. R* **2014**, *77*, 1–34. [CrossRef]
3. Brar, H.S.; Platt, M.O.; Sarntinoranont, M.; Martin, P.I.; Manuel, M.V. Magnesium as a biodegradable and bioabsorbable material for medical implants. *JOM* **2009**, *61*, 31–34. [CrossRef]
4. Levy, G.; Aghion, E. Effect of diffusion coating of Nd on the corrosion resistance of biodegradable Mg implants in simulated physiological electrolyte. *Acta Biomater.* **2013**, *9*, 8624–8630. [CrossRef]
5. Staiger, M.P.; Pietak, A.M.; Huadmai, J.; Dias, G. Magnesium and its alloys as orthopedic biomaterials: A review. *Biomaterials* **2006**, *27*, 1728–1734. [CrossRef]
6. Ezechieli, M.; Ettinger, M.; König, C.; Weizbauer, A.; Helmecke, P.; Schavan, R.; Lucas, A.; Windhagen, H. Biomechanical characteristics of bioabsorbable magnesium-based (MgYREZr-alloy) interference screws with different threads. *Knee Surg. Sports Traumatol. Arthrosc.* **2016**, *24*, 3976–3981. [CrossRef]
7. Åhman, H.N.; Thorsson, L.; Mellin, P.; Lindwall, G.; Persson, C.J.M. An Enhanced Understanding of the Powder Bed Fusion–Laser Beam Processing of Mg-Y$_{3.9wt\%}$-Nd$_{3wt\%}$-Zr$_{0.5wt\%}$ (WE43) Alloy through Thermodynamic Modeling and Experimental Characterization. *Materials* **2022**, *15*, 417. [CrossRef]
8. Esmaily, M.; Zeng, Z.; Mortazavi, A.; Gullino, A.; Choudhary, S.; Derra, T.; Benn, F.; D'Elia, F.; Müther, M.; Thomas, S.J.A.M. A detailed microstructural and corrosion analysis of magnesium alloy WE43 manufactured by selective laser melting. *Addit. Manuf.* **2020**, *35*, 101321. [CrossRef]
9. Li, H.; Xu, K.; Yang, K.; Liu, J.; Zhang, B.; Xia, Y.; Zheng, F.; Han, H.; Tan, L.; Hong, D. The degradation performance of AZ31 bioabsorbable magnesium alloy stent implanted in the abdominal aorta of rabbits. *J. Interv. Radiol.* **2010**, *19*, 315–317.
10. Straumal, P.; Martynenko, N.; Kilmametov, A.; Nekrasov, A.; Baretzky, B.J.M. Microstructure, microhardness and corrosion resistance of WE43 alloy based composites after high-pressure torsion. *Materials* **2019**, *12*, 2980. [CrossRef]
11. Jia, G.-L.; Wang, L.-P.; Feng, Y.-C.; Guo, E.-J.; Chen, Y.-H.; Wang, C.-L. Microstructure, mechanical properties and fracture behavior of a new WE43 alloy. *Rare Met.* **2021**, *40*, 2197–2205. [CrossRef]
12. Ghorbanpour, S.; McWilliams, B.A.; Knezevic, M. Effect of hot working and aging heat treatments on monotonic, cyclic, and fatigue behavior of WE43 magnesium alloy. *Mater. Sci. Eng. A* **2019**, *747*, 27–41. [CrossRef]
13. Agnew, S.; Yoo, M.; Tome, C. Application of texture simulation to understanding mechanical behavior of Mg and solid solution alloys containing Li or Y. *Acta Mater.* **2001**, *49*, 4277–4289. [CrossRef]
14. Argade, G.; Panigrahi, S.; Mishra, R. Effects of grain size on the corrosion resistance of wrought magnesium alloys containing neodymium. *Corros. Sci.* **2012**, *58*, 145–151. [CrossRef]
15. Li, Z.; Gu, X.; Lou, S.; Zheng, Y. The development of binary Mg–Ca alloys for use as biodegradable materials within bone. *Biomaterials* **2008**, *29*, 1329–1344. [CrossRef] [PubMed]
16. Wang, T.; Zhou, X.; Li, Y.; Zhang, Z.; Le, Q. Effect of Rolling Temperature on the Microstructure and Mechanical Properties of Mg-2Zn-0.4 Y Alloy Subjected to Large Strain Rolling. *Metals* **2018**, *8*, 937. [CrossRef]
17. Sabat, R.; Samal, P. Effect of strain path on the evolution of microstructure, texture and tensile properties of WE43 alloy. *Mater. Sci. Eng. A* **2018**, *715*, 348–358. [CrossRef]
18. Zhu, S.; Yan, H.; Chen, J.; Wu, Y.; Liu, J.; Tian, J. Effect of twinning and dynamic recrystallization on the high strain rate rolling process. *Scr. Mater.* **2010**, *63*, 985–988. [CrossRef]
19. Dvorský, D.; Kubásek, J.; Hosová, K.; Čavojský, M.; Vojtěch, D.J.M. Microstructure, Mechanical, Corrosion, and Ignition Properties of WE43 Alloy Prepared by Different Processes. *Metals* **2021**, *11*, 728. [CrossRef]
20. ASTM E8/E8M-09; Standard Test Methods for Tension Testing of Metallic Materials. ASTM: West Conshohocken, PA, USA, 2011.
21. Haynes, G.S. *Laboratory corrosion tests and standards: a symposium by ASTM Committee G-1 on Corrosion of Metals, Bal Harbour, FL, 14–16 Nov. 1983*; ASTM International: Conshohocken, PA, USA, 1985.
22. Cui, Z.; Ge, F.; Lin, Y.; Wang, L.; Lei, L.; Tian, H.; Yu, M.; Wang, X. Corrosion behavior of AZ31 magnesium alloy in the chloride solution containing ammonium nitrate. *Electrochim. Acta* **2018**, *278*, 421–437. [CrossRef]
23. Wang, B.; Xu, D.; Xin, Y.; Sheng, L.; Han, E. High corrosion resistance and weak corrosion anisotropy of an as-rolled Mg-3Al-1Zn (in wt.%) alloy with strong crystallographic texture. *Sci. Rep.* **2017**, *7*, 16014. [CrossRef] [PubMed]
24. Li, H.; Xie, X.; Zheng, Y.; Cong, Y.; Zhou, F.; Qiu, K.; Wang, X.; Chen, S.; Huang, L.; Tian, L.; et al. Development of biodegradable Zn-1X binary alloys with nutrient alloying elements Mg, Ca and Sr. *Sci. Rep.* **2015**, *5*, 10719. [CrossRef] [PubMed]
25. Yue, R.; Huang, H.; Ke, G.; Zhang, H.; Pei, J.; Xue, G.; Yuan, G.J.M.C. Microstructure, mechanical properties and in vitro degradation behavior of novel Zn-Cu-Fe alloys. *Mater. Charact.* **2017**, *134*, 114–122. [CrossRef]
26. Christian, J.W.M. Subhash, Deformation twinning. *Prog. Mater. Sci.* **1995**, *39*, 1–157. [CrossRef]
27. Bakhsheshi-Rad, H.; Abdul-Kadir, M.; Idris, M. Relationship between the corrosion behavior and the thermal characteristics and microstructure of Mg–0.5 Ca–xZn alloys. *Corros. Sci.* **2012**, *64*, 184–197. [CrossRef]
28. Yan, Y.; Gang, Z.; REN, F.-J.; DENG, H.-J.; WEI, G.-B. Effect of rolling reduction and annealing process on microstructure and corrosion behavior of LZ91 alloy sheet. *Trans. Nonferrous Met. Soc. China* **2020**, *30*, 1816–1825.

MDPI
St. Alban-Anlage 66
4052 Basel
Switzerland
www.mdpi.com

Materials Editorial Office
E-mail: materials@mdpi.com
www.mdpi.com/journal/materials

Disclaimer/Publisher's Note: The statements, opinions and data contained in all publications are solely those of the individual author(s) and contributor(s) and not of MDPI and/or the editor(s). MDPI and/or the editor(s) disclaim responsibility for any injury to people or property resulting from any ideas, methods, instructions or products referred to in the content.

www.ingramcontent.com/pod-product-compliance
Lightning Source LLC
LaVergne TN
LVHW070448100526
838202LV00014B/1685